Smart Electronic Devices

The book analyzes the use of smart medical devices that use artificial intelligence and machine learning to analyze medical images, detect diseases, and assist in diagnosis. It further focuses on real-world applications of artificial intelligence and machine learning in smart electronic devices, demonstrating how these technologies are being used in various industries, such as healthcare, automotive, finance, and consumer electronics.

Features:

- Explores how cloud and edge computing work together to enhance the capabilities and performance of smart devices, enabling a seamless user experience and facilitating the growth of the Internet of Things ecosystem.
- Discusses the use of smart devices within a smart home system, exploring the seamless connectivity, interoperability, and centralized control.
- Explains the advancements in smart traffic management and smart parking systems, which leverage cutting-edge technologies to address the growing challenges of urban mobility.
- Surveys the growing importance of smart energy management and the integration of renewable energy sources in the pursuit of a sustainable and eco-friendly energy landscape.
- Covers the dynamic relationship between the adoption of smart devices and artificial intelligence technologies and the diverse regulatory frameworks governing these innovations.

It is primarily written for senior undergraduates, graduate students, and academic researchers in the fields including electrical engineering, electronics and communications engineering, computer science and engineering, and biomedical engineering.

Smart Electronic Devices

Artificial Intelligence, Machine Learning, and the Future

Edited by

Yogesh Kumar Verma

Manoj Singh Adhikari

Varun Mishra

Suman Lata Tripathi

Manoj Kumar Shukla

CRC Press

Taylor & Francis Group

Boca Raton London New York

CRC Press is an imprint of the
Taylor & Francis Group, an **informa** business

Designed cover image: Getty Images

First edition published 2026
by CRC Press
2385 NW Executive Center Drive, Suite 320, Boca Raton FL 33431

and by CRC Press
4 Park Square, Milton Park, Abingdon, Oxon, OX14 4RN

CRC Press is an imprint of Taylor & Francis Group, LLC

ISBN: 978-1-032-85406-9 (hbk)
ISBN: 978-1-041-07988-0 (pbk)
ISBN: 978-1-003-64317-3 (ebk)

DOI: 10.1201/9781003643173

Typeset in Sabon
by SPi Technologies India Pvt Ltd (Straive)

Contents

Contributors

Manoj Singh Adhikari
School of Electronics and Electrical Engineering
Lovely Professional University
Phagwara, India

Kadambri Agarwal
Department of CSE
ABES Engineering College
Ghaziabad, India

Vinay Anand
School of Electronics and Electrical Engineering
Lovely Professional University
Phagwara, India

Shaik Vaseem Akram
Department of ECE, School of Engineering
SR University
Warangal, India

Krishan Arora
Lovely Professional University
Punjab, India

Harwant Singh Arri
Department of Computer Science and Engineering
Lovely Professional University
Phagwara, India

Sandhya Avasthi
Department of Computer Science and Engineering
ABES Engineering College
Ghaziabad, India

Aditya
Department of Computer Science and Engineering
ABES Engineering College
Ghaziabad, India

Indu Bala
School of Electronics and Electrical Engineering
Lovely Professional University
Punjab, India

Harpreet Singh Bedi
School of Electronics and Electrical Engineering
Lovely Professional University
Phagwara, India

Shakti Raj Chopra
School of Electrical and Electronics Engineering
Lovely Professional University
Phagwara, India

Deepika Ghai
School of Electronics and Electrical Engineering

Lovely Professional University
Phagwara, India

Sudip Ghosh
School of VLSI Technology
Indian Institute of Engineering Science and Technology (IIEST)
Howrah, India

Sachin Gupta
School of Electronics and Electrical Engineering
Lovely Professional University
Phagwara, India

Aditya Pai H
MIT – School of Computing, MIT Art
Design and Technology University
Pune, India

Shahid Hamid
Physics Wallah
Uttar Pradesh, India

S. K. HimaBindu
Department of Electronics and Communication Engineering
Marri Laxman Reddy Institute of Technology and Management
Hyderabad, India

Anita Gehlot
Uttaranchal Institute of Technology
Uttaranchal University
Dehradun, India

Praveen Gujjar J
Faculty of Management Studies
JAIN (Deemed-to-be University)
Bengaluru, India

Amith K Jain
Department of Electronics and Communication Engineering
S D M Institute of Technology
Ujire, India

Sandeep Kumar
School of Computer Science and Artificial Intelligence
SR University
Warangal, India

K. Aruna Kumari
Department of CSE
SRKR Engineering College
Bhimavaram, India

Mandeep Kaur
Department of Computer Science and Engineering
Chitkara University Institute of Engineering and Technology
Chitkara University
Punjab, India

Lavpreet Kaur
Journalism and Mass Communication Department
Punjabi University
Patiala, India

Simarpreet Kaur
Baba Banda Singh Bahadar Engineering College
Fatehgarh Sahib, India

Archana Kanwar
Department of Applied Engineering
Chitkara University
Rajpura, India

Arunava Majumder
Department of Mathematics
Lovely Professional University
Phagwara, India

Praveen Kumar Malik
School of Electronics and Electrical Engineering
Lovely Professional University
Phagwara, India

Varun Mishra
Department of Electronics and Communication Engineering
Madhav Institute of Technology and Science (Deemed University)
Gwalior, India

Sunpreet Kaur Nanda
Department of Electronics and Telecommunication Engineering
P. R. Pote Patil College of Engineering and Management
Amravati, India

Naveen Kumar H N
Department of Electronics and Communication Engineering
Vidyavardhaka College of Engineering
Mysuru, India

M. Nagalakshmi
Department of Information Technology
Marri Laxman Reddy Institute of Technology and Management
Hyderabad, India

Raju Patel
School of Electronics Engineering
Vellore Institute of Technology
Chennai, India

Aman Pandey
Department of Computer Science and Engineering
ABES Engineering College
Ghaziabad, India

Suman Rani
Assistant Professor, Electronics & Communication Engineering
Lovely Professional University
Punjab, India

M. Nagabhushana Rao
Department of Computer Science and Engineering
Marri Laxman Reddy Institute of Technology and Management
Hyderabad, India

Shanky Saxena
School of Electronics & Electrical Engineering
Lovely Professional University
Phagwara, India

Balraj Singh
School of Engineering and Technology
Babasaheb Bhimrao Ambedkar University
Lucknow, India

Department of Computer Science and Engineering
Lovely Professional University
Phagwara, Punjab

Gurmukh Singh
School of Computer Applications
Lovely Professional University
Phagwara, India

Manoj Sindhwani
School of Electronics and Electrical Engineering
Lovely Professional University
Phagwara, India

Shippu Sachdeva
School of Electronics and Electrical Engineering
Lovely Professional University
Phagwara, India

Tanishk Singhal
School of Electronics and Electrical Engineering
Lovely Professional University
Phagwara, India

Manoj Kumar Shukla
Robotics and Automation
Symbiosis Institute of Technology
Symbiosis International (Deemed University)
Pune, India

Chandrappa S
Department of Computer Science and Engineering
Nagarjuna College of Engineering and Technology
Bengaluru, India

Guru Prasad M S
Department of Computer Science and Engineering
Graphic Era (Deemed to be University)
Dehradun, India

Jaibir Singh
Computer Science & Engineering
Lovely Professional University
Punjab, India

Tanushree Sanwal
Department of Management
KIET Group of Institution
Delhi-NCR, India

A. Leela Sravanthi
Department of Information Technology
Marri Laxman Reddy Institute of Technology and Management
Hyderabad, India

Himanshu Sharma
Department of Electrical Engineering
G.H. Raisoni Skilltech University
Nagpur, India

Shallu Sharma
Department of Computer Applications
Lovely Professional University
Phagwara, Punjab

Ravindra Sharma
Swami Rama Himalayan University
Dehradun, India

Rajesh Singh
Uttaranchal Institute of Technology
Uttaranchal University
Dehradun, India

Shailendra Kumar Tripathi
Department of Physics and Material Science
Jaypee University
Anoopshahr, India

Suman Lata Tripathi
Department of Electronics and Telecommunication Engineering
Symbiosis Institute of Technology (SIT), Symbiosis International (Deemed University)
Pune, India

Neha Thakur
Durham College
Oshawa, Canada

Pavan Thimmavajjala
Economic Policy Analyst
Nikore Associates
Hyderabad, India

Yogesh Kumar Verma
School of Electronics and Electrical Engineering
Lovely Professional University
Jalandhar, India

Editors

Yogesh Kumar Verma is working as Associate Professor in the School of Electronics & Electrical Engineering, Lovely Professional University, Phagwara, Punjab, India. He received his B.Tech. from Sharda Group of Institutions, Mathura, UP, India, in the year 2009. He received his M.Tech. degree in Digital Systems from Madan Mohan Malaviya Engineering College, Gorakhpur, UP, India, in 2012. He received his Ph.D. degree from Motilal Nehru National Institute of Technology, Allahabad, Prayagraj, UP, India, in 2020. His research interests are simulation and modeling of semiconductor devices. He had published more than 25 research papers in SCI indexed journals. He has seven years of teaching experience. He has supervised 2 Ph.D. Thesis in Microelectronics and VLSI design.

Manoj Singh Adhikari is working as an Assistant Professor in the School of Electronics & Electrical Engineering, Lovely Professional University, Phagwara, Punjab, India. He received his M.Tech. degree in Digital Signal Processing Engineering from the G. B. Pant Institute of Engineering and Technology (formerly known as G. B. Pant Engineering College), Pauri Garhwal, India, in 2013. He received his Ph.D. degree (Specialization – Solid State Devices) from the G. B. Pant Institute of Engineering and Technology (formerly known as G. B. Pant Engineering College), Pauri Garhwal, India, in 2019. His research interests are the simulation and modeling of power semiconductor devices. He has six years of teaching experience. He is serving as a reviewer for IEEE Transactions on Electron Devices, AEU – International Journal of Electronics and Communications, Measurement Science and Technology in international science journals.

Varun Mishra has completed his Ph.D. in the area of microelectronics and VLSI from MNNIT, Allahabad. He did his M.Tech in VLSI Design at Uttarakhand Technical University, Dehradun, and his B.Tech in Electronics and Communication Engineering at Graphic Era University, Dehradun. He is associated with Graphic Era (Deemed to be University) as an Assistant Professor with more than two years of experience in academics. He has published more than 15 research papers in SCI-indexed journals and more than 10 research papers in international conference proceedings. His area of expertise includes modeling and simulation of novel semiconductor devices such as Tunnel FETs, Negative Capacitance FETs, HEMTs, etc., and low-power VLSI circuit design.

Suman Lata Tripathi is working as a Professor at Department of Electronics and Communication Engineering, Symbiosis Institute of Technology (SIT), Symbiosis International (Deemed University) more than 23 years of experience in academics and research. She has completed her Ph.D. in the area of microelectronics and VLSI Design from MNNIT, Allahabad. She completed her remote post-doc from Nottingham Trent University, London, UK, in the year 2022–2023. She has published more than 164 research papers in international science journals, conference proceedings, and e-books. She has also published 20 Indian patents and 4 copyrights. She has been listed in the world's top 2% of scientists published by Stanford University for the year 2024. Her area of expertise includes microelectronics device modeling and characterization, low-power VLSI circuit design, VLSI design of testing, and advanced FET design for IoT, Embedded System Design, reconfigurable architecture with FPGAs, and biomedical applications.

Manoj Kumar Shukla is currently working as an Assistant Professor in the School of Electronics and Electrical Engineering at Symbiosis Institute of Technology (SIT), Symbiosis International (Deemed University), Pune. He completed his M.Tech and Ph.D. in Electrical Engineering from NIT, Hamirpur (Himachal Pradesh), India. He has 7 years of teaching and research experience. His research interests include control systems, fractional calculus, robotics, automation, artificial intelligence, sensor technology, etc. He has published several research papers in reputed international journals and conferences.

Preface

Artificial intelligence (AI) and machine learning (ML) are critical technologies in the development of smart devices. Smart devices can use AI and ML algorithms to learn and recognize individual users' preferences and habits, allowing them to personalize the user experience. AI and ML algorithms can help smart devices predict when maintenance is required and notify users before a failure occurs. This can reduce downtime and prevent expensive repairs. AI and ML are used in smart speakers to recognize and respond to voice commands accurately. The more the device is used, the better it can understand the user's speech patterns and language. AI and ML algorithms can analyze usage patterns and optimize the device's power consumption to reduce energy consumption. Smart devices can use AI and ML algorithms to automate routine tasks, such as setting the thermostat or turning on lights, based on the user's habits and preferences. AI and ML can help smart devices detect unusual activity and alert users of potential security threats. Overall, AI and ML play a critical role in making smart devices more intuitive, efficient, and secure. As these technologies continue to advance, we can expect smart devices to become even more capable and user-friendly.

The aim of incorporating AI and ML in smart devices is to create intelligent devices that can adapt and learn from their environment, provide personalized experiences to users, and automate routine tasks to improve efficiency. The scope of AI and ML in smart devices is vast and can be seen in various applications, including voice assistants like Siri, Alexa, and Google Assistant use natural language processing (NLP) and machine learning algorithms to understand and respond to users' voice commands. Smart homes use AI and ML to automate routine tasks, such as turning on/off lights, adjusting thermostats, and controlling home appliances based on user behavior. Wearable devices like smartwatches and fitness trackers use AI and ML to track users' physical activity, monitor their health, and provide personalized recommendations based on their habits. Autonomous vehicles use AI and ML to analyze sensor data from cameras and other sources to navigate and make decisions on the road. AI and ML are used in industrial automation to optimize processes, reduce waste, and increase productivity. Overall, the integration of AI and ML in smart devices enables devices to become more intelligent and intuitive, providing users with a better and more personalized experience.

The future scope of smart electronic devices using artificial intelligence and machine learning is immense. With advancements in technology, these devices are becoming more sophisticated and capable of performing complex tasks. Here are some potential future developments in this field: Smart electronic devices are likely to become more interconnected, with the ability to communicate and share data seamlessly. This could lead to more efficient and effective systems. With AI and ML, smart devices will be able to learn from their surroundings and make autonomous decisions. This could result in more accurate and timely responses to changes in the environment. Smart devices will be able to recognize individual users and adapt their behavior accordingly. This could lead to more personalized experiences for users. With smart energy management systems, smart devices will be able to optimize their energy consumption based on usage patterns and other factors. This could lead to significant energy savings. Smart healthcare devices using AI and ML will be able to analyze vast amounts of data to diagnose and treat diseases. This could lead to more accurate and personalized healthcare. Smart agriculture devices using AI and ML will be able to monitor crops, soil, and weather conditions to optimize yields and reduce waste. This could lead to more sustainable farming practices. Smart electronic devices in robotics will be able to perform more complex tasks, such as surgical procedures and autonomous vehicles. This could lead to significant advancements in these fields. Overall, the future of smart electronic devices using AI and ML is bright. These devices are likely to become more integrated, automated, personalized, and energy-efficient. They will also have expanded applications in healthcare, agriculture, and robotics.

Chapter 1

AI for smart agriculture

Raju Patel, Sachin Gupta, Balraj Singh and Shailendra Kumar Tripathi

1.1 INTRODUCTION

In the modern era, the smart farming is one of the best alternatives to improve the scenario of agriculture sector in terms of productivity, management, and improvement of quality of the crop [1–4]. From the beginning to harvesting, the crop has certain kinds of situations in form of data to understand the nature and make decisions to improve the production of a crop [5, 6]. The integration of Internet of Things (IoT), Artificial Intelligence (AI), and Machine learning tools helps to capture the data. Moreover, based on trends of the data, certain decisions can be taken by the machine to manage the overall field. It has been observed that the IoT helps to access the data from certain remote locations to main node for processing and computing processes [7–9]. The Machine learning tools contain certain kinds of algorithms to classify and predict the trends. It unbundles very important and delicate trends of crops to sustain the progress [10–14]. The machine learning algorithms have state-of-the-art methodologies to predict the trends. One of the main advantages of these techniques is to process the data in real time [15, 16]. The implementation of these tools helps to originate certain kinds of applications for the stakeholders in terms of time and money [17, 18]. These tools help the users from local to global levels, in which the geo-tagging is one of the marvelous tools. It provides efficient management of fields. The crops are very sensitive to spatiotemporal changes due to which they can be affect by certain diseases and fungal infections [19, 20]. The AI-based tools can have the ability to monitor the inch-to-inch progress of a crop to feed the nutrition's and fertilizer. As per the need of the hour, the precision and accuracy can be attained by the machine learning tools. As per research data, every year million tons of resources are wasted due to lack of knowledge fault analysis [21–26]. The IoT, and machine learning tools are the best alternatives to improve the agriculture sector. In upcoming years, the experimental results of crops will improve the profits in local and global markets [27–30]. The IoT and AI will also assist for policies of agriculture reforms. Various countries are investing in these technologies to improve in the yield management [31, 32]. These cutting-edge technologies have versatility to

DOI: 10.1201/9781003643173-1

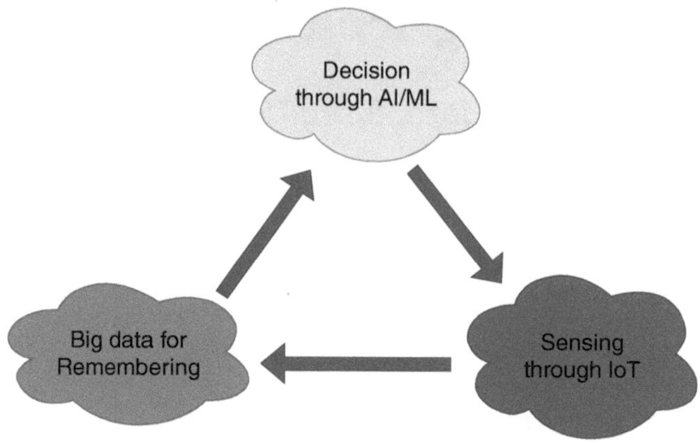

Figure 1.1 Latest trends of agriculture using AI and IoT.

make agriculture sector smarter and profitable. Figure 1.1 represents the latest trends in agriculture sector. Different kinds of sensors are available, which can process the collected chunk of data-based information to give on-the-spot decisions to certain kinds of actuators. The integration of these technologies may improve the decision taking capability in terms of feeding water, management of temperature, and humidity. Overall, the IoT, AI, and ML are very advanced levels of tools which are helpful in transforming the agriculture sector for betterment of humankind [33–36].

1.2 CROP MONITORING AND DISEASE DETECTION

Precision and smart farming require crop monitoring and disease detection. Early crop disease detection and timely crop health monitoring can have a major impact on yield. It reduces the need for heavy pesticides. Crop monitoring and disease detection use a variety of technologies:

Sensors: Temperature, humidity, and soil moisture sensors all offer useful information for tracking environmental conditions. These sensors aid in ensuring that crops are grown in ideal conditions of humidity and temperature, as well as the proper amount of water.

Satellite imagery: Satellites take high-resolution pictures of fields, enabling extensive crop health monitoring. The data from satellite images can reveal changes in plant development that may be due to disease or nutrient deficiency. The smart agriculture AI model is depicted in Figure 1.2.

Thermal imaging drones, multi-spectral drones, and unmanned aerial vehicles can provide comprehensive information about crop health and

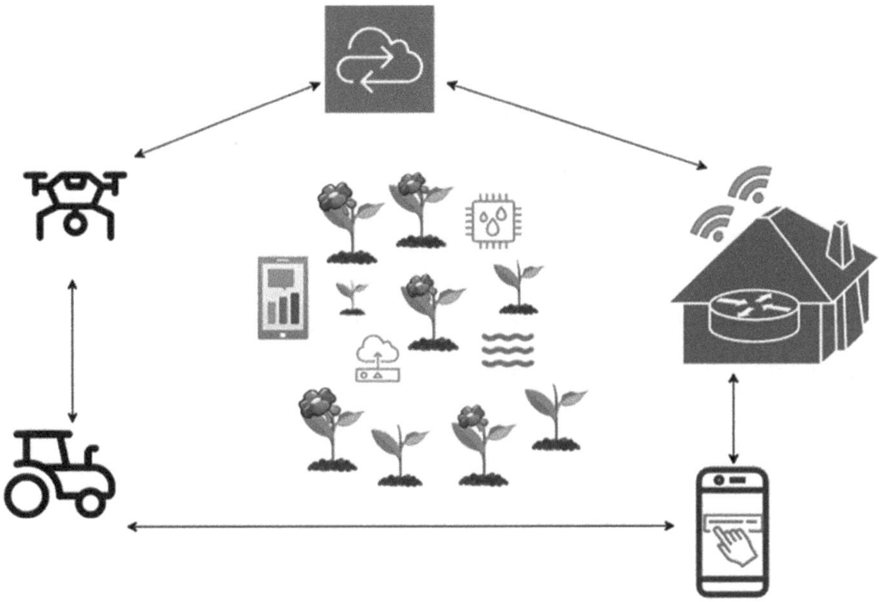

Figure 1.2 Smart AI-based agriculture.

abnormalities. Analysis of aerial photos and data can help find the problems in the field. Machine learning and AI systems have been trained to analyze the data from a wide range of sources, which includes IoT sensors, drones, and satellite photography, to find the pattern related to both good and ill harvests.

By using sensor data, machine learning algorithms are able to identify flaws and give an alert. The Normalized Difference Vegetation Index (NDVI) and spectral imaging measure how much light plants reflect at various wavelengths. Stress or illness can change a plant's spectral signature. Imaging and leaf analysis technologies enable the collection of high-resolution images of individual leaves, which can then be analyzed to gain a comprehensive understanding of the plant's overall health. Real-time temperature, humidity, and precipitation data can be integrated into disease prediction models, and weather data can be used to detect the conditions that contribute to disease epidemics. Ground-based sensors monitor the microclimate that is closest to crops to provide information on specific conditions in the area. The sensors are frequently used with meteorological data to assess the risk of illness.

Efficient crop monitoring and early disease diagnosis are essential for minimizing crop losses and maintaining sustainable methods of farming. With these tools and methods, farmers can now make data-driven decisions, make the most of their resources, and protect their crops from diseases and pests.

1.3 WEATHER FORECASTING AND RISK MITIGATION

In contemporary agriculture, weather speculation and risk alleviation are crucial features, assisting cultivators in the allocation of resources, making well-informed decisions, and adjusting to weather patterns. Figure 1.3 illustrates the parts of the smart agricultural model. Here is a brief summary of how the weather prediction is used in agriculture and how it plays a role in avoiding risks.

1.3.1 Accurate weather prediction in agriculture

By directing farmers' decisions regarding sowing, watering, reaping, and pest management, meteorological forecasts play a vital role. With access to up-to-date data, they may boost the quantity of produce, mitigate probable dangers, and work in a more efficient manner. Meteorological forecasts have become immensely detailed, providing precise data that is tailored to each cultivator's field. Accurate data on temperature, humidity, and precipitation

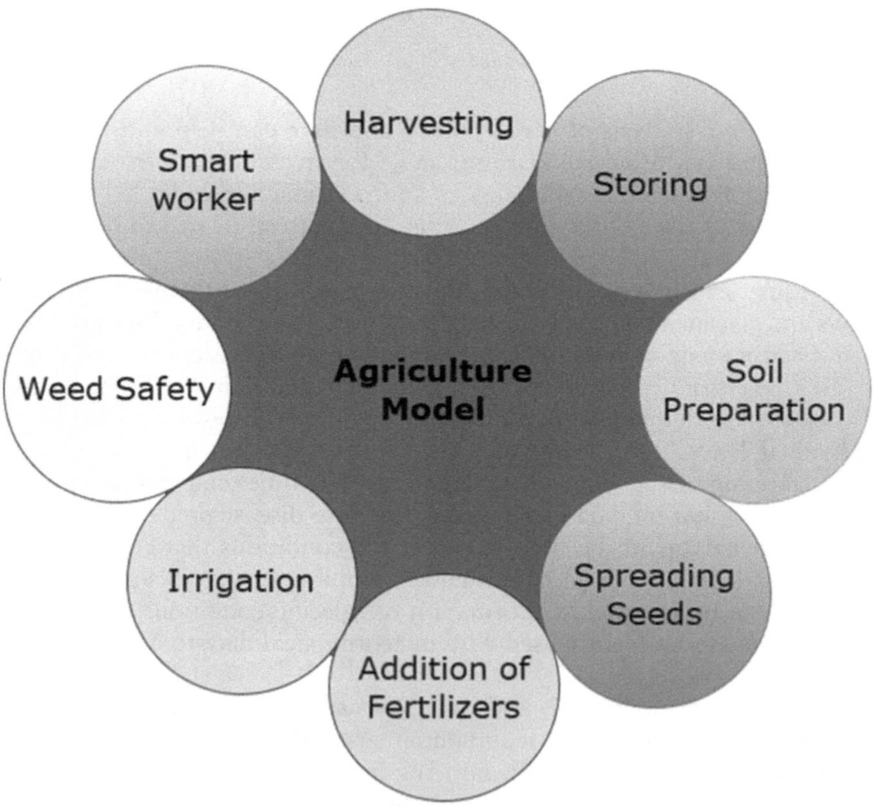

Figure 1.3 Component parts of the model used for smart agriculture.

for specific locations is procured through microclimate forecasting. For the entirety of the cultivation period, predictions aid farmers in working out strategies, including the selection of suitable crops and irrigation methods in the long term. Farmers may effectively revise their plan of action by closely monitoring variable conditions and employing real-time weather information.

1.3.2 Factors for reduction in agriculture

Farmers should consider drought mitigation to ensure that they can promptly implement water-saving measures, such as efficient irrigation techniques, when early warnings of an approaching drought are released. This will help minimize crop stress and wastage.

Farmers can take the necessary steps for probable flooding occurrences by using flood risk evaluations based on weather predictions. Implementing raised beds or altering planting schedules can be accomplished to prepare for potential flooding incidents. Weather predictions can benefit treatments for illnesses and pests. Forecasts can also point to favorable circumstances for the spread of infections, illnesses, or infestations. Farmers who implement early intervention strategies or take preventive action can reduce their reliance on chemical administration.

It is the duty of the farmer to make sure that the crops are provided with the right amount of moisture. This involves using cutting-edge technologies that are connected to weather forecasts. Excess irrigation is protected, which preserves water resources. Farmers can reduce the adverse effects of frost by covering crops beforehand and using frost protection equipment.

Forecasts during times of extreme heat help farmers in decision-making regarding irrigation schedules and whether to provide shade to crops that are vulnerable to damage caused by the heat. Conditions: We refer to this method as "extreme temperature management."

To be able to protect from hail, meteorological radar devices are crucial because they offer warnings in advance of oncoming hailstorms. The impact of the hailstorm can be minimized by using protective measures like utilizing netting.

Insurance policies related to weather are quite significant. Farmers can be compensated when they suffer losses due to adverse weather conditions. Market analysis is affected by the way weather impacts the agriculture pricing.

By adjusting their planting and harvesting schedules in line with price expectations, farmers may maximize their profits. Precision agricultural systems employ weather data to maximize resource usage. Applications of variable rate technology (VRT) can be modified in response to anticipated weather. By limiting environmental impact and protecting resources, accurate weather forecasts, risk mitigation, and environmental conservation all contribute to ecologically responsible agricultural practices.

1.4 LIVESTOCK MONITORING

Keeping an eye on the well-being, contentment, and conduct of livestock raised on farms or ranches is an essential part of modern agriculture. A close eye is necessary to prevent disease outbreaks, maintain the welfare of livestock, and increase productivity. Figure 1.4 displays the most recent advancements in smart farming innovation. four points. There are many tools and methods used in livestock monitoring, including.

Sheep and cattle, for instance, are equipped with wearable sensors that are affixed to them in order to collect data on their movements, body temperatures, and activity levels. Information collected by wearable sensors can be utilized to deduce specific aspects of an animal's behavior and well-being. GPS and radio-frequency identification (RFID) tags are used to track animals' activities and locations. They are particularly useful in regulating grazing patterns and preventing animal theft.

Remote monitoring cameras use live video feeds from cameras positioned strategically in barns, pastures, and feeding areas to enable farmers to keep a close eye on the behavior and health of their livestock.

It is beneficial for the steps involved in foaling or calving. Animal facilities are equipped with IoT sensors for environment tracking. These sensors

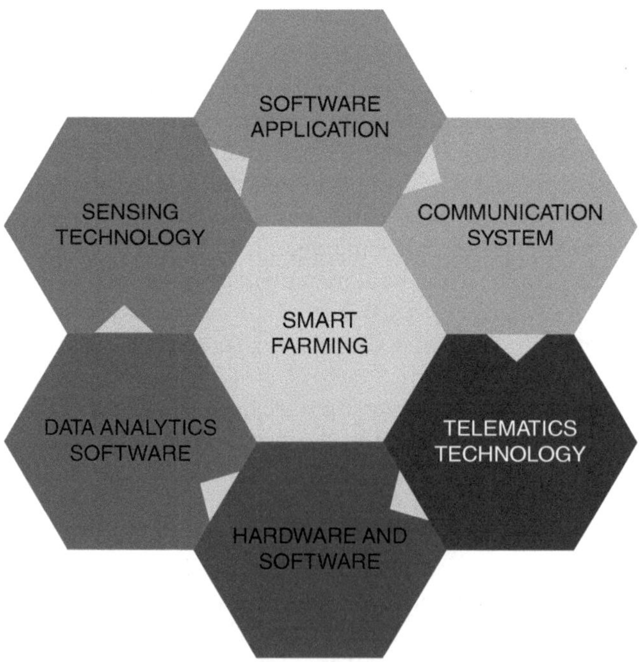

Figure 1.4 Smart farming trends.

detect factors like moisture content, humidity, temperature, and air quality. IoT sensors improve animals' well-being and cleanliness.

Wearable sensor data is analyzed utilizing AI and machine learning addresses to assess cow's behavior. The goal of these sensors is to find any unusual behavior patterns. The abnormal activities of animals may be an indication of a disease or physical distress. Sensors and AI-enabled devices can be used to track and assess vital statistics and healthcare data.

Variations in body temperature or heart rate might indicate a disease or injury. Feeding systems with sensors for nutrition and feeding management make sure that animals receive the correct amount of food. The eating habits of animals might provide details regarding their specific nutritional requirements.

Cows are equipped with sensors to enhance the quantity and quality of milk they produce. These sensors provide information to identify health issues and adjust feeding and milking schedules. Livestock movements, births, and deaths could be monitored by using RFID tagging and tracking devices, and this enables more accurate stock management. Livestock monitoring systems that use health data and pest and disease detection tools can facilitate the identification of disease outbreaks at early stages. Spread of infections can be prevented by taking quick actions.

Environmental monitoring systems can assess the condition of a livestock region using variables such as dust levels and gas concentrations. This benefits both farmworkers and animals. Keeping track of livestock is crucial for maintaining animal welfare and also for improving farm efficiency and productivity. It aids farmers in reducing disease outbreaks, optimizing breeding and feeding programs, and promptly identifying and resolving issues.

1.5 BENEFITS OF AI AND IOT INTEGRATION

The convergence of IoT with AI provides numerous advantages across diverse sectors such as manufacturing, healthcare, agriculture, and smart cities. Figure depicts an IoT-enabled smart agricultural cycle. There are five points. The primary benefits of integrating AI and the IoT are as follows: Data-driven decision-making is the use of AI to analyze the large volumes of real-time data collected by IoT sensors and transform it into practical and useful insights. Businesses can utilize this tool to make educated decisions that optimize operational efficiency and facilitate data-driven decision-making. This automation leads to increased productivity and a decrease in breakdowns, resulting in improved efficiency. AI-driven agricultural automation enhances irrigation and improves resource utilization. Figure 1.5 is a cycle of smart farming that is based on the IoT. Improved Security and Safety: IoT sensors and AI-powered systems can enhance public safety in smart cities through the implementation of real-time surveillance, predictive

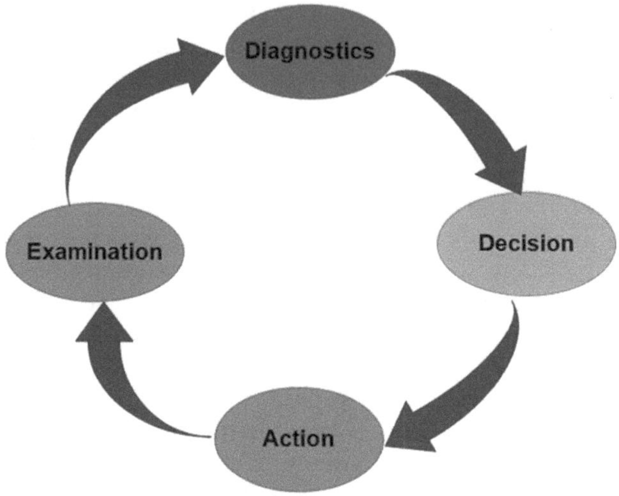

Figure 1.5 IoT smart farming cycle framework.

policing, and video analytics. AI has the capability to take appropriate action in industrial settings.

Environmental sustainability: Sustainable practices can also be implemented through AI and IoT technologies. For instance, by optimizing energy usage, carbon emissions are reduced. Smart irrigation in agriculture minimizes the impact on the environment and wastes less water.

Cost savings: Through process and resource optimization, IoT and AI can lower operational costs. For example, predictive maintenance can save maintenance costs and budget on medical equipment used in healthcare.

More output quality and productivity are ensured by AI-driven quality control systems, which can identify defects immediately. Machine uptime and productivity in manufacturing are increased by predictive maintenance.

Resource optimization: AI can assist in optimizing these resources, lowering waste and expenses. IoT sensors can track resource usage, such as energy and water consumption. IoT-based smart energy-efficient platform for crop monitoring is shown in Figure 1.6. AI can improve inventory optimization and demand forecasting in supply chain management, saving waste and increasing effectiveness.

Healthcare and patient monitoring: Personalized treatment plans and remote patient monitoring are made possible by IoT sensors and AI systems, which enhance healthcare outcomes and lower readmission

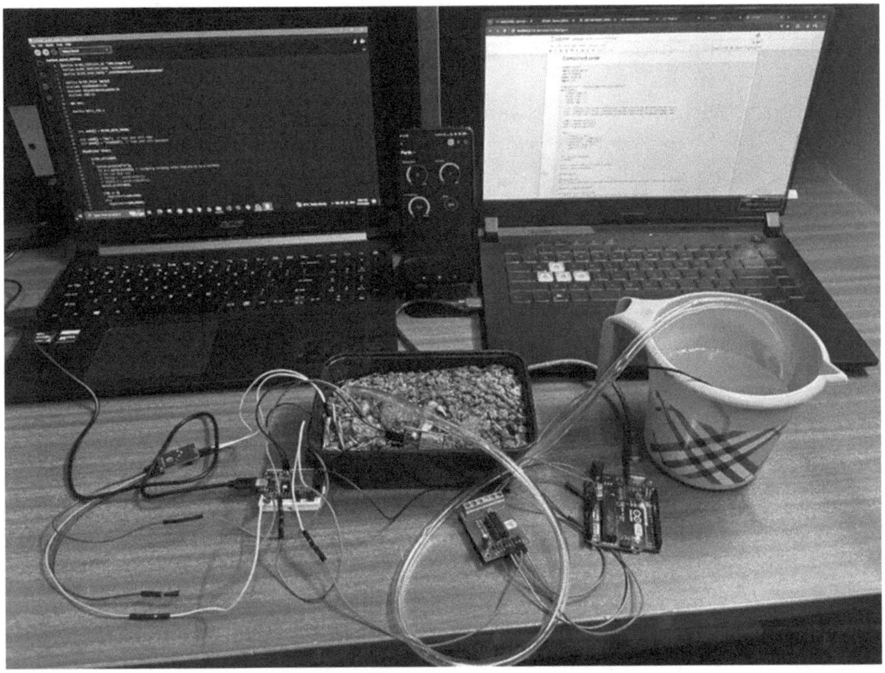

Figure 1.6 Smart energy-efficient platform for crop monitoring.

rates to hospitals. Wearable technology can monitor health indicators and give early alerts for medical problems.

Supply chain visibility: AI analytics and IoT tracking systems offer end-to-end supply chain visibility, assisting businesses in locating bottlenecks and optimizing logistics. Real-time tracking lowers the chance of product loss or theft while increasing transparency.

Predictive analytics: AI can foresee illness outbreaks, traffic jams, equipment failures, and more, enabling businesses to take preventive action. Demand forecasting and inventory management are aided by predictive analytics.

Customization and personalization: AI and IoT facilitate customized user experiences in e-commerce, smart homes, and content recommendations. Crop and resource management in agriculture can be customized to regions of a field with precision agriculture.

Scalability: Because IoT and AI solutions are so scalable, they can be used for both large industrial operations and small farms or smart homes.

Competitive advantage: Businesses that use IoT and AI to their advantage are able to provide data-driven, creative, and effective solutions.

Public services and smart cities: AI and IoT are used in smart city applications to improve waste collection, traffic management, and emergency response times, which raises people's quality of life in general.

Health and Well-Being: Telemedicine, telehealth, and remote patient monitoring are made easier and more accessible in the healthcare industry by IoT and AI.

Real-time Monitoring: IoT sensors enable real-time monitoring, which enables businesses to react quickly to emergencies and changing conditions.

Speedier Problem Resolution: AI and IoT technologies have the ability to automatically identify problems and notify the appropriate parties, which speeds up the process of solving problems.

1.6 INCREASED CROP YIELDS

A key objective of agriculture is raising crop yields in order to satisfy the rising demand for food on a worldwide scale. To increase crop productivity, farmers and other agricultural professionals can use a variety of tactics and procedures. Process of precision agriculture is shown in Figure 1.7. The following are some essential elements and techniques for raising crop yields:

Better crop selection: You can greatly increase yields by choosing high-yielding crop varieties that are compatible with the local soil, climate, and pest situation. Modern methods of crop breeding, such as genetic modification, have produced crops with higher yields.

Soil management: Crop growth depends on the health of the soil. Determining pH levels and nutrient deficits can be achieved through soil testing and analysis. Plant nutrient availability can be increased by applying the appropriate number of fertilizers and soil conditioners in accordance with the findings of soil tests.

Precision agriculture: Accurate planting, fertilization, and irrigation are made possible by the application of precision farming techniques, such as VRT and GPS-guided equipment. By doing this, resource waste is reduced, and crop yields are increased.

Optimized irrigation: Soil moisture sensors and drip irrigation are two efficient strategies that ensure crops receive the appropriate amount of water at the right time. Both excessive and insufficient irrigation can lead to decreased crop production.

Effective treatment of both diseases and pests depends on early diagnosis. Effective methods of lowering disease rates include integrated pest management (IPM) techniques, including biological control, resistant crop varieties, and sparing use of pesticides. The life cycles of pests and illnesses can be efficiently disrupted by implementing crop rotation and improving crop diversity.

Effective weed control is necessary to keep weeds from outcompeting crops for resources. There are several ways to do this, including mulching,

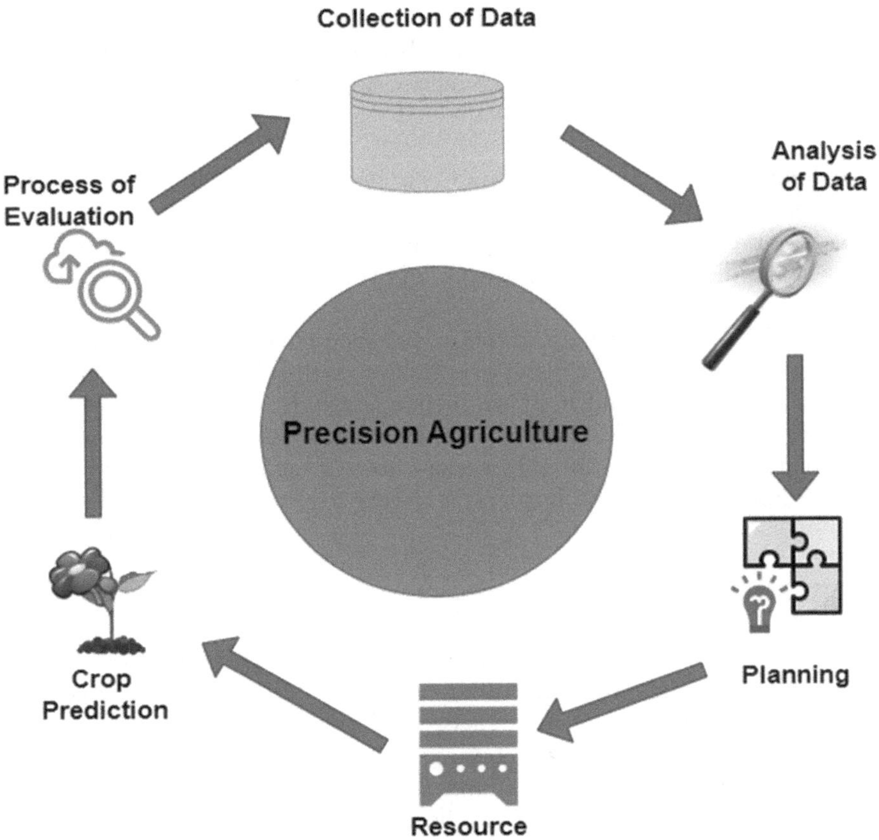

Figure 1.7 Process of precision agriculture.

mechanical cultivation, and the use of herbicides. Reduce the detrimental impact of weeds through the use of cover crops and crop rotation. To guarantee that every plant receives the proper quantity of sunshine, moisture, and nutrients, plants must be spaced optimally. Crop output can be decreased by both overcrowding and underplanting.

Seasonal planning: Increasing agricultural output requires making the most of crop production and harvesting. If farmers have access to optimal growth conditions and seasonal patterns, they should arrange their planting schedules accordingly. Utilizing Cover Crops: Planting cover crops in the soil during periods of low agricultural output improves the soil's structure, lowers erosion, and increases the amount of organic matter in the soil. Agricultural production will rise in the next years with the application of these tactics. Providing fertilizers at right time and with right amounts can ensure crops receive the nutrients for optimum development of crops. In this case, soil testing and the use of nutrient management strategies could be beneficial. Climate-resilient crops are chosen specifically to be resistant to drought or other adverse impacts of climate change. Using this strategy the agricultural productivity will remain unaffected regardless of challenges faced by changing climate.

Pollination management: Pollinators are essential for crops whose yields rely on them.

Post-harvest management: To minimize post-harvest losses, crops must be handled, stored, and transported properly.

Farmers should be educated and trained in modern technology and advanced farming techniques, which result in good crop yields. Landowners can make data-based decisions and improve their operations by using modern technologies like AI, the IoT, and precision farming. By promoting environmentally conscious and sustainable agriculture through constant monitoring and method improvement, farmers can boost crop yields.

1.7 RESOURCE EFFICIENCY

Resource optimization is the process of utilizing resources more wisely in order to increase production and reduce waste. This idea significantly impacts several areas, including energy, industry, agriculture, and environmental management. Resource optimization saves costs, lessens environmental impact, and enhances resource use.

To achieve resource efficiency, the following policies and practices should be:

Efficient Distribution of Resources: The efficient use of natural resources, such as raw materials, energy, land, and water, is necessary for resource efficiency. Minimizing waste and losses at every stage of a resource's life cycle is crucial.

Adopting the circular economy approach offers a ray of hope: By promoting conserving assets through effective reuse, recycling, and refurbishment, this approach drastically reduces waste and diminishes our dependency on resources.

Energy efficiency, which reduces energy consumption through the use of energy-efficient devices and methods, is an essential part of resource efficiency. It's important to stress the impact of modernizing lighting systems, along with utilizing energy-efficient appliances and streamlining industrial procedures, in achieving energy efficiency.

In order to preserve water resources, water conservation entails using water-saving techniques in all facets of life, including everyday tasks, industrial operations, and agricultural practices. Water recycling, rainwater collecting, and drip irrigation are common techniques.

To reduce material waste, material efficiency is essential in the manufacturing and construction sectors. This may be achieved by lowering material waste in the production process and producing durable, easily recyclable products. The goal of the clean manufacturing concept is the elimination of waste from the fabrication processes. It is pivotal to decrease excess inventory, streamline procedures, and increase yield to achieve the proposed results.

Precision agriculture minimizes waste generation and adverse effects on the environment, while precise sowing, irrigation, and fertilization approaches maximize resource utilization. Precision agriculture will use more inputs in accordance with certain requirements. It is possible to reduce greenhouse gases remarkably by shifting from fossil fuels like hydrocarbons to renewable energy sources like solar power, wind energy, and hydroelectricity. The process of extracting valuable materials and chemicals from waste streams in order to increase resource efficiency is known as resource recovery. Two cases of resource recovery are recycling metals and turning organic waste into compost or biogas. Using long-lasting, limiting packaging, and adopting a zero-waste mentality are some of the strategies to cut waste. By following these steps, the amount of garbage that ends up in landfills can be reduced, and resources can be protected.

Electric cars and various low-emission industrial processes are a few examples of eco-friendly technologies. Whose idea is it to reduce environmental harm and resource waste?

1.8 REDUCED ENVIRONMENTAL IMPACT

Preventing environmental harm is an important objective in numerous sectors, including energy, transportation, production, and agriculture. These sectors need to lessen the damage they cause to the environment and natural resources.

Sustainable practices

Agriculture: Sustainable farming techniques protect soil, lessen water pollution, and maintain habitats. These techniques include crop rotation, organic farming, and agroforestry. High-quality agriculture reduces harm to the environment while improving resource efficiency.

Energy: Switching from fossil fuels to renewable energy sources like wind, solar, and hydropower can reduce air pollution and greenhouse gas emissions. Enhancing transportation, commercial, and building energy efficiency reduces energy use and associated environmental impact.

Waste management: Recycling, composting, and conscious consumerism all help reduce waste and conserve resources by keeping less rubbish in landfills. Under Extended Producer Responsibility (EPR), manufacturers are held responsible for product disposal.

Industry: Acquiring clean technologies and green environment industries can reduce pollution in the environment.

Sustainable transportation: Carbon gases and pollution can be decreased by advertising electric vehicles, bicycles, and public transportation uses.

Water conversation: Water scarcity can be reduced by developing rainwater storage management and introducing pit holes.

Eco-friendly products and production: RRR should be followed: reduce, reuse, and recycle. By following this, the environment will be in safe hands.

Awareness regarding environmental education: Industries can advertise regarding environmental education and be responsible toward it.

Rules and regulations: Public authorities will ensure that the public follows rules and regulations to protect them from hazardous gases and pollution.

Carbon management: Carbon offsetting and reducing carbon emissions through reforestation and renewable energy initiatives are vital for mitigating climate change.

Resource management: Sustainable practices in forestry and fisheries ensure the long-term availability of these valuable resources while protecting ecosystems.

Collaboration: Addressing global environmental challenges requires cooperation among nations and stakeholders. International agreements and partnerships are essential for achieving shared goals. By implementing these strategies, we can work toward a healthier planet and a sustainable future for generations to come.

1.9 CHALLENGES AND CONSIDERATIONS

To ensure a sustainable future, it's imperative to address environmental challenges. Mitigating the harmful impacts of human activities on the planet requires focused efforts to resolve these issues.

Key environmental challenges

Climate change: The escalating buildup of greenhouse gases threatens global stability. Transitioning to renewable energy, energy efficiency, and sustainable land use are critical steps to mitigate its impacts.

Loss of biodiversity: Declining biodiversity jeopardizes ecosystems and their essential services. Conservation efforts and sustainable land management are crucial to protect our planet's rich variety of life.

Resource scarcity: Finite resources like minerals and fossil fuels are diminishing. Adopting circular economy principles, recycling, and responsible resource management are essential for future generations.

Pollution of the air and water: Finite resources like minerals and fossil fuels are diminishing. Adopting circular economy principles, recycling, and responsible resource management are essential for future generations.

Waste management: Improper waste disposal harms the environment. Promoting recycling, reducing waste generation, and exploring waste-to-energy options are vital for sustainable waste management.

Environmental degradation: Unsustainable land use, deforestation, and habitat loss threaten ecosystems. Sustainable land management practices and reforestation can help restore and protect our natural environment.

Water scarcity: Many regions face severe water shortages. Water recycling, efficient wastewater treatment, and water conservation are essential for ensuring adequate water supplies.

Fishery depletion: Overexploitation of marine resources endangers marine ecosystems. Implementing sustainable fishing practices and robust fishery management is crucial for ocean health.

Degradation of the soil: Deteriorating soil quality impacts food production and ecosystem health. Soil conservation and sustainable farming methods are essential for maintaining soil fertility.

Hazardous waste disposal: Improper disposal of hazardous waste contaminates land and water. Strict regulations and proper handling practices are necessary to protect human health and the environment.

Regulatory compliance: Commitment to environmental laws and standards is crucial for businesses and industries. Non-compliance can lead to severe financial and reputational consequences.

Environmental education and awareness: Promoting moral conduct and decision-making requires educating people and organizations about environmental challenges. The goal of sustainable development is to take these factors into account.

Governments, corporations, communities, and individuals must all play a part in the multifaceted approach needed to address these environmental issues. Given the interdependence of environmental and social issues, sustainability initiatives should place a high priority on responsible resource management, environmental conservation, and the welfare of future generations.

1.10 CONCLUSION

AI, ML, and IoT are the future generation technologies to upscale the revolution in the agriculture field. There is still scarcity in the development of such applications and their implementation on the micro level. In future, these technologies help the nations to procure the stock of raw food for any kind of situation. AI techniques are used to analyze the massive volumes of data produced by IoT devices in smart agriculture with machine learning and deep learning. AI can recognize trends, forecast outcomes, and provide useful information. AI, for instance, can optimize irrigation schedules, forecast crop disease outbreaks, and suggest the best times to plant and harvest crops. So, the research communities have diverse areas to explore and create the opportunities for progression in the agriculture sector. The integration of IoT, AI, and Machine learning tools helps to capture the data. Further, based on trends of the data, certain decisions can be taken by the machines to manage the overall field. The IoT helps to access the data from certain remote locations to the main node for processing and computing processes. The Machine learning tools contain certain kinds of algorithms and methods to classify and predict the trends and sustain the progress.

REFERENCES

1. Guo, Q., Ren, M., Wu, S., Sun, Y., Wang, J., Wang, Q., Ma, Y., Song, X., Chen, Y., "Applications of artificial intelligence in the field of air pollution: a bibliometric analysis," *Front. Public Health*, vol. 10, pp. 933665, 2022.
2. Adli, H. K., Remli, M. A., Salihin, W., Wong, K. N. S., Ismail, N. A., González-Briones, A., Corchado, J. M., Mohamad, M. S., "Recent Advancements and Challenges of AIoT Application in Smart Agriculture: A Review," *Sensors*, vol. 23, pp. 3752, 2023.
3. Adhikari, M. S., Majumder, A., Sheifudeen, A. M., Reddy, K. H., Sidhwani, M., Verma, Y. K., "High Gain Farming and Data Analytics using IoT", *IEEE International Conference on Devices for Integrated Circuit (DevIC)*, pp. 158–162, 2023.
4. Davenport, T., Kalakota, R., "The potential for artificial intelligence in healthcare," *Future Health J.*, vol. 6, pp. 94–98, 2019.
5. Kaloxylos, A., Wolfert, J., Verwaart, T., Terol, C. M., Brewster, C., Robbemond, R., Sundmaker, H., "The use of future internet technologies in the agriculture and food sectors: integrating the supply chain," *Proc. Technol* vol. 8, pp. 51–60, 2013.
6. Tripathi, S. L., Mahmud, M., *Explainable Machine Learning Models and Architecture*, Wiley & Scrivener Publishing LLC, USA, 2023.
7. Rajak, P., Roy, S., Pal, A. K., Paramanik, M., Dutta, M., Podder, S., Sarkar, S., Ganguly, A., Mandi, M., Dutta, A., Das, K., Ghanty, S., Khatun, S., "In silico study reveals binding potential of rotenone at multiple sites of pulmonary surfactant proteins: a matter of concern," *Curr. Res. Toxicol*, vol. 2, pp. 411–423, 2021.
8. Joshi, V., Adhikari, M. S., *Advanced Agricultural-Based IoT Technology*, Intelligent Communication and Automation Systems, CRC Press, pp. 239–254, 2021.
9. Joshi, V., Adhikari, M. S., "IOT-based technology for smart farming", *Electronic Devices and Circuit Design: Challenges and Applications in the Internet of Things*, CRC Press, pp. 223–242, 2022.

10. Ghanty, S., Mandi, M., Ganguly, A., Das, K., Dutta, A., Nanda, S., Biswas, G., Rajak, P., "Lung surfactant proteins as potential targets of prallethrin: an in silico approach," *Toxicol. Environ. Health Sci*, vol. 14, pp. 89–100, 2022.
11. Navarro, E., Costa, N., Pereira, A., "A systematic review of IoT solutions for smart farming," *Sensors*, vol. 20, pp. 4231, 2020.
12. Gao, D., Sun, Q., Hu, B., Zhang, S., "A framework for agricultural pest and disease monitoring based on internet-of-things and unmanned aerial vehicles," *Sensors*, vol. 20, pp. 148, 2020.
13. Lan, L., Le, X., Dong, H., Xie, J., Ying, Y., Ping, J., "One-step and large-scale fabrication of flexible and wearable humidity sensor based on laser-induced graphene for real-time tracking of plant transpiration at bio-interface," *Biosens. Bioelectron*, vol. 165, pp. 112360, 2020.
14. Yang, L. B., "Application of artificial intelligence in electrical automation control," *Procedia Comput. Sci.*, vol. 166, pp. 292–295, 2020.
15. Jha, K., Doshi, A., Patel, P., Shah, M., "A comprehensive review on automation in agriculture using artificial intelligence," *Artif. Intell. Agric*, vol. 2, pp. 1–12, 2019.
16. Ashwinkumar, S., Rajagopal, S., Manimaran, V., Jegajothi, B., "Automated plant leaf disease detection and classification using optimal MobileNet based convolutional neural networks," *Mater. Today Proc.*, vol. 51, pp. 480–487, 2022.
17. Taneja, P., Vasava, H. K., Daggupati, P., Biswas, A., "Multi-algorithm comparison to predict soil organic matter and soil moisture content from cell phone images," *Geoderma*, vol. 385, pp. 114863, 2021.
18. Patel, R., Adhikari, M. S., Tripathi, S. K., Sahu, S., "Design, optimization and performance assessment of single port film bulk acoustic resonator through finite element simulation," *Sensors*, vol. 23, pp. 8920, 2023.
19. Khaki, S., Wang, L., Archontoulis, S. V., "A CNN-RNN framework for crop yield prediction," *Front. Plant Sci.*, vol. 10, pp. 1750, 2020.
20. Lu, B., Dao, P., Liu, J., He, Y., Shang, J., "Recent advances of hyperspectral imaging technology and applications in agriculture," *Remote, Sensors*, vol. 12, pp. 2659, 2020.
21. Lippitt, C. D., Stow, D. A., Riggan, P. J., "Application of the remote-sensing communication model to a time-sensitive wildfire remote-sensing system," *Int. J. Rem. Sens.*, vol. 37, pp. 3272–3292, 2016.
22. Shepherd, M., Turner, J. A., Small, B., Wheeler, D., "Priorities for science to overcome hurdles thwarting the full promise of the 'digital agriculture' revolution," *J. Sci. Food Agric.*, vol. 100, pp. 5083–5092, 2020.
23. Yarlagadda, N., Verma, Y. K., Adhikari, M. S., and Mishra, V., "Opto-VLSI Devices and Circuits for Biomedical and Healthcare Applications," *A Review of Multi-material, Multi-gate MOSFET Structures*, CRC Press, pp. 29–40, 2023.
24. Shafi, U., Mumtaz, R., Iqbal, N., Zaidi, S. M. H., Hussain, I., Mahmood, Z., "A Multimodal approach for crop health mapping using low altitude remote sensing. Internet of Things (IoT) and machine learning," *IEEE Access*, vol. 8, pp. 112708–112724, 2020.
25. Ikram, A., Aslam, W., Aziz, R. H. H., Noor, F., Mallah, G. A., Ikram, S., Ahmad, M. S., Abdullah, A. M., Ullah, I., "Crop yield maximization using an IoT-based smart decision", *J. Sens.*, vol. 22, pp. 2022923, 2020.

26. Rezvani, S. M., Abyaneh, H. Z., Shamshiri, R. R., Balasundram, S. K., Dworak, V., Goodarzi, M., Sultan, M., Mahns, B., "IoT-based sensor data fusion for determining optimality degrees of microclimate parameters in commercial greenhouse production of tomato," *Sensors*, vol. 20, pp. 6474. 2020.
27. Sagheer, A., Mohammed, M., Riad, K., Alhajhoj, M., "A cloud based IoT platform for precision control of soilless greenhouse cultivation," *Sensors*, vol. 21, pp. 223, 2021.
28. Kiani, F., Seyyedabbasi, A., "Wireless sensor network and internet of things in precision agriculture," *Int. J. Adv. Comput. Sci. Appl.*, vol. 9, pp. 220–226, 2018.
29. Umapathy, P., Hemanth, J., Khera, S., Inbamani, A., Tripathi, S. L., *Machine Learning and the Internet of Things in Solar Power Generation*, CRC Press Boca Raton, UK, 2023.
30. O'Malley, L., Bronson, A., van der Burg, K., Klerkx, S. L., "The future (s) of digital agriculture and sustainable food systems: an analysis of high-level policy documents," *Ecosyst. Serv.*, vol. 45, pp. 101183, 2020.
31. Baghel, S. S., Rawat, P., Singh, R., Akram, S. V., Pandey, S., Baghel, A. S., "AI, IoT and cloud computing based smart agriculture," *International Conference on Contemporary Computing and Informatics (IC3I)*, Uttar Pradesh, India, pp. 1658–1661, 2022.
32. Adhikari, M. S., Alkhayyat, A. H., "Prediction and classification for smart grid applications," *Data Analytics for Smart Grids Applications-A Key to Smart City Development*. Intelligent Systems Reference Library, Springer, vol. 247, pp 87–102, 2023.
33. Adhikari, M. S., Thakur, N., Malik, P. K., "Advanced digital twin technology: opportunity and challenges," *Data Analytics for Smart Grids Applications—A Key to Smart City Development*. Intelligent Systems Reference Library, Springer, vol. 247, pp 233–250, 2023.
34. Wolfert, S., Goense, D., Sørensen, C. A. G., "A future internet collaboration platform for safe and healthy food from farm to fork," *Annual SRII Global Conference*, pp. 266–273, 2014.
35. Dutta, M., Rajak, P., Roy, S., "Determination of chronic median lethal concentration of sodium fluoride in Drosophila melanogaster and exploring effect of sub-lethal concentrations on differential hemocyte count," *Proc. Zool. Soc.*, vol. 72, 2019.
36. Tripathi, S. L., Ganguli, S., Kumar, A., Magradze, T. *Intelligent Green Technologies for Sustainable Smart Cities*, Wiley, USA, 2022, doi: 10.1002/9781119816096

Chapter 2

IoT empower

Your one-stop device manager

Shanky Saxena and Gurmukh Singh

2.1 INTRODUCTION

With the ability to connect a vast array of objects and systems, the Internet of Things (IoT) has become a disruptive force that is enabling a smarter and more efficient world. To achieve seamless automation and control in this ever-expanding IoT world, device coordination and administration are essential. In this centralized IoT system, a Raspberry Pi (RPi) takes on the central role of controlling a suite of diverse devices. In essence, this research contributes significantly to the evolution of IoT technologies [1–6]. By presenting a robust and adaptable solution for centralized device management, it underscores the influential role of the Raspberry Pi 4 as a master device, poised to shape the future landscape of IoT ecosystems [6–8]. In a world where interconnectedness is key, our approach stands as a promising step toward a smarter, more seamlessly integrated future. To create an IoT Empower: Your One-Stop Device Manager using the Raspberry Pi as the master device, equipped with multiple sensors for enhanced and precise data collection, contributing to the advancement of home and industrial automation. This system efficiently combines cutting-edge technology with IoT capabilities to offer a dependable solution [9–10].

2.2 LITERATURE REVIEW

This section gives a complete literature review for the home automation system given in Table 2.1 below.

2.3 RESEARCH METHODOLOGY

Methodology for Building a Centralized IoT Hub using the Raspberry Pi 4: The development of a Centralized IoT Hub with the Raspberry Pi as the master device revolves around a series of key components and processes,

DOI: 10.1201/9781003643173-2

Table 2.1 Literature survey and gaps for home automation system

Paper title	*Authors name*	*Year*	*Methodology*	*Limitations*
A smart TV viewing and web access interface based on video indexing techniques. IEEE Link	H. Kuwano, Y. Taniguchi, K. Minami, M. Morimoto, and H. Kojima by. (n.d.)	2002	The functionality of this project, Joy TV, is to enhance the TV viewing experience by automatically detecting scene changes, music, and superimposed texts in live TV broadcasts in real time. It then uses these indexes to enable users to browse and retrieve TV content more efficiently. Additionally, Joy TV offers a novel method for web access through TV, linking TV content to web pages using image features, all without the need for manually created electronic program guides (EPG).	It faces accuracy, compatibility, resource, and privacy challenges. Superimposed text recognition and web linking effectiveness vary.
Memory access scheduling for a smart TV. IEEE Link		2016	The functionality of this project is to design a specialized memory access scheduler tailored for the demanding computational and memory requirements of smart TV System-on-Chip (SoC) devices. This scheduler optimizes memory utilization for various real-time graphics and user-responsive tasks, ensuring high memory throughput and priority handling, even under heavy memory traffic conditions. Through innovative future prediction and priority management techniques, the project significantly enhances memory performance, achieving up to 98% of the ideal upper bound throughput for smart TVs.	It faces challenges in scenarios with exceptionally intense memory traffic or unforeseen system demands.
Smart TV interaction system using face and hand gesture recognition. IEEE Link	Sang-Heon Lee, Myoung-Kyu Sohn, Dong-Ju Kim, Byungmin Kim, and Hyunduk Kim	2013	The functionality of this project is to create a system for smart TV interaction by recognizing human faces and natural hand gestures. It enables viewer authentication through face recognition and control of smart TV functions, such as adjusting volume and changing channels, through hand gesture recognition. Additionally, the system allows for personalized services like recommending favorite channels or providing parental guidance based on face recognition. The project demonstrates high accuracy in both face and hand gesture recognition, making it effective for enhancing the smart TV user experience.	The limitations include distance sensitivity, lighting variations, potential false detections, and privacy concerns with face recognition for viewer authentication.

When smart devices are stupid: negative experiences using home smart devices IEEE Link	Weijia He, Jesse Martinez, Roshni Padhi, and Lefan Zhang	2019	The functionality of this project is to investigate and understand the negative experiences that users encounter with household smart devices, including issues related to power outages, network failures, false alarms, and user programming concerns. Through an online survey-based study, the project aims to identify these challenges and proposes a research agenda focused on enhancing the transparency and usability of smart devices to provide a safer and more user-friendly experience for device owners.	Limitations include a small sample size, self-reported data, potential response bias, and specificity to certain devices and demographics.
SEED smart pixel devices IEEE Link	A. L. Lentine (n.d.)	1993	The functionality of this project revolves around the development and utilization of optoelectronic processing devices known as "smart pixels." These smart pixels, particularly the quantum well self electro-optic device (SEED), are designed to perform a range of advanced functions beyond basic logic gates. They integrate quantum wells within p-i-n diode structures to enable both light detection and modulation. These devices, including transistor-diode SEEDs like the field effect transistor SEED (FET-SEED), aim to enhance functionality, reduce optical energy requirements, and achieve complex digital and analog operations, thereby advancing optoelectronic processing capabilities.	Limitations include complexity, sensitivity, compatibility, energy consumption, integration challenges, speed constraints, cost, and limited commercial availability.
Smart pixel devices and free-space digital optics applications IEEE Link	A. A. Sawchuk	1995	The functionality of this project involves the integration of optics and electronics in Free-Space Digital Optics (FSDO) systems. These systems use smart pixel nodes with optical detectors and modulators to create high-capacity, parallel interconnections. They employ fixed optical components to establish intricate 3-D interconnection topologies, enhancing data bandwidth, spatial density, and information spatial channel density. FSDO is especially effective for applications requiring high data capacity, spatial channel density, and parallel data transfer, such as communication switching, photonic space switching, parallel computing, and optical neural systems.	Challenges include complex integration, varying error correction efficiency, compatibility with data formats, size constraints, cost considerations, scalability issues, and potential real-world interference.
A Responsive Web Design Approach to Enhancing the User Web Browsing Experience on Smart TVs IEEE Link	E. Perakakis and Ghinea		The functionality of this project is to enhance the user experience of viewing websites on smart TV devices by implementing responsive web design techniques. This approach aims to optimize website layouts and interactions for larger screens, improving usability and task completion times without the need for creating separate TV-specific websites or apps.	It encounters issues with browser support, remote control input, and performance. Content adaptation, standardization, testing, and accessibility are crucial for an effective TV-responsive website.

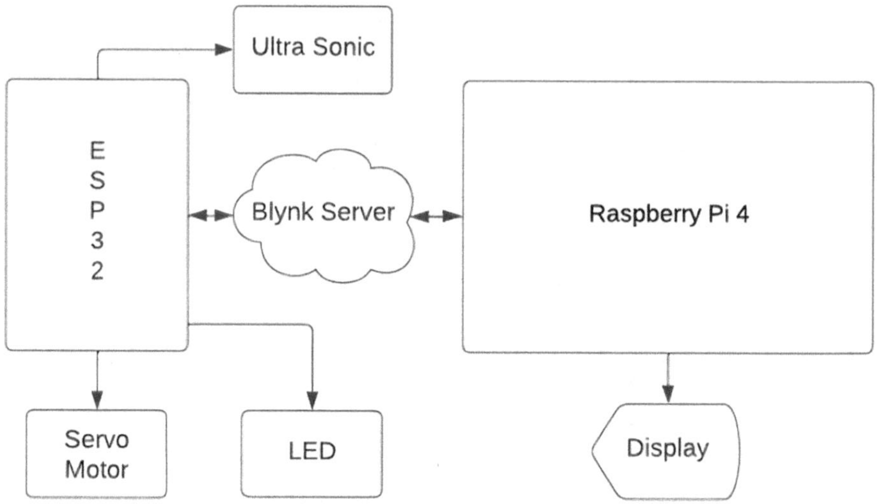

Figure 2.1 Block diagram.

with a primary focus on the master device itself, rather than slave devices, as shown in Figure 2.1.

1. Light-dependent resistor (LDR) for bulb control: The LDR, also referred to as a photoresistor, continuously monitors ambient light levels. The Raspberry Pi periodically reads LDR values through analog or digital interfaces. When the light level crosses a predefined threshold (e.g., transitioning from daylight to darkness), the Raspberry Pi triggers a command to control indoor lighting, either by activating a relay or interfacing with a smart lighting control system. Conversely, when the LDR detects ample light (e.g., daytime conditions), the Raspberry Pi sends a command to turn off the lighting.
2. Servo motor for door control: The servo motor is connected to the Raspberry Pi through an ESP32. The Raspberry Pi receives commands from the user interface or predefined automation rules. For instance, when the user issues a command such as "Open Door," the Raspberry Pi transmits instructions to the ESP32, which, in turn, communicates with the servo motor to rotate and open the door to the desired angle. Similarly, a command like "Close Door" prompts the Raspberry Pi to relay instructions to the servo motor to close the door.
3. Ultrasonic sensor for water tank level monitoring: The ultrasonic sensor is responsible for measuring the distance to the water's surface in a tank. Periodically, the Raspberry Pi 4 activates the ultrasonic sensor to emit an ultrasonic pulse. The time taken for the pulse to bounce back is used to calculate the distance to the water surface. This distance measurement is then translated into a water level reading, which can

be displayed on the user interface or used to trigger predefined actions based on specified thresholds.

4. Users can interact with the system through the user interface to issue commands for controlling the door or checking sensor statuses. The system maintains a comprehensive log of sensor data, which proves valuable for historical analysis and troubleshooting. In summary, the Raspberry Pi 4 serves as the core intelligence of the Centralized IoT Hub, coordinating the functioning of multiple sensors and devices, responding to user inputs, and ensuring the seamless integration of these components to create a smart and automated environment as shown in Figures 2.2–2.4.

2.4 DISCUSSION

The implementation of the centralized IoT hub using the Raspberry Pi as the master device yielded promising results. Throughout our project, we successfully integrated multiple sensors and devices into a cohesive ecosystem, allowing for efficient centralized control. Taking Raspberry Pi to the next level, making Raspberry Pi multipurpose use at the same time. The output images of Master device (Raspberry Pi) and the Salce Device (IoT devices: Servo Motor, Ultra Sonic, LED) are shown from Figures 2.5–2.10.

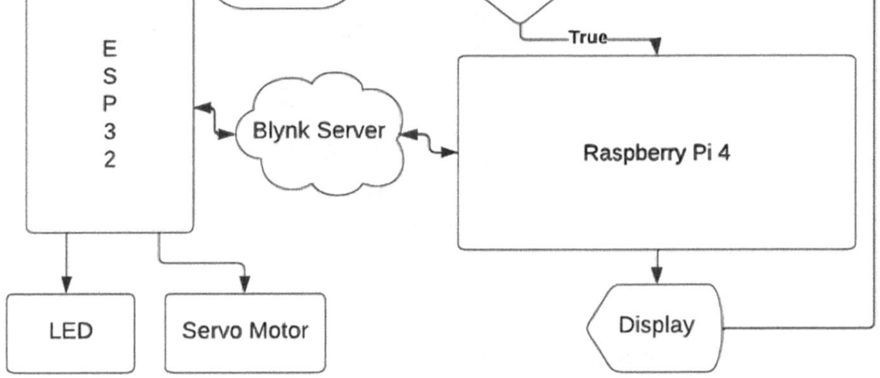

Figure 2.2 Data flow diagram.

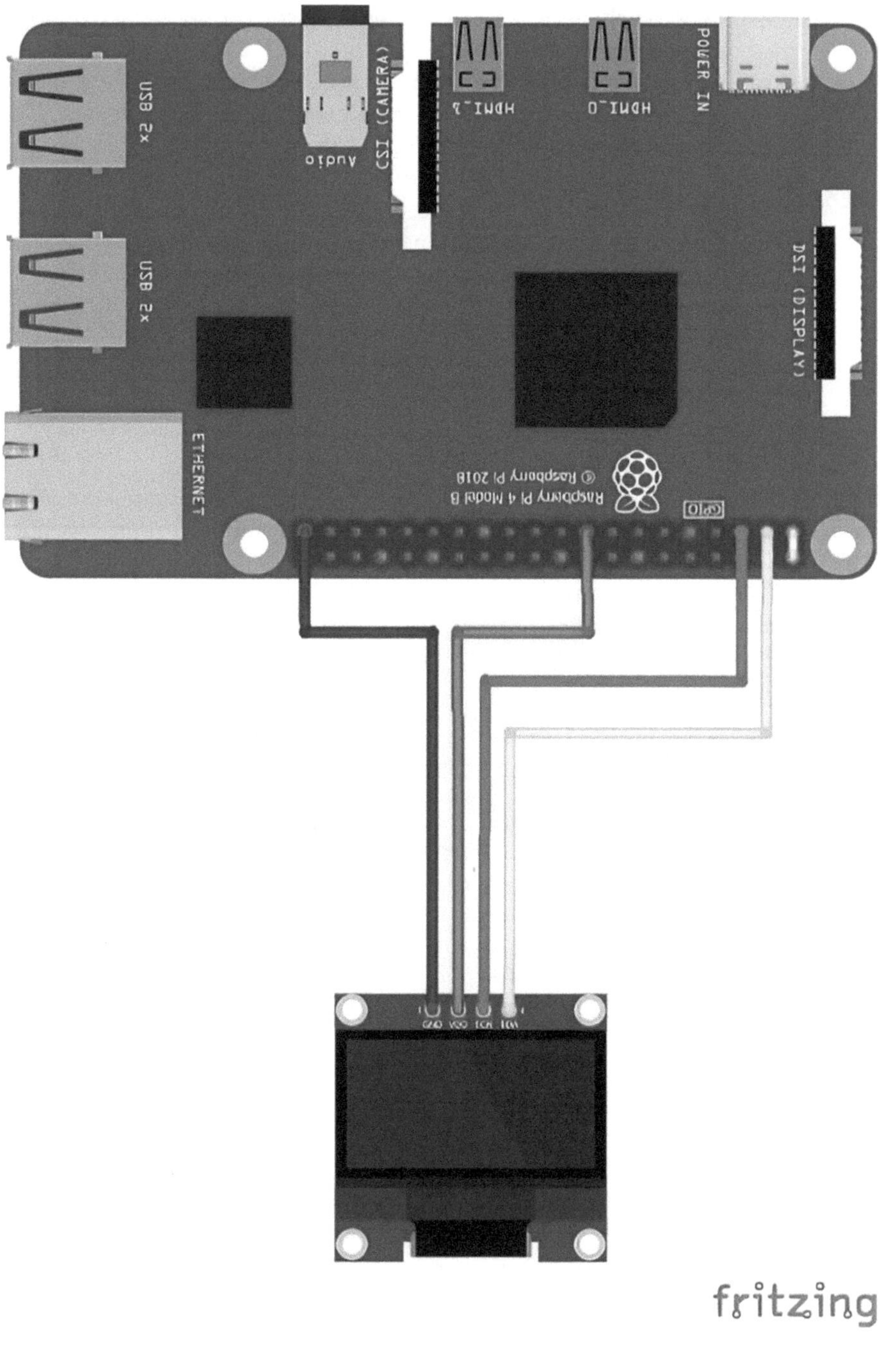

Figure 2.3 Master device (Rpi) circuit diagram.

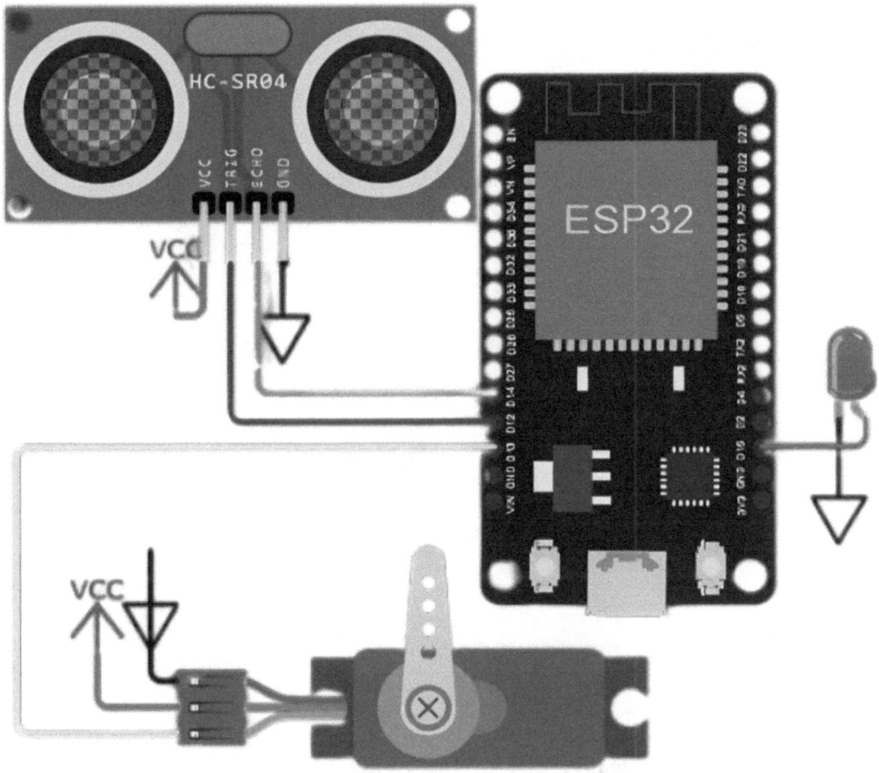

Figure 2.4 Slave device ESP32 circuit diagram (example IoT device).

These positive outcomes underscore the immense potential of the Raspberry Pi as a master device in shaping the future of IoT ecosystems [11, 12] and its exceptional capacity to deliver effective centralized device management.

2.5 CONCLUSION

In summary, the creation of a centralized IoT hub utilizing the Raspberry Pi as the master device marks a significant milestone in the realm of IoT technologies. Our project successfully showcased the hub's ability to effectively oversee and govern a diverse array of IoT devices, establishing a seamless and interconnected ecosystem that can even double as a smart TV and personal computer. The incorporation of sensors like the servo motor for door control and the ultrasonic sensor for water tank level monitoring underscored the adaptability and reliability of our system. The servo motor flawlessly managed door operations, elevating security and convenience, while the

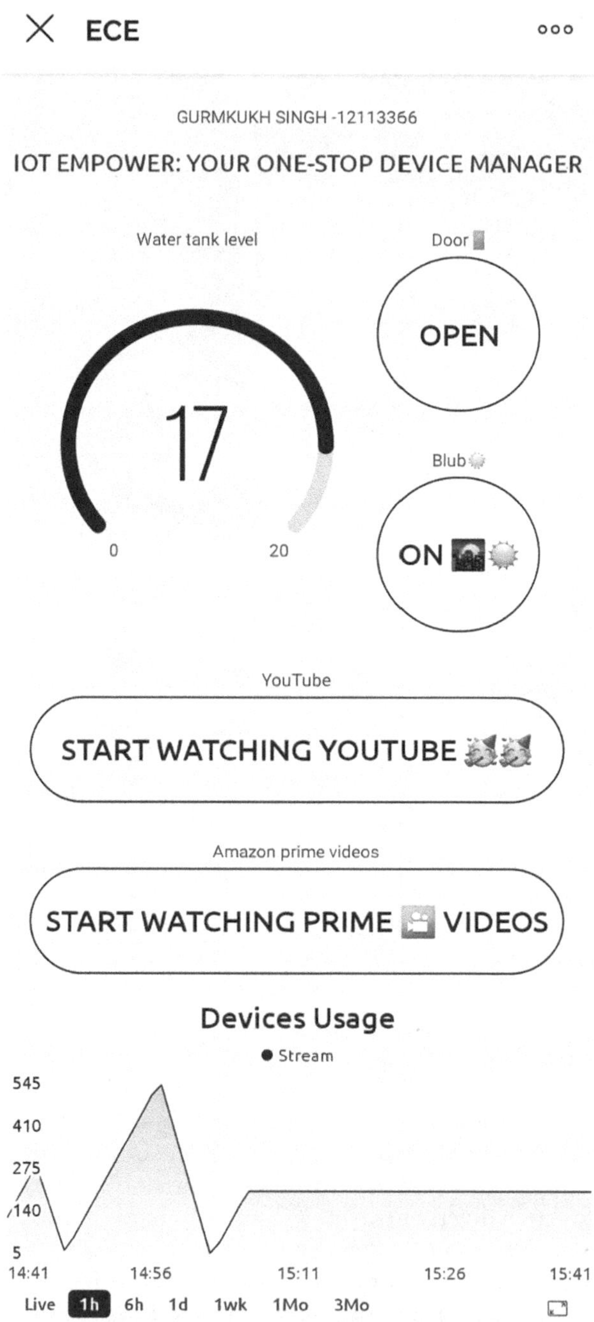

Figure 2.5 Blynk app controlling and showing data.

Figure 2.6 Water tank level (ultra sonic).

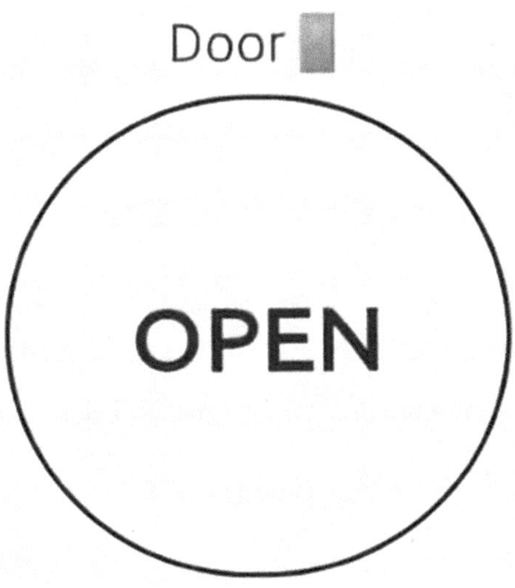

Figure 2.7 Door open and close control (servo motor).

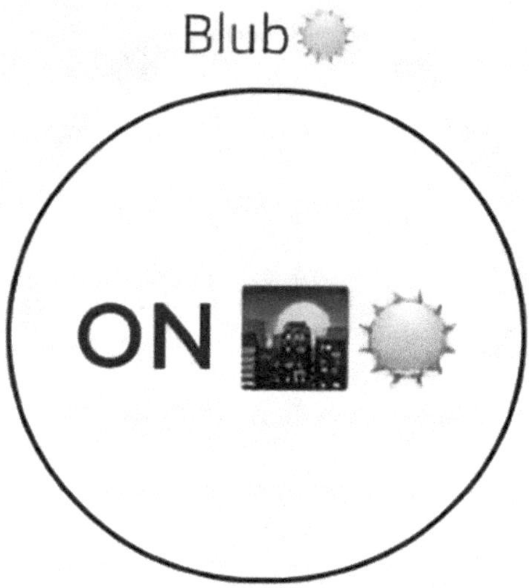

Figure 2.8 Blub on and off control (LED).

Figure 2.9 Watch movies and videos from Amazon Prime and YouTube.

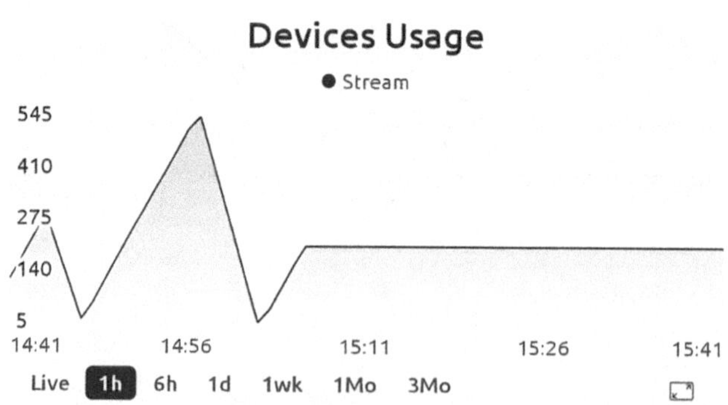

Figure 2.10 Monitor device usage.

ultrasonic sensor delivered precise data for water management. Additionally, we harnessed the Raspberry Pi for entertainment purposes, enabling users to watch movies and videos from platforms such as Amazon Prime Video and YouTube, taking their Raspberry Pi experience to the next level.

REFERENCES

1. Perakakis, E., & Ghinea, G. (2017). Smart enough for the web? A responsive web design approach to enhancing the user web browsing experience on smart TVs. *IEEE Transactions on Human-Machine Systems*, 47(6), 860–872.
2. Kuwano, H., Taniguchi, Y., Minami, K., Morimoto, M., & Kojima, H. (n.d.). A smart TV viewing and Web access interface based on video indexing techniques. *2002 Digest of Technical Papers. International Conference on Consumer Electronics.*
3. Hahm, C.-H., Lee, S., Lee, T., & Yoo, S. (2016). Memory access scheduling for a smart TV. *IEEE Transactions on Circuits and Systems for Video Technology*, 26(2), 399–411. doi:10.1109/tcsvt.2015.2389414
4. Lee, S.-H., Sohn, M.-K., Kim, D.-J., Kim, B., & Kim, H. (2013). Smart TV interaction system using face and hand gesture recognition. *2013 IEEE International Conference on Consumer Electronics (ICCE)*. doi:10.1109/icce.2013.6486845
5. He, W., Martinez, J., Padhi, R., Zhang, L., & Ur, B. (2019). When smart devices are stupid: negative experiences using home smart devices. *2019 IEEE Security and Privacy Workshops (SPW)*. doi:10.1109/spw.2019.00036
6. Sawchuk, A. A. (n.d.). Smart pixel devices and free-space digital optics applications. LEOS '95. *IEEE Lasers and Electro-Optics Society 1995 Annual Meeting. 8th Annual Meeting. Conference Proceedings*. doi:10.1109/leos.1995.484697
7. Kim, J., Kim, S., Park, S., & Hong, J. (2013). Home appliances controlling through Smart TV set-top box with screen-mirroring remote controller. *2013 International Conference on ICT Convergence (ICTC)*. doi:10.1109/ictc.2013.6675541
8. Lee, S. Y., & Park, S. T. (2014). Implementation of cloud server for home screen based on smart TV. *The 18th IEEE International Symposium on Consumer Electronics (ISCE 2014)*. doi:10.1109/isce.2014.6884295
9. Anufrienko, A. (2019). Appliances of smart TV as an IoT device for Industry 4.0. *2019 IEEE 21st Conference on Business Informatics (CBI)*. doi:10.1109/cbi.2019.10087
10. Yang, S.-H., Liu, X.-W., & Lo, Y.-C. (2016). A design framework for smart TV: Case study of the TaipeiTech smart TV system. *2016 IEEE International Conference on Consumer Electronics-Taiwan (ICCE-TW)*. doi:10.1109/icce-tw.2016.7521053
11. Tripathi, S. L., Ganguli, S., Kumar, A., & Magradze, T. (2022). *Intelligent Green Technologies for Sustainable Smart Cities*, Wiley, USA. doi:10.1002/9781119816096
12. Goyal, N., Sharma, S., Kumar Rana, A., & Tripathi, S.L. (Eds.). (2022). *Internet of Things: Robotic and Drone Technology* (1st ed.). CRC Press, Boca Raton UK. doi:10.1201/9781003181613

Chapter 3

AI and IoT integration for smart agriculture

Manoj Singh Adhikari, Arunava Majumder, Manoj Sindhwani and Shippu Sachdeva

3.1 INTRODUCTION

The need for efficient and sustainable agricultural practices has led to the concept of smart agriculture. This fact has motivated research in many different areas and, particularly, the possibility of complementing Internet of Things (IoT) with artificial intelligence (AI), which makes this combination potentially more interesting [1–4]. AI- and IoT-based technologies have also enabled farmers to collect, and analyze data, which can be used in taking smarter decisions [5–7]. The present study contributes to the current research field by exploring AI and IoT interactions on precision agriculture (smart farming) uses, in which we discussed benefits, challenges, along with future implications – sustainability issues or better yield production [8–10]. The incorporation of AI and IoT in agriculture field has extended the horizon toward a smarter solution to address contemporary farming struggles prevailing [11–14]. Smart agriculture refers to the use of advanced data and technology within this business, which can help maximize a few elements: resource efficiency in agricultural production or practices. Additionally, precision farming manages environmental sustainability. Apart from that, animal monitoring and control are other features, along with smart crop management are essential tools for a stronger yield ratio by estimating productivity measures through automated techniques. Along with automation, real-time monitoring, and data-driven decision-making, AI and IoT integration in agriculture add-on cultivation as well as the total efficiency [15–21]. Humidity, temperature, and soil moisture are all useful environmental conditions to which IoT sensors can provide depth. These sensors aid in ensuring that crops are grown in ideal conditions of humidity and temperature, as well as the proper amount of water.

The latest trends in AI, ML, and IoT are shown in Figure 3.1. Precision agriculture, sometimes referred to as smart agriculture, is a cutting-edge farming method that uses technology to improve every step of the farming process. Its objectives are to increase yields, reduce resource waste, and enhance the sustainability and quality of agricultural output.

DOI: 10.1201/9781003643173-3

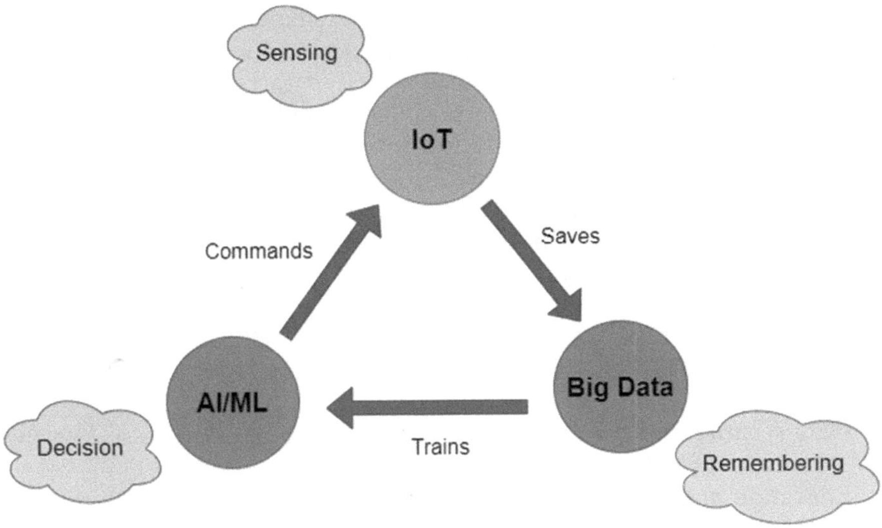

Figure 3.1 Latest trends in AI, ML, and IoT.

IoT in smart agriculture: To collect data in real time from the agricultural environment, IoT devices like sensors, drones, and cameras are essential. They collect information about crop health, weather patterns, livestock welfare, and soil condition. All that data can be received and analyzed by a central system through the internet for making decisions [22–26].

AI in smart agriculture: AI – Machine Learning and Deep learning – is used to analyze the large volumes of data generated by various IoT devices. This AI can identify patterns, predict future scenarios, and help you with useful insights. Irrigation schedules can be optimized, crop disease outbreaks anticipated, and the most suitable time for planting and harvesting crops recommended using AI [27–35].

3.2 MAIN ADVANTAGES

The devices collect information about the health of crops, weather patterns, and even whether livestock are distressed, or soil is in poor condition. The data is then sent through the internet to a central system, which processes it and makes decisions based on this analysis.

Improved resource management: An IoT integration allows farmers to optimize water, fertilizer, and pesticide consumption, thereby reducing their wastage and consequent damage to the environment.

Enhanced productivity: Smart agriculture systems will firstly be able to detect and solve some of the problems (related to disease or nutrient shortages) in early stages, which will lead to better availability for their livestock as well as crop-raising bottom lines.

Saves money: Smart Agriculture leads to the automation of procedures hence reducing the labor costs.

Environmental sustainability: Precision agriculture helps in sustainable farming by minimizing harmful environmental impacts and promoting proper use of resources.

3.3 DIFFICULTIES

Data security and privacy: Agriculture is drowning in data, so keeping that valuable and sensitive info secure.

Education and training: Data generated using these state-of-the-art systems is only as good as the farmers' abilities to use it correctly.

Future prospects: AI and IoT in the arena where Smart agriculture, a part are not limited to it, as this field is still growing into its space.

As technology progresses, we should expect even more advanced solutions with self-governing farm machinery, better detection of pests and disease control methods, as well as increased connectivity between every segment of agriculture.

3.4 AI AND IoT IN AGRICULTURE

The traditional concept of agricultural approach is undergoing a sea change with the application of IoT and AI. The convergence of these new technological solutions, or "AgTech," provides answers to questions the agriculture industry is facing as a result.

Due to issues that include global food production needing to more than double in order to feed our growing world population, optimize resource consumption, and maintain ongoing sustainability. Figure 3.2 presents the method used in precision farming. This also gives a peek into how AI and IoT are applied in agriculture:

AI in agriculture: AI is a broad term covering various technologies, including machine learning and deep learning, which allow machines to perform tasks that typically necessitate human intelligence. In agriculture, AI is used to process vast amounts of data and make predictions based on trends and other insights.

IoT network: An IoT network is a group of connected devices and sensors that communicate over the internet to share information or act upon

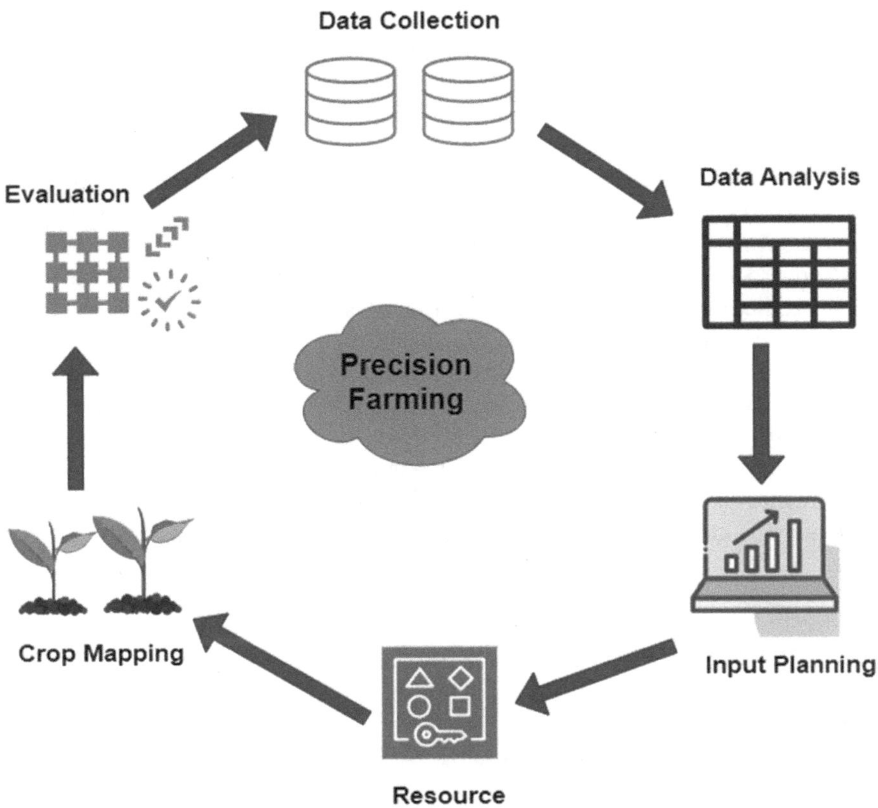

Figure 3.2 Method used in precision farming.

it. They are used in agriculture to collect field data live and provide details about soil content, the status of crops, etc.

AI and IoT elements for agriculture: Sensors and Devices IoT sensors to monitor soil moisture, temperature, humidity, and the growth of crops.

Data connectivity: This is the transmission of data from the IoT devices to a central system where it can be analyzed, and a decision taken.

AI algorithms help to find patterns, anomalies, and trends in the collected data, which is known as Data Analysis.

Automation-engineered AI systems can automate livestock rearing, extermination of organisms that are pests to the crops, and watering.

Decision support: AI offers real-time advanced decision support to help farmers decide on where and how much of their resources (labor, fertilizer) can be used for more productivity at lesser risks.

AI and IoT's advantages for agriculture: These technologies provide the ability to manage resources significantly more accurately by reducing wastage while increasing crop yield. Farmers could monitor their fields from a distance, receiving alerts if the system detected potential problems such as outbreaks of disease or unseasonably cold weather. Through saving water, minimizing the usage of chemicals, and negatively affecting less on environment AI and IoT have made sustainable agriculture possible to a new extent. Farmers will increase productivity and thus profitability by using data to guide their decisions.

AI and IoT's difficulties for agriculture: It can often be expensive for some farmers to implement AI and IoT systems. Privacy of farm data is paramount. Farmers and other agricultural professionals may need training to use the technologies properly for field monitoring, crop yield prediction, and mean examination of sensor data from these tools. Technological development in agriculture, specifically the use of AI and IoT, are, as expected, always growing segments. 1 – Autonomous agricultural machinery: more efficient as the need for human management of field-based systems is eliminated; 2 – Crop and pest monitoring system: better accuracy as pests are identified swiftly by drones rather than a foot patrol, and much can be done through biotechnologies to control them. AI and IoT are very well tackling these global issues as food security and climate change.

3.5 IoT SENSORS AND DATA COLLECTION

Sensors in the IoT touch on a fundamental data collection challenge that plagues agriculture, and many other business domains, which are physical as well as environmental parameters. These sensors are meant to harvest real-time data, which is sent via the internet or other communication mediums for later analysis and decision-making. Figure 3.3 module is responsible

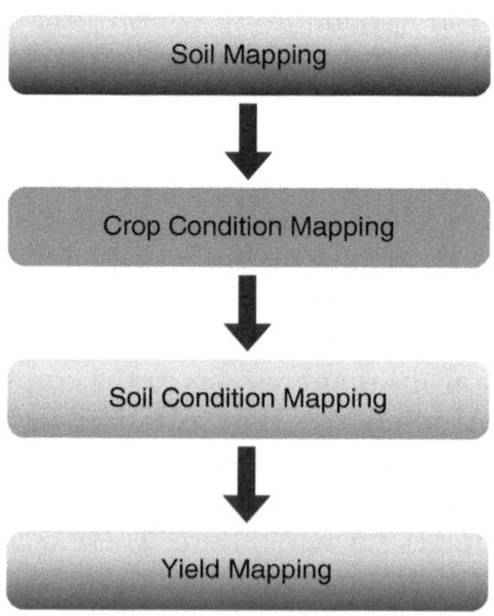

Figure 3.3 Data collection module.

for data collection. In terms of IoT sensors and data collection in agriculture, the following are highlights:

IoT sensor used in agriculture:

Soil moisture sensors: These can be used to detect the level of soil moisture, which assists farmers in knowing when to apply water and how much they require. They are repositioned at different depths, to see the soil moisture levels of each root level.

Weather sensors: Such devices store the data about temperature, humidity, and other meteorological characteristics around them. Knowledge decisions for planting, harvest, and control. Crop Health Sensors: These sensors assess canopy density, leaf temperature, and chlorophyll content to monitor crop health. They help the early diagnosis of plant diseases, nutritional deficiencies, and stress.

Livestock monitoring sensors: IoT devices can attach to your livestock and monitor its location, health, and behavior via Livestock Monitoring Sensors. They can provide details on animal movement, body temperature, and even detect stress signals.

Asset tracking sensors: These sensors also improve asset management and act as an anti-theft by tracking the location and states of vehicles, farm equipment, or anything else in this role.

3.6 DATA COLLECTION PROCESS

IoT sensors record data, every moment as per the variable they are measuring. Soil moisture sensor tracks how much moisture is present in the soil regularly. In Figure 3.4 the data interpretation module is demonstrated.

Data transmission: The collected data is sent to a central DBMS system for processing and storage. To achieve this, there is a wide variety of communication concepts, from cellular networks and Wi-Fi to low-power wide area networks (LPWAN), that can be used.

Data aggregation: The data collected by different sensors is organized and aggregated in a central system. This step refers to collecting and storing data from multiple sensors in order to create a large dataset.

Data analysis: Using AI and machine learning algorithms, the data is then processed in aggregate form to identify patterns, trends, or anomalies. For instance, if soil moisture drops very much, that might be an indication of the need for irrigation – something an AI system can pick up on.

Visualization and reporting: Analyzed data are usually reported to the farmers or stakeholders using reports, user-friendly interfaces for better decision-making. This data can be accessed by farmers through mobile apps and web applications.

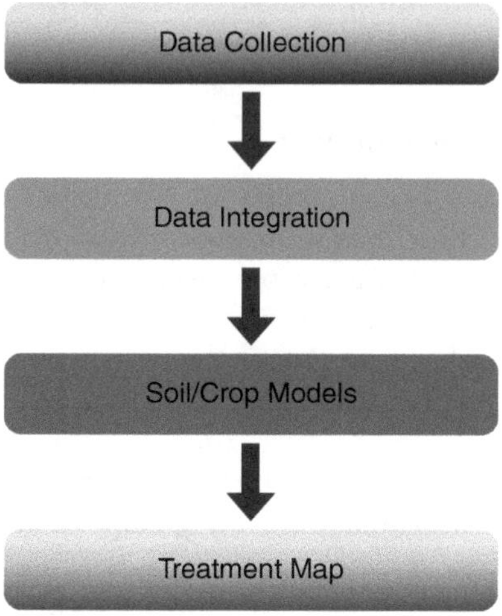

Figure 3.4 Data interpretation module.

3.7 BENEFITS OF IOT SENSORS AND DATA COLLECTION IN AGRICULTURE

Precision farming: IoT sensors provide accurate resource management, reducing waste and increasing production efficiently.

Real-time monitoring: Farmers are able to monitor the status and identify any issues that may arise quickly.

Resource optimization: Better decision-making because of data, meaning less use of resources (fertilizer and water) from an economical point.

Early problem detection: Replace the previous, non-functional example: By equipping a home with sensors, it can detect diseases or pests and other problems at their nascence to allow preventive action.

3.8 AI IN AGRICULTURE

Opposite of agricultural diversity, AI is empowering the modern agriculture sector through data-driven solutions for myriad opportunities and challenges. AI is playing a significant role in agriculture. AgTech, also known as AI in agriculture, provides advanced solutions for crop management and various farming activities. Figure 3.5 demonstrates the smart precision agriculture model. A brief on AI in agriculture is given below:

Precision agriculture and crop management: They can predict crop yields, disease outbreaks, and suggest planting and harvesting times based on the analysis of past/current data such as weather conditions, soil characteristics/ratios, or even plant health. Finding the most competitive

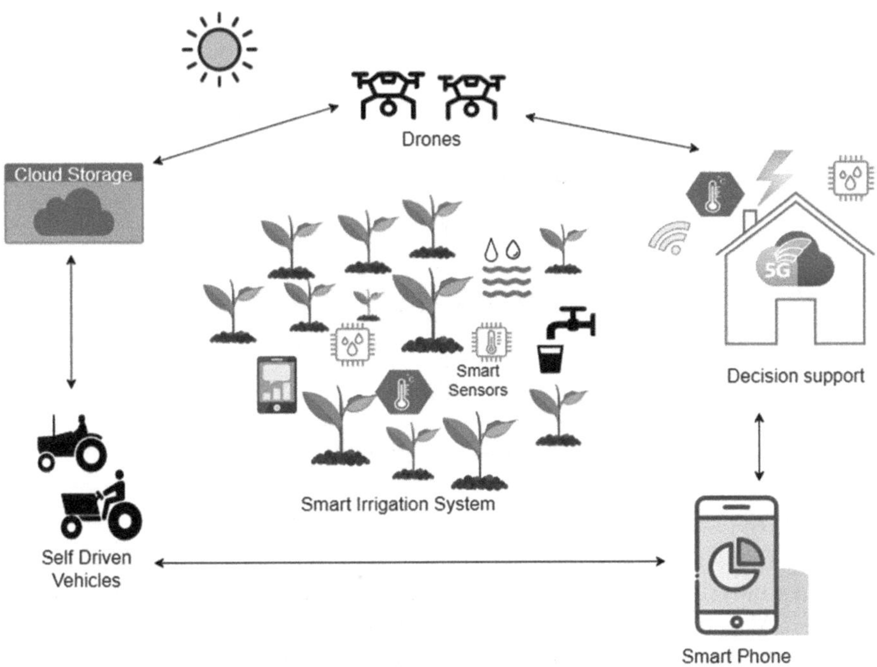

Figure 3.5 Proposed model of smart precision agriculture.

crop can be a fine gauge, and AI systems that are so driven to control irrigation by evaluating moisture data in soil or weather readings help reduce wastewater while bringing hope for resource sustainability. AI can detect weeds and pests with high accuracy, so pesticide use that harms the environment is minimized. Figure 3.6 illustrates the application module of the proposed model.

Livestock monitoring: AI cameras and sensors operate by gathering knowledge from that database on animal habits, thermal input, and the exact surroundings of wildlife for a large number of times each day. Detecting diseases and stress early on will contribute to improving animal welfare, thereby increasing production output.

Supply chain administration: Robots powered by AI and systems have the ability to collect and organize crops efficiently. This leads to savings on labor expenses and improved quality checks. AI assists in devising transportation routes, keeping an eye on inventory, and reducing food wastage by anticipating demands as well as supply chain needs.

Detection of disease and pests: AI-based software, which recognizes images, helps in identifying diseases, pests, and lack of nutrients in crops at an early stage. With the help of drones, we can enhance crops and connect to IoT devices. So, reducing the crop loss. AI drones monitor the crops in a large area.

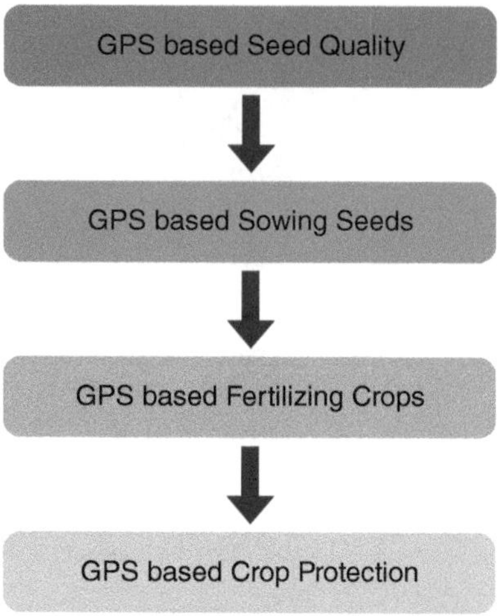

Figure 3.6 Application module of the proposed model.

Data analysis: As the technology changes, the farming is managed by AI. AI is helpful for farming operations. It is useful for the farmers to make data-driven decisions. With support of AI, it analyzes market trends and data to determine which crops, grow easily and the suitable time for planting.

Sustainability: The new AI technology pays attention to environmental sustainability in agriculture. This technology improves the resource use, reduces the requirement for chemicals, and declines destructive impacts. It inspires agriculturalists to implement sustainable practices.

Challenges: Nowadays, the data collection is in a huge amount, so it creates a key challenge in managing it. The private farming data remains secure. But for the poor farmers, the technology costs can be an obstacle. For the effective operation and understanding of data, farmers must develop the necessary skills.

Future outcomes: Now farming has become easier with the new tools and technology, with trends including improved disease, self-driving farming equipment, pest detection, and increased connectivity among components of the agricultural. With the use of AI, we can solve global issues like climate change, food security, and sustainable farming methods.

3.9 APPLICATIONS IN SMART AGRICULTURE

Precision agriculture, sometimes referred to as smart agriculture, uses technology to optimize several farming processes, such as resource management, livestock monitoring, and crop management. It makes use of a variety of applications to boost productivity and sustainability. Figure 3.7 shows the

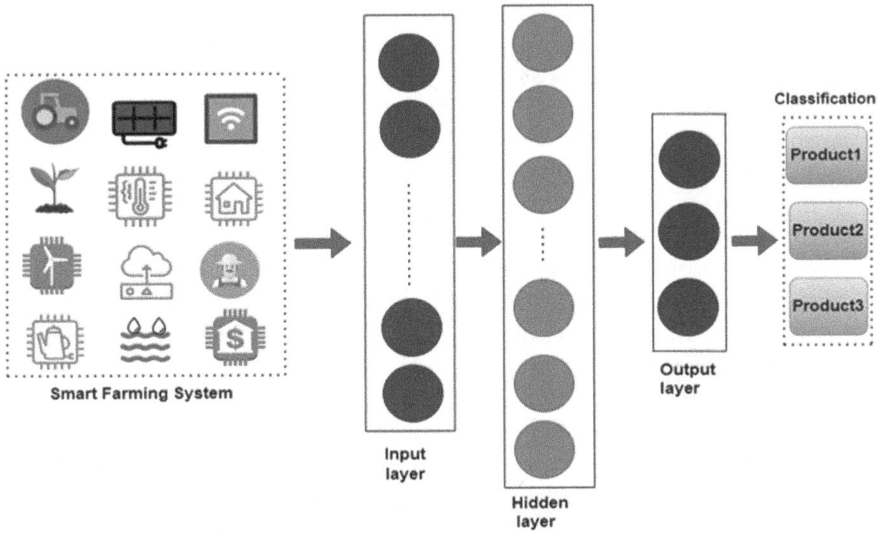

Figure 3.7 AI and IoT for smart farming management system.

AI and IoT for smart farming management system. The following are some crucial smart agriculture applications:

Accurate farming: Variable Rate Technology (VRT) applies inputs like water, pesticides, and fertilizers exactly where they are needed by using data from sensors and GPS. This raises crop yields while cutting waste.

Sensors based on IoT: IoT soil sensors track temperature, moisture content, and nutrient levels to help with fertilization and irrigation schedule selections. Weather sensors help with crop management and weather forecasting by providing real-time data on temperature, humidity, wind, and precipitation. These sensors monitor the condition of plants by finding diseases and shortfalls of nutrients. They do this by gauging leaf temperature, the thickness of foliage, and levels of chlorophyll.

Water directive: In smart agriculture, managing water is very important. It consists of fundamentals like weather conditions, crop type, and soil moisture to assist in managing water resources better.

Crop management: AI is very supportive for prediction of pest attacks and crop illnesses. It gives an immediate response.

Food outlining: Various tools are used for tracking food items. The interesting tools as blockchain and IoT, are used to track food quality. It provides help to enhance food safety standards.

Smart irrigation: These setups reduce power costs and water loss by changing the way water is provided using current details. With IoT-based drip irrigation strategies, we can adjust water use based on soil moisture levels, improving water efficiency.

Unmanned aerial vehicles: AI-based software is used for the crop and pest control. Crop diseases are easily found out with new technology. Autonomous drones provide the solution by minimizing weeds, so the requirement for herbicides is reduced.

Farm monitoring: Different types of farm management apps now exist. So, now the farming planning becomes easier. It provides useful information about crop and livestock performance.

Field software: Latest tools available in the market provide help to farmers for farming. It helps the farmer to provide correct information about the routine of the financial management of plants and animals.

Supply control: AI robots are working perfectly in the supply management system. So, with the help of this technology, harvesting and dispersing crops efficiently. Now, labor costs reduce and the quality of the product increases. AI helps the supply chain work in a smooth way.

Environmental preservation: In environmental preservation, AI plays an important role. It enables farmers to use capital effectively. Less use of harmful chemicals. So, reducing the damage to ecosystem.

Market analysis: AI is also playing a good role in market analysis. It is helpful for farmers to look at market news trends. Trends may be when the crops should be grown or when is the best time to sell to get the maximum benefit.

3.10 PRECISION AGRICULTURE

Precision agriculture is also called intelligent farming. It is a good method that serves the use of technology and information. It is also used to estimate how to recover diverse agricultural actions. The main goal of precision agriculture is to boost harvests. It also cuts resource wastage and improves farm performance efficiency. To get all these, so the need for fertilizers, water, labor, and pesticides. The important components of precision agriculture include:

Data collection: The collection of data is important. For that, sensors and IoT devices are used. It is worked in real time to send the data and analyze it. IoT devices are used to find out the soil properties, temperature, and humidity. So, all these data are easily analyzed and necessary action will be taken for the further farming.

Data administration: Software is used for gathering all the data. All data is kept in one main system. It's like a personal assistant for making decisions. Further, it is used for analyzing in a very simple manner.

Variable rate technology: It is important for farming. VRT customizes the usage of inputs. So, the amounts of pesticides, fertilizers, and water are provided in an appropriate way to the fields.

Seeding and planting: The crop's growth is important in terms of the right seeding. It is related to depths and gaps so that we can get the uniform growth of all plants.

Sustainability: Environmental sustainability is important for modern farming. The awareness about less resource consumption is important to maintain sustainability. We can perform less environmental harm via farming practices. AI tools provide the support for environmental sustainability.

Exactness irrigation: Water conservation is very important for land farming. So, the need for good irrigation systems. AI-based irrigation systems work accurately. It automatically provides the needed water amount and saves the wastage of water.

Control bugs: As the data is sensed via the sensor, the data is processed with the help of AI. So, rapidly identify the bug and control it. The modern ML technique is helpful to maintain the accuracy for crop farming. GPS trackers keep an eye on livestock. These devices keep track of location and use data to boost productivity. Aerial surveillance robots are used in farming to distribute pesticides and fertilizers as per the requirement. They are also used for crop analysis.

3.11 CONCLUSION

Modern farming has long made use of AI and IoT for eco-friendly farming. The agricultural revolution is a bipartisan movement that has spread methods of high-yield farming across an array of different crops and has been one of the greatest contributors to human suffering prevention in all eras. In today's life, IoT technologies are used, which will cover and keep under control entire agricultural ecosystems to provide us with goods of the highest quality. AI- and IoT-based technologies have also enabled farmers to collect and analyze data, which can be used in taking smarter decisions.

Farmers have used AI and IoT alongside other advanced computer science methods for a long time. Lately, however, there has been an emphasis on finding the best ways to apply these modern technologies. Agriculture has been vital for food production for humans for many centuries. Its most notable achievement is its popularization of efficient methods for different crops. Modern IoT technology is now emerging that allows for the observation of agricultural ecosystems, ensuring high-quality goods are produced. However, adopting agricultural practices like deploying and managing IoT and AI devices, administering and sharing data, ensuring interoperability, analysis, and storage of huge amounts of data provides significant challenges to achieve SSA. As such, the identification of architectural components that could help in building smart, sustainable agricultural platforms is crucial.

REFERENCES

[1] H. K. Adli, M. A. Remli, K. N. S. Wan Salihin Wong, N. A. Ismail, A. González-Briones, J. M. Corchado, M. S. Mohamad, "Recent advancements and challenges of aiot application in smart agriculture: a review," *Sensors*, vol. 23, p. 3752, 2023.

[2] Q. Guo, M. Ren, S. Wu, Y. Sun, J. Wang, Q. Wang, Y. Ma, X. Song, Y. Chen, "Applications of artificial intelligence in the field of air pollution: a bibliometric analysis," *Front. Public Health*, vol. 10, p. 933665, 2022.

[3] T. Davenport, R. Kalakota, "The potential for artificial intelligence in healthcare," *Future Health J.*, vol. 6, pp. 94–98, 2019.

[4] L. O'Malley, A. Bronson, K. van der Burg, S. L. Klerkx, "The future (s) of digital agriculture and sustainable food systems: an analysis of high-level policy documents," *Ecosyst. Serv.*, vol. 45, p. 101183, 2020.

[5] A. Kaloxylos, J. Wolfert, T. Verwaart, C. M. Terol, C. Brewster, R. Robbemond, H. Sundmaker, "The use of future internet technologies in the agriculture and food sectors: integrating the supply chain," *Proc. Technol.* vol. 8, pp. 51–60, 2013.

[6] M. S. Adhikari, A. Majumder, A. M. Sheifudeen, K. H. Reddy, M. Sidhwani, Y. K. Verma, "High gain farming and data analytics using IoT", *IEEE International Conference on Devices for Integrated Circuit (DevIC)*, pp. 158–162, 2023.

[7] S. Wolfert, D. Goense, C. A. G. Sørensen, "A future internet collaboration platform for safe and healthy food from farm to fork," *Annual SRII Global Conference*, pp. 266–273, 2014.

[8] M. Dutta, P. Rajak, S. Roy, "Determination of chronic median lethal concentration of sodium fluoride in Drosophila melanogaster and exploring effect of sub-lethal concentrations on differential hemocyte count," *Proc. Zool. Soc.*, vol. 72, pp. 111–117, 2019.

[9] S. Ghanty, M. Mandi, A. Ganguly, K. Das, A. Dutta, S. Nanda, G. Biswas, P. Rajak, "Lung surfactant proteins as potential targets of prallethrin: an in-silico approach," *Toxicol. Environ. Health Sci*, vol. 14, pp. 89–100, 2022.

[10] P. Rajak, S. Roy, A. K. Pal, M. Paramanik, M. Dutta, S. Podder, S. Sarkar, A. Ganguly, M. Mandi, A. Dutta, K. Das, S. Ghanty, S. Khatun, "In silico study reveals binding potential of rotenone at multiple sites of pulmonary surfactant proteins: a matter of concern," *Curr. Res. Toxicol.*, vol. 2, pp. 411–423, 2021.

[11] E. Navarro, N. Costa, A. Pereira, "A systematic review of IoT solutions for smart farming," *Sensors*, vol. 20, p. 4231, 2020.

[12] D. Gao, Q. Sun, B. Hu, S. Zhang, "A framework for agricultural pest and disease monitoring based on internet-of-things and unmanned aerial vehicles," *Sensors*, vol. 20, p. 148, 2020.

[13] L. Lan, X. Le, H. Dong, J. Xie, Y. Ying, J. Ping, "One-step and large-scale fabrication of flexible and wearable humidity sensor based on laser-induced graphene for real-time tracking of plant transpiration at bio-interface," *Biosens. Bioelectron.*, vol. 165, p. 112360, 2020.

[14] L. B. Yang, "Application of artificial intelligence in electrical automation control," *Proc. Comput. Sci.*, vol. 166, pp. 292–295, 2020.

[15] R. Patel, M. S. Adhikari, S. K. Tripathi, S. Sahu, "Design, optimization and performance assessment of single port film bulk acoustic resonator through finite element simulation," *Sensors*, vol. 23, p. 8920, 2023.

[16] K. Jha, A. Doshi, P. Patel, M. Shah, "A comprehensive review on automation in agriculture using artificial intelligence," *Artif. Intell. Agric*, vol. 2, pp. 1–12, 2019.

[17] S. Ashwinkumar, S. Rajagopal, V. Manimaran, B. Jegajothi, "Automated plant leaf disease detection and classification using optimal MobileNet based convolutional neural networks," *Mater. Today Proc.*, vol. 51, pp. 480–487, 2022.
[18] P. Taneja, H. K. Vasava, P. Daggupati, A. Biswas, "Multi-algorithm comparison to predict soil organic matter and soil moisture content from cell phone images," *Geoderma*, vol. 385, p. 114863, 2021.
[19] S. Khaki, L. Wang, S. V. Archontoulis, "A CNN-RNN framework for crop yield prediction," *Front. Plant Sci.*, vol. 10, p. 1750, 2020.
[20] B. Lu, P. Dao, J. Liu, Y. He, J. Shang, "Recent advances of hyperspectral imaging technology and applications in agriculture," *Remote, Sensors*, vol. 12, p. 2659, 2020.
[21] C. D. Lippitt, D. A. Stow, P. J. Riggan, "Application of the remote-sensing communication model to a time-sensitive wildfire remote-sensing system," *Int. J. Rem. Sens.*, vol. 37, pp. 3272–3292, 2016.
[22] J. R. Robles, A. Martin, S. Martin, J. Ruiperez-Valiente, M. Castro, "Autonomous sensor network for rural agriculture environments, low cost, and energy self-charge," *Sustainability*, vol. 12, p. 5913, 2020.
[23] N. Yarlagadda, Y. K. Verma, M. S. Adhikari, V. Mishra, "Opto-VLSI devices and circuits for biomedical and healthcare applications," *A Review of Multi-material, Multi-gate MOSFET Structures*, pp. 29–40, CRC Press, 2023.
[24] U. Shafi, R. Mumtaz, N. Iqbal, S. M. H. Zaidi, I. Hussain, Z. Mahmood, "A multimodal approach for crop health mapping using low altitude remote sensing. Internet of Things (IoT) and machine learning," *IEEE Access*, vol. 8, pp. 112708–112724, 2020.
[25] A. Ikram, W. Aslam, R. H. H. Aziz, F. Noor, G. A. Mallah, S. Ikram, M. S. Ahmad, A. M. Abdullah, I. Ullah, "Crop yield maximization using an IoT-based smart decision", *J. Sens.*, vol. 2022, p. 2022923, 2020.
[26] V. Joshi, M. S. Adhikari, "Advanced agricultural-based IoT technology," *Intelligent Communication and Automation Systems*, CRC Press, pp. 239–254, 2021.
[27] V. Joshi, M. S. Adhikari, "IOT-based technology for smart farming", *Electronic Devices and Circuit Design: Challenges and Applications in the Internet of Things*, CRC Press, pp. 223–242, 2022.
[28] F. Kiani, A. Seyyedabbasi, "Wireless sensor network and internet of things in precision agriculture," *Int. J. Adv. Comput. Sci. Appl.*, vol. 9, pp. 220–226, 2018.
[29] P. Umapathy, J. Hemanth, S. Khera, A. Inbamani, S. L. Tripathi, *Machine Learning and the Internet of Things in Solar Power Generation*, CRC Press, Boca Raton, UK, 2023.
[30] S. L. Tripathi, M. Mahmud, *Explainable Machine Learning Models and Architecture*, Wiley & Scrivener Publishing LLC, USA, 2023.
[31] S. S. Baghel, P. Rawat, R. Singh, S. V. Akram, S. Pandey, A. S. Baghel, "AI, IoT and cloud computing based smart agriculture," *International Conference on Contemporary Computing and Informatics (IC3I)*, Uttar Pradesh, India, pp. 1658–1661, 2022.
[32] M. Shepherd, J. A. Turner, B. Small, D. Wheeler, "Priorities for science to overcome hurdles thwarting the full promise of the 'digital agriculture' revolution," *J. Sci. Food Agric.*, vol. 100, pp. 5083–5092, 2020.
[33] M. S. Adhikari, A. H. Alkhayyat, "Prediction and classification for smart grid applications," *Data Analytics for Smart Grids Applications—A Key to Smart City Development*. Intelligent Systems Reference Library, Springer, vol. 247, pp. 87–102, 2023.

[34] S. M. Rezvani, H. Z. Abyaneh, R. R. Shamshiri, S. K. Balasundram, V. Dworak, M. Goodarzi, M. Sultan, B. Mahns, "IoT-based sensor data fusion for determining optimality degrees of microclimate parameters in commercial greenhouse production of tomato," *Sensors*, vol. 20, p. 6474. 2020.
[35] A. Sagheer, M. Mohammed, K. Riad, M. Alhajhoj, "A cloud based IoT platform for precision control of soilless greenhouse cultivation," *Sensors*, vol. 21, p. 223, 2021.

Chapter 4

Hardware prototype of RIW algorithm on FPGA and P-SoC platform

A trade-off between resource and time constraint

Sudip Ghosh and Suman Lata Tripathi

4.1 INTRODUCTION

Reversible Image Watermarking (RIW) [1], meets the purpose of copyright protection [2] and authentication. It can also be categorized as robust, semi-fragile, or fragile, visible/invisible [3]. Again, it can be blind [4] as well as non-blind, depending on whether the extraction mechanism requires the cover image or not. The watermark embedding and decoding could be done in spatial domain or in transform domain, where transforms like Fourier-Mellin [5], DCT [6], Discrete Fast Walsh-Hadamard Transform (DFWHT) [7], and Discrete Wavelet Transform (DWT) [8] are widely used.

Reversible Watermarking (RW) [9–11] is one type that finds potential application in recent times on telemedicine and tele-surgery for its high security. RW is lossless. Essential requirements of RW on digital images are to meet low imperceptibility at high capacity, blind decoding, and less-complex operations for real-time implementation.

Various RW techniques in the literature that include Difference Expansion (DE) [9], Histogram Bin Shifting [12], Wavelet Transform [8], Prediction Error Expansion (PEE) [13], Reversible Contrast Mapping (RCM) [10], Integer DCT [14], Improved Rhombus Interpolation (IRI) using DE [11], etc.

PEE- and RCM-based RIW offer improved performance than DE. RCM involves non-complex computational operations and offers better perceptual quality, robustness against cropping, and high embedding capacity. Furthermore, RCM, due to its mathematical simplicity, can be easily mapped into hardware and is considered in this work.

The rest of the chapter is organized as follows: Next section describes the related works. Then the next section briefly introduces the mathematical preliminaries of RCM, followed by RCM-based RIW embedding and decoding algorithm. After that section, describes the embedding algorithm. Next, the proposed decoding algorithm is described. The experimental results and the analysis are presented after that. Finally, chapter concludes with the scope of future work.

DOI: 10.1201/9781003643173-4

4.2 RELATED WORKS

VLSI realization could be done on Field Programmable Gate Array (FPGA) board [15–20], Trimedia processor board [21], Graphics Processing Unit (GPU) [22] or Application Specific Integrated Circuit (ASIC) [23].

The hardware platforms are FPGA [23, 24], in GPU [22], in ASIC [23], etc. Furthermore, VLSI realization of RCM-based RIW, according to the research findings in the literature, was first proposed by Ghosh et al. [15] in which a multiplier-less FPGA-based architecture for invisible and semi-fragile watermarking has been implemented on Xilinx Spartan 3E FPGA with a decoder clock frequency of 45 MHz. Maity et al. in [16, 17] developed an architecture for a modified RCM-based RIW algorithm, where [16] is implemented on Xilinx Virtex (xcv3200efg1156-6) FPGA and requires 67.98 MHz for embedding. The architecture proposed in [17] has been implemented on Spartan 3E (XC3S1600E-4FG320) FPGA running at 98.7 MHz. Ghosh et al. [19] proposed another FPGA, P-SoC, and Ultra-Scale FPGA-based architecture for invisible semi-fragile DE-RIW based on IRI. The implementation results for [19] have been shown in Table 4.14. In [20], DE-based RIW was proposed on three different architectures, namely data-flow, pipelined, and modified-pipelined, where the last one takes 8 cycles for embedding and 8 cycles for decoding. Among the three architectures proposed in [20], it is reported that the modified-pipelined architecture consumes the least number of hardware resources. The timing cycle in [20] takes 8-cycles each for embedding and decoding and takes 76 number of adders, 140 subtractors, 45 D Flip-flops, 10 XOR gates, and 2 inverters. The architecture reported in this work takes only 1 adder, 1 subtractor, along with 4 registers, 2 multiplexers for the entire watermark embedding. The decoder also needs 3 adders, 1 subtractor, and 1 divider for the entire decoding operation, along with 4 registers and 2 multiplexers. This shows that the architecture proposed in this paper presents improved results and outperforms the architectures proposed in [16, 17] in terms of both timing cycles and resources utilized.

4.3 RCM-BASED RIW ALGORITHM

Let [0, L] be the image graylevel range (L = 255 for eight-bit graylevel images), and let (x,y) be a pair of pixel intensity values. The forward RCM transforms the pair of intensity values into the respective pair of values as [10]

$$\acute{x} = 2x - y, \acute{y} = 2y - x \tag{4.1}$$

Domain, $D \subset [0, L] \times [0, L]$ is considered for non-underflow and overflow, defined by the equations

$$0 \leq 2x - y \leq L, 0 \leq 2y - x \leq L \tag{4.2}$$

D is rhombic and forms the diagonal of $[0, L] \times [0, L]$. The reverse transform is:

$$x = \lceil (2/3)(\acute{x}) + (1/3)(\acute{y}) \rceil, y = \lceil (1/3)(\acute{x}) + (2/3)(\acute{y}) \rceil \tag{4.3}$$

where $\lceil a \rceil$ is the ceiling of a.

Mathematical relations shown in (4.1)–(4.3) are used in RCM-based reversible watermark embedding and decoding algorithm proposed by Coltuc et al. in [10] for which the corresponding VLSI architecture has been proposed in Sections 4.4 and 4.5, respectively.

4.4 WATERMARK EMBEDDING

STEP 1: Converting multiple into single assignment

$$\acute{x} = 2x - y \tag{4.4}$$

$$\acute{y} = 2y - x \tag{4.5}$$

$$a = 2x \tag{4.6}$$

$$\acute{x} = a - y \tag{4.7}$$

$$b = 2y \tag{4.8}$$

$$\acute{y} = b - x \tag{4.9}$$

STEP 2: Single assignment program code to DFG
Then we construct a graph as shown in Figures 4.1–4.6.
STEP 3: Scheduling (Figure 4.7)
STEP 4: Binding (Figures 4.8–4.11)
STEP 5: Datapath (Figure 4.12–4.15)
STEP 6: Controller (Figures 4.16–4.19)

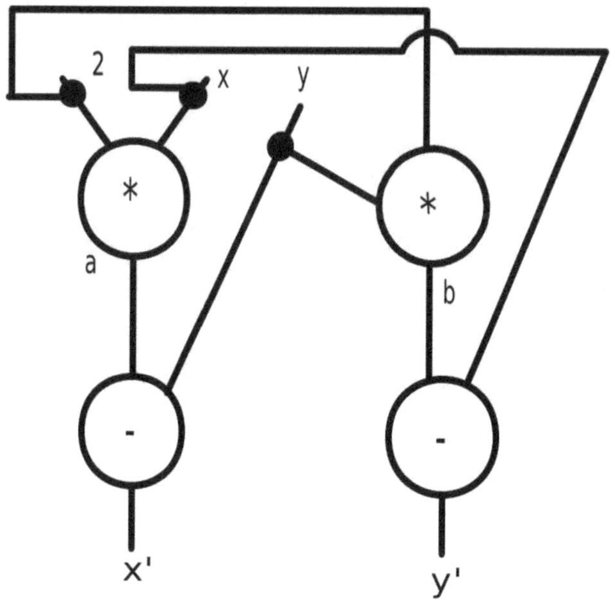

Figure 4.1 DFG of RCM embedding equations for Scheme 1.

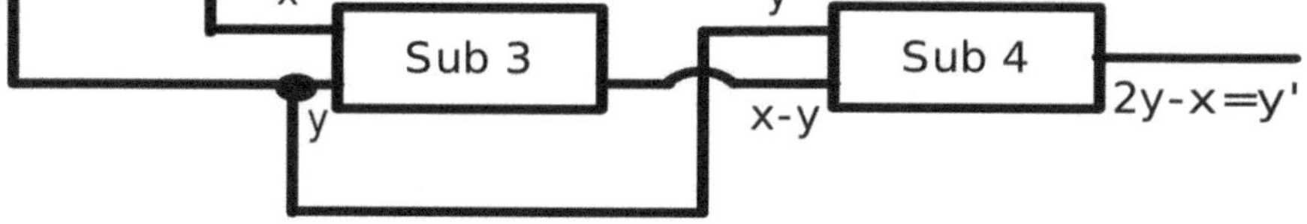

Figure 4.2 Block Schematic Representation of RCM embedding equations for Scheme 2.

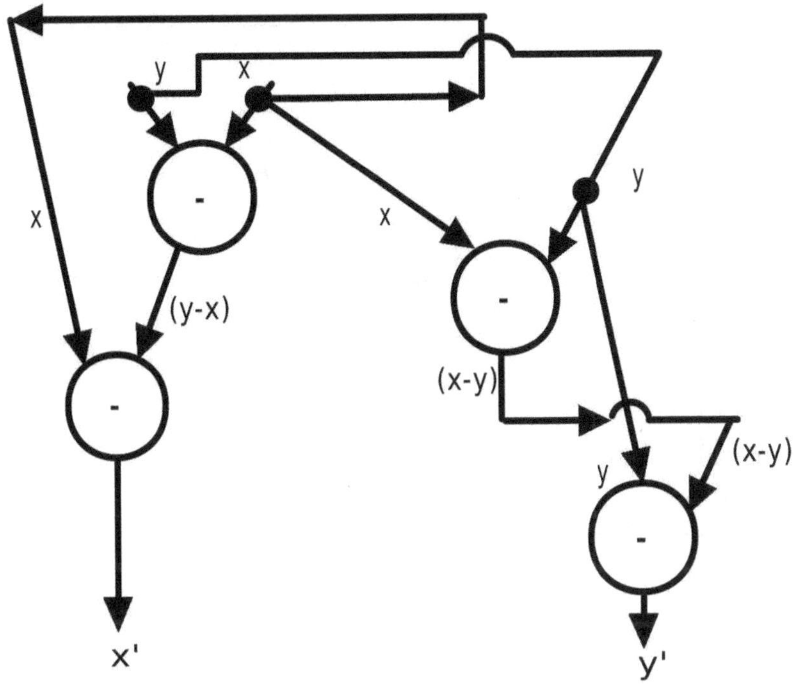

Figure 4.3 DFG of RCM embedding equations for Scheme 2.

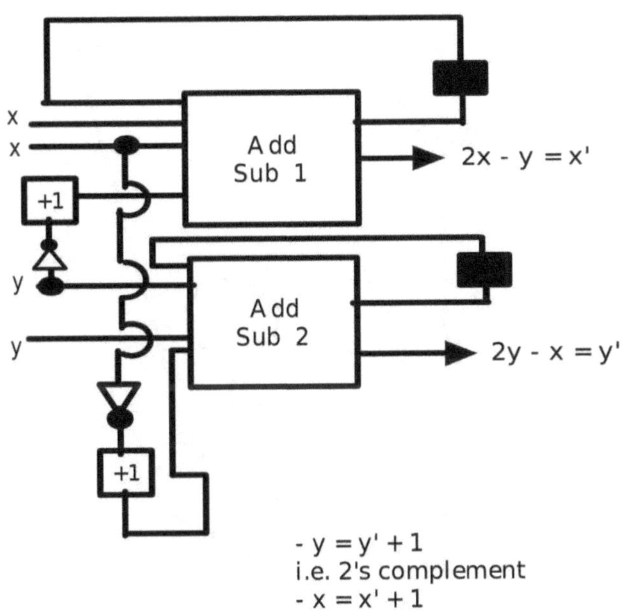

Figure 4.4 Block Schematic Representation of RCM embedding equations for Scheme 3.

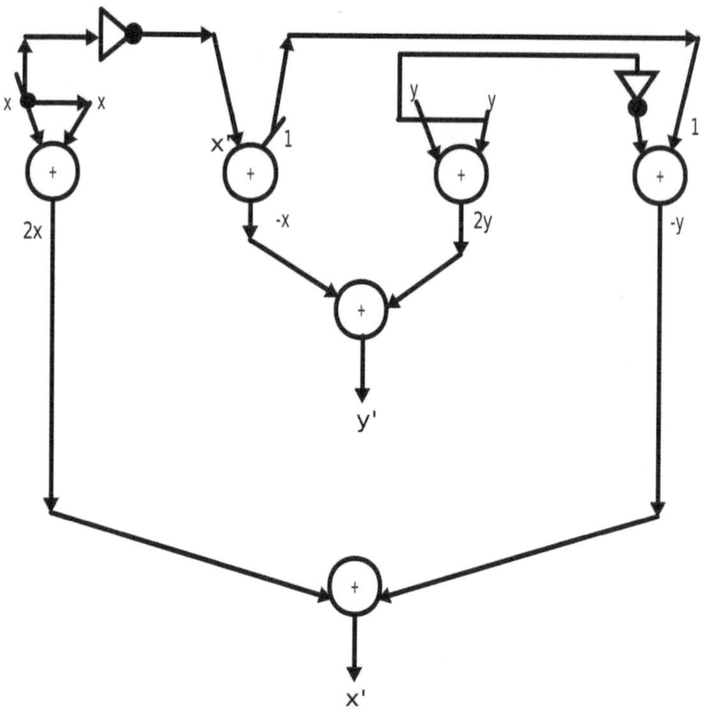

Figure 4.5 DFG of RCM embedding equations for Scheme 3.

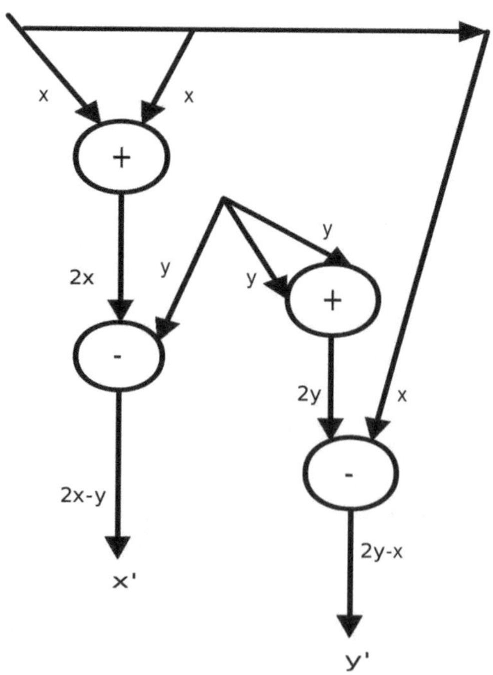

Figure 4.6 DFG of RCM embedding equations for Scheme 4.

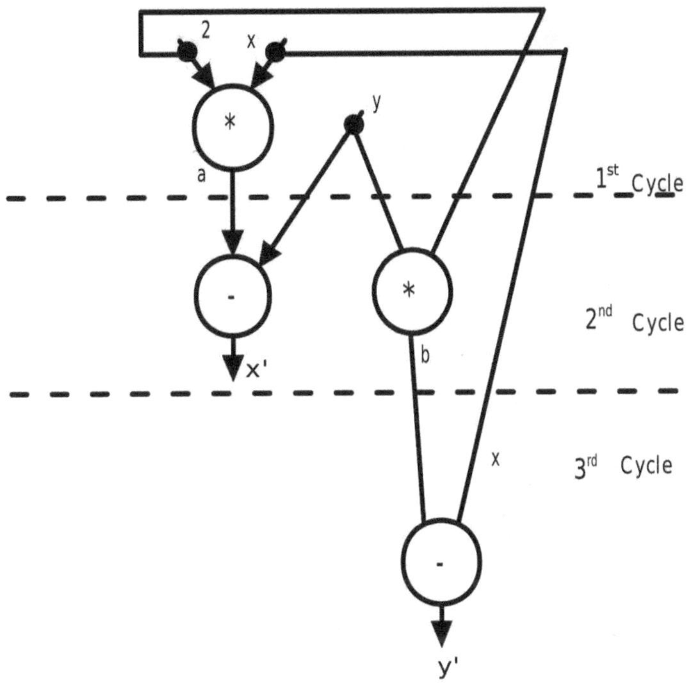

Figure 4.7 Scheduling of RCM embedding equations for Scheme 1.

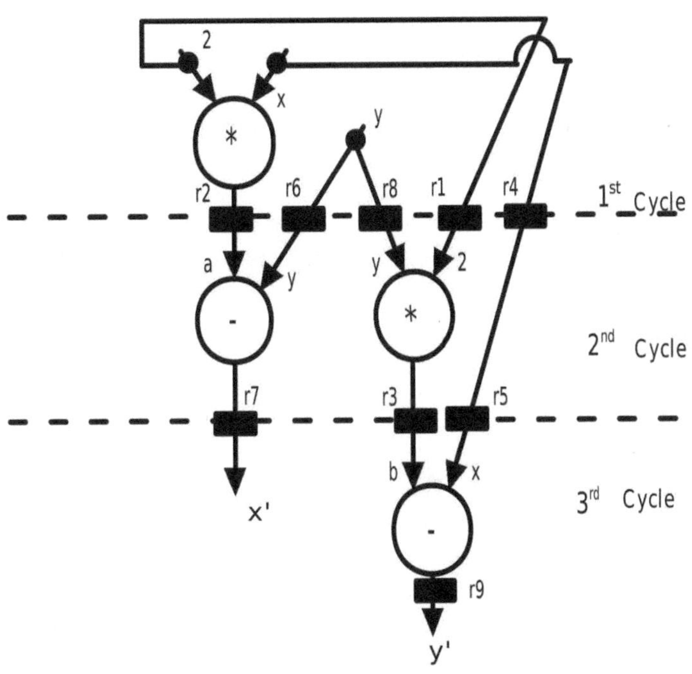

Figure 4.8 Binding and Register Allocation of RCM embedding equations for Scheme 1.

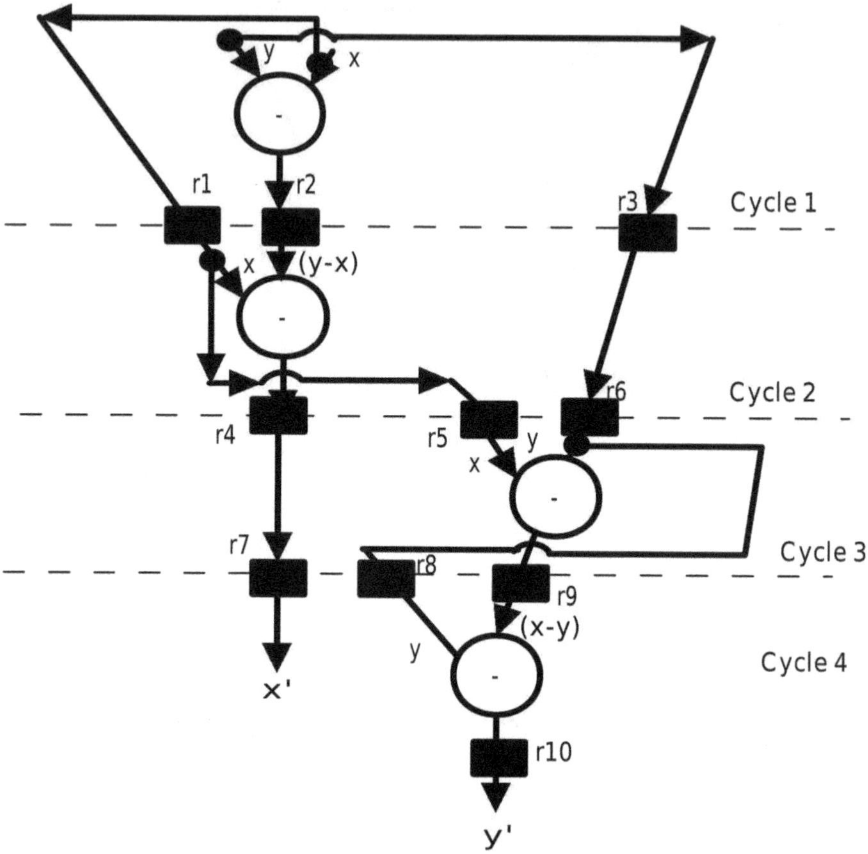

Figure 4.9 Scheduling, Binding, and Register Allocation of RCM embedding equations for Scheme 2.

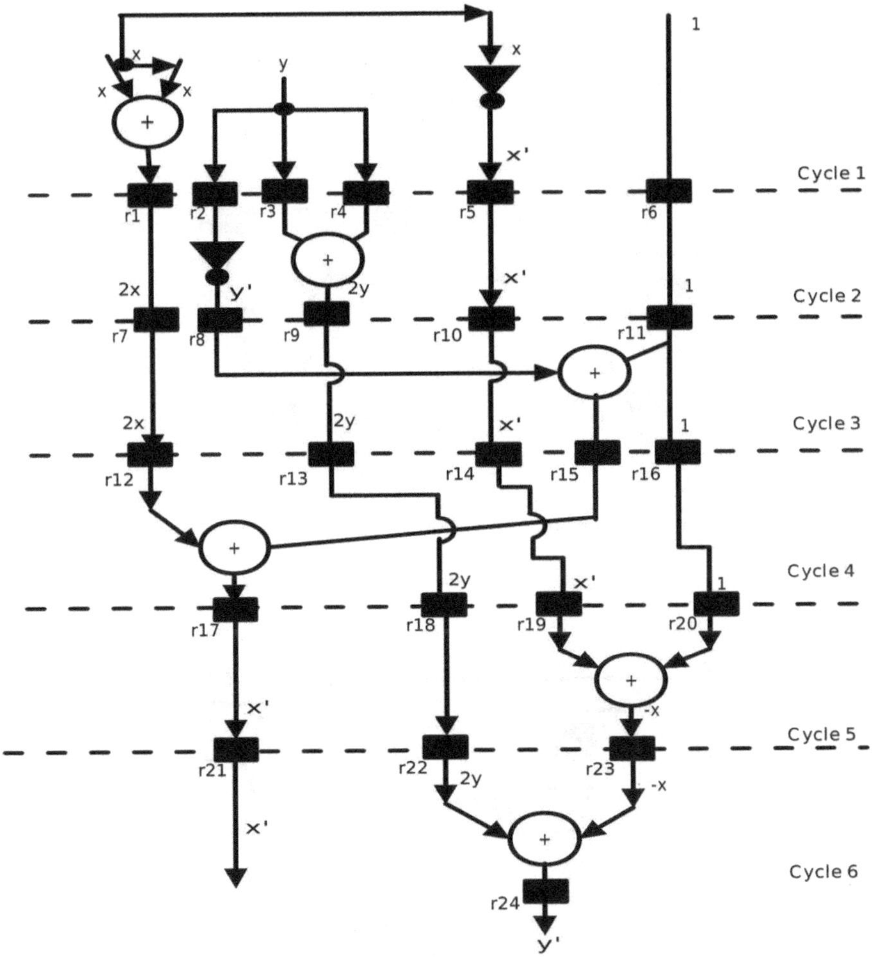

Figure 4.10 Scheduling, Allocation, and Binding of RCM embedding equations for Scheme 3.

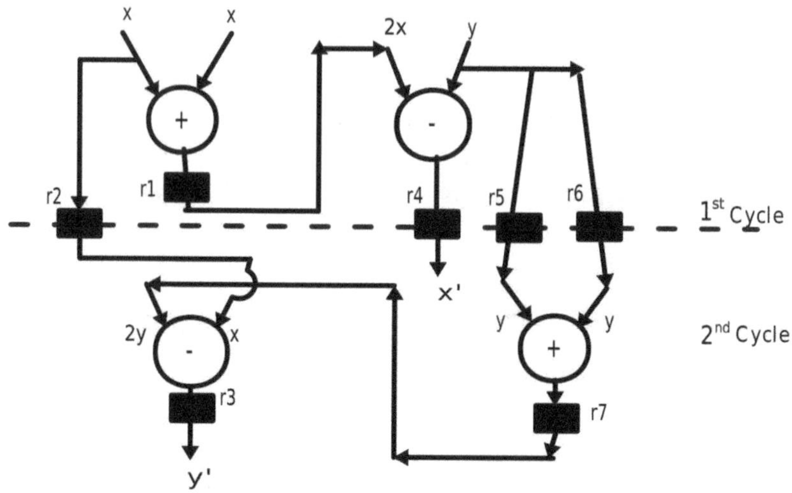

Figure 4.11 Scheduling with Allocation and Binding of RCM embedding equations for Scheme 4.

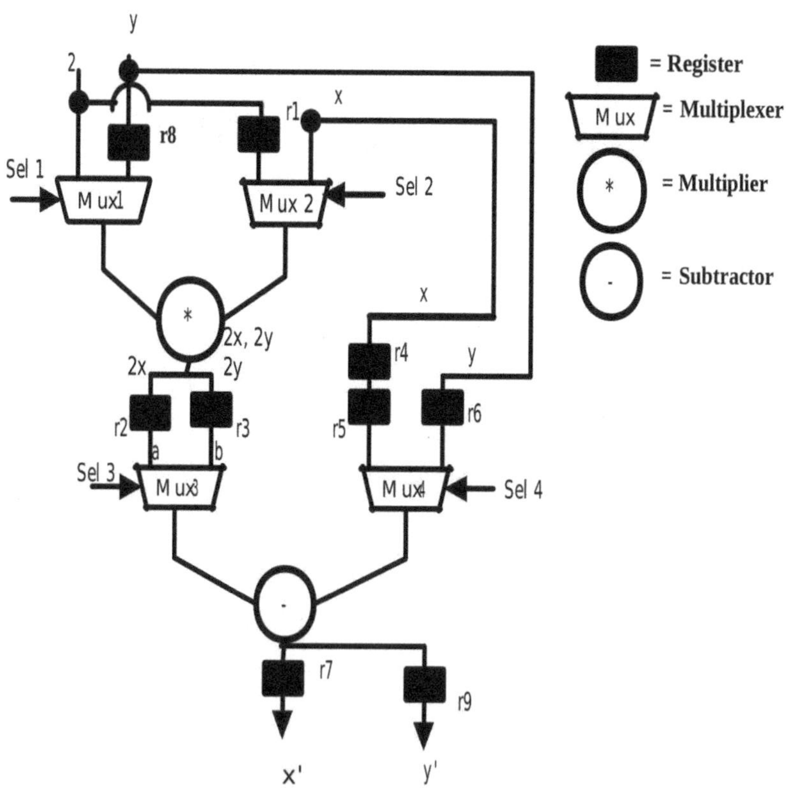

Figure 4.12 Datapath of RCM embedding equations for Scheme 1.

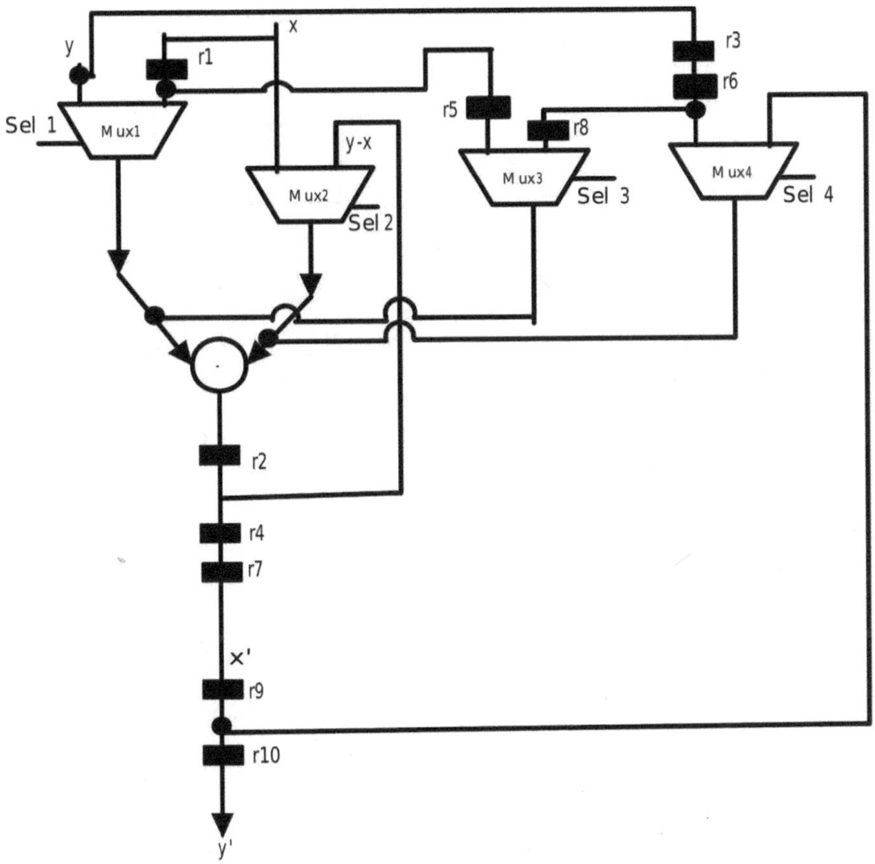

Figure 4.13 Datapath of RCM embedding equations for Scheme 2.

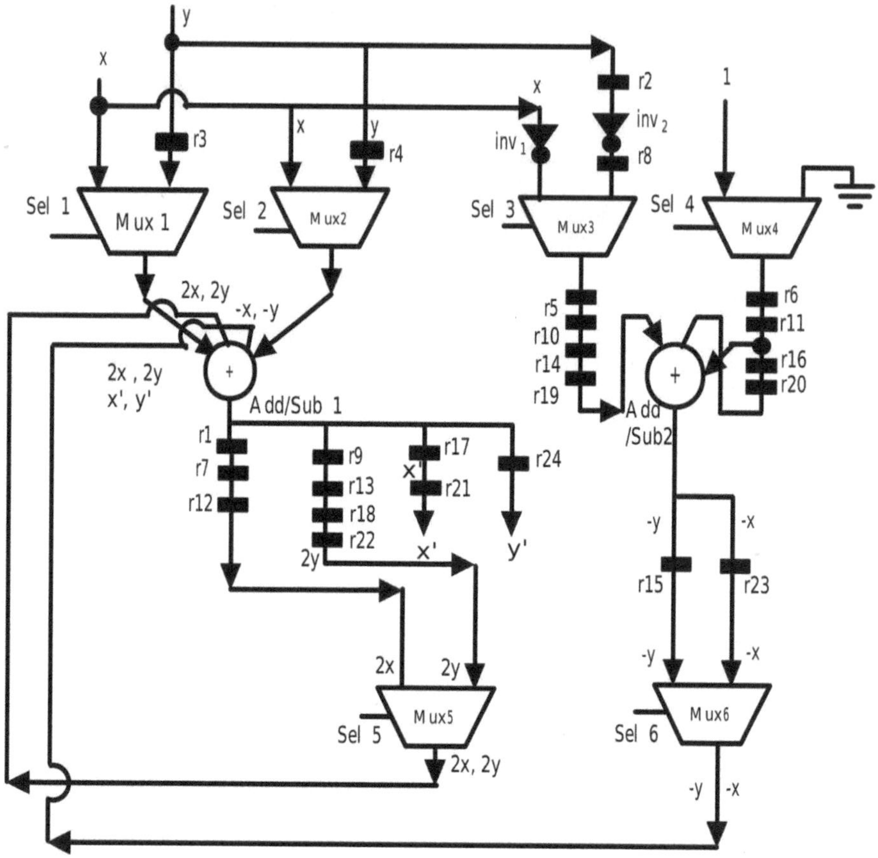

Figure 4.14 Datapath of RCM embedding equations for Scheme 3.

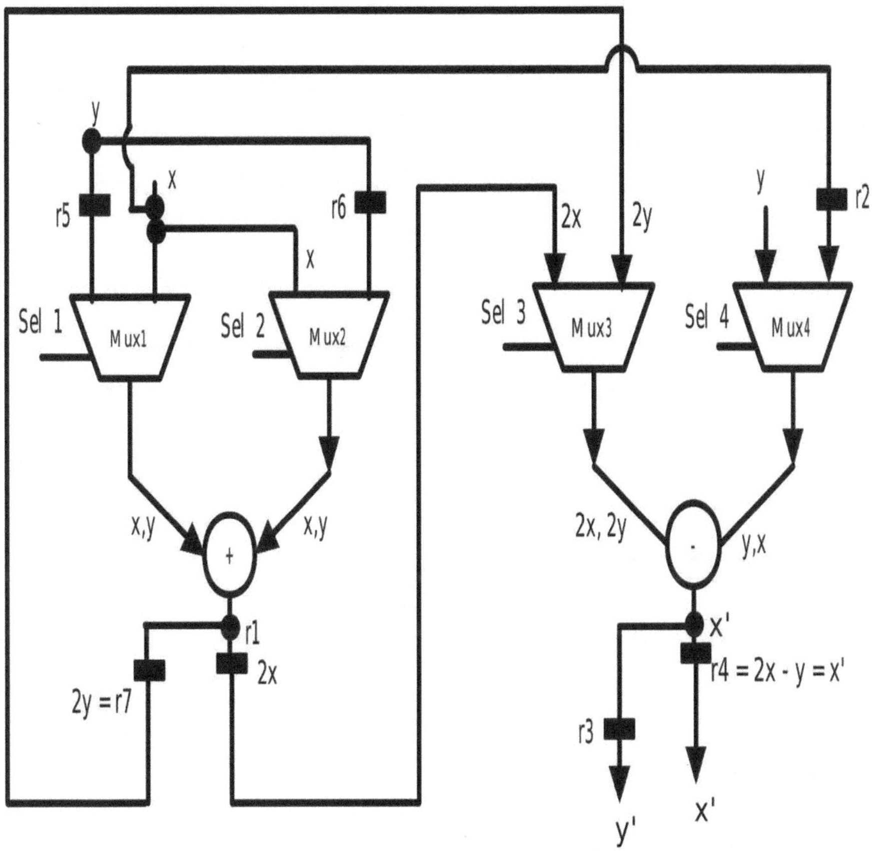

Figure 4.15 Datapath of RCM embedding equations for Scheme 4.

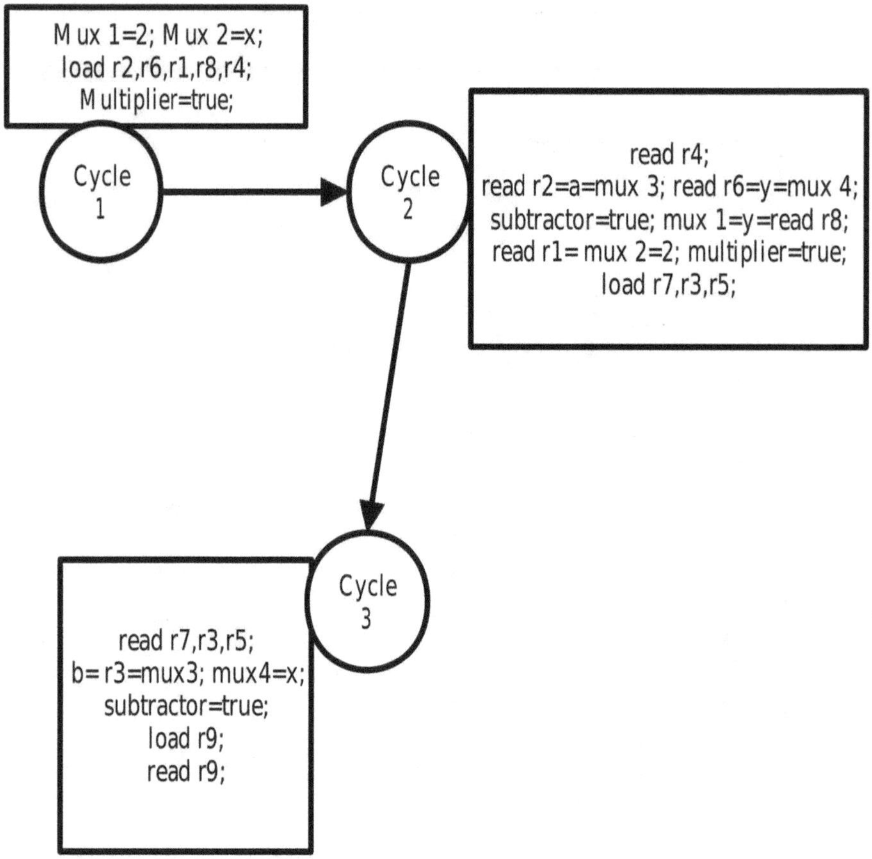

Figure 4.16 Controller of RCM embedding equations for Scheme 1.

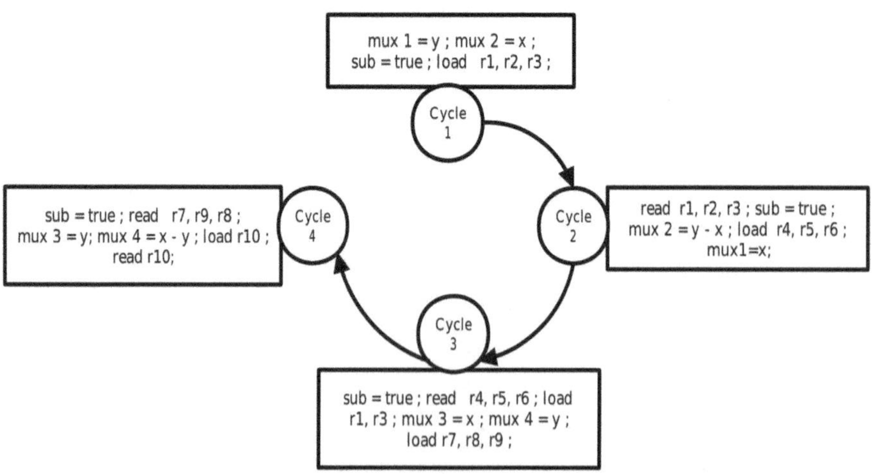

Figure 4.17 Controller of RCM embedding equations for Scheme 2.

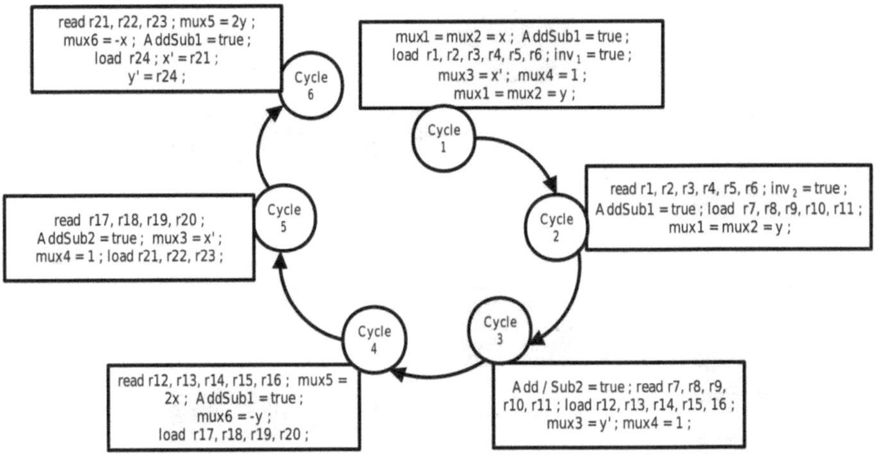

Figure 4.18 Controller of RCM embedding equations for Scheme 3.

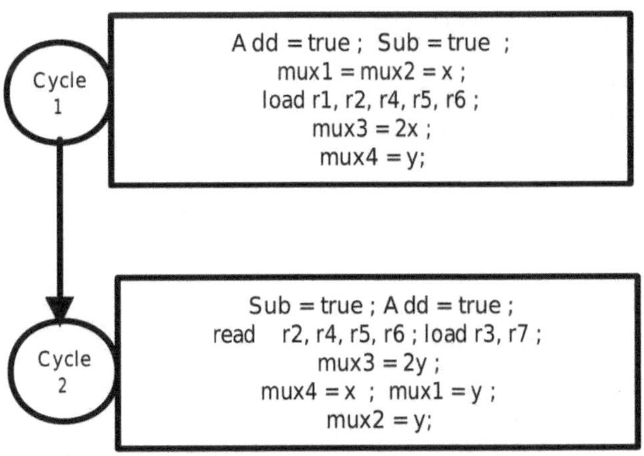

Figure 4.19 Controller of RCM embedding equations for Scheme 4.

4.5 REAL-TIME IMPLEMENTATION OF RCM-BASED DECODING

Equations for inverse RCM transform to decode the same as the forward ones with two additional operations: an additional division operation, specifically a divide-by-3 operation, followed by a ceiling function. So here, the best hardware scheme, with respect to the hardware resource utilization of forward transform, is selected with additional adders, dividers, and ceiling functions. Figure 4.20 shows the inverse RCM block for decoding pixel pair (x, y). The division and the ceiling function modules are shown in Figure 4.21.

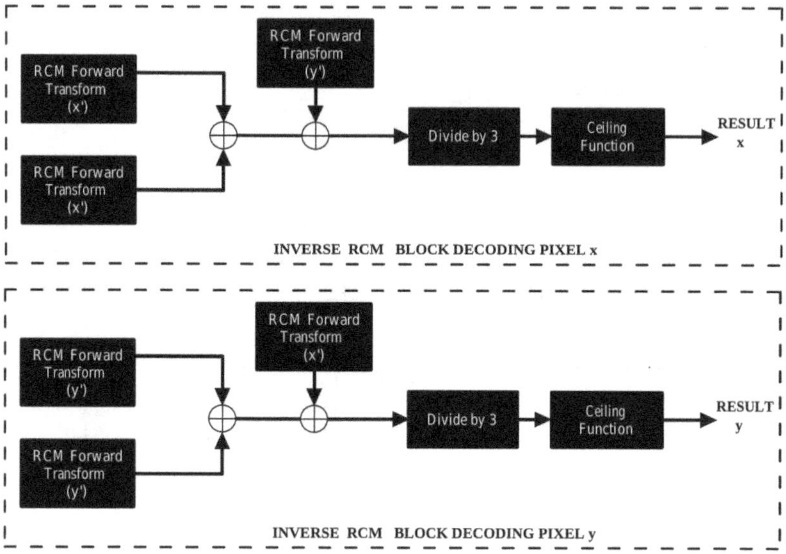

Figure 4.20 Inverse RCM Block for decoding pixel pair (x, y).

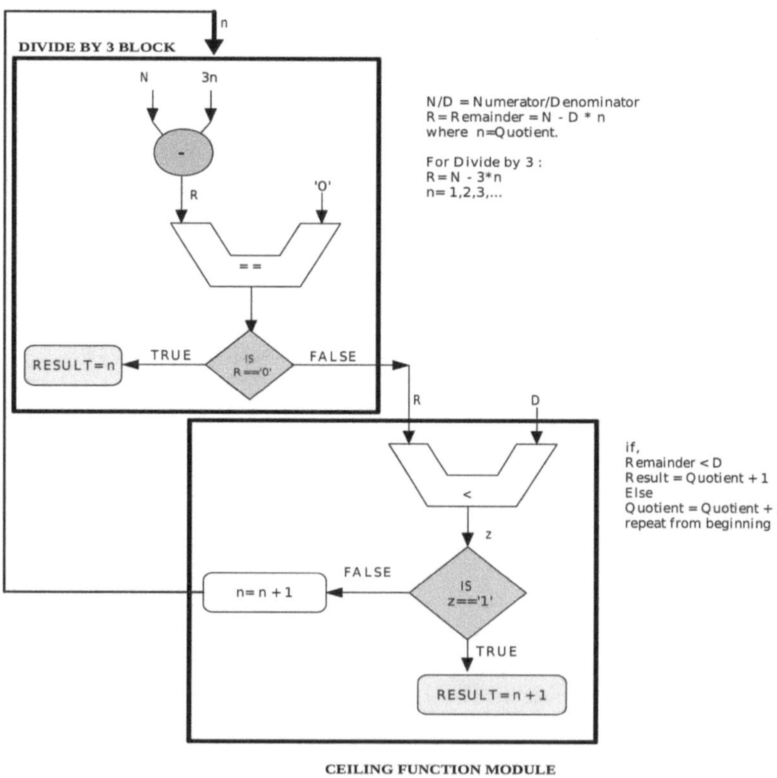

Figure 4.21 Divide by 3 and ceiling function modules.

4.6 EXPERIMENTAL RESULTS, PERFORMANCE ANALYSIS, AND DISCUSSIONS

Here, architectures are developed considering 8 bpp gray cover image of size (32 × 32) to (512 × 512).

4.7 SOFTWARE IMPLEMENTATION

OS used is MS Windows7 64-bit architecture environment with 4GB RAM and Intel Corei5 processor of 3.20 GHz using Matlab R2013b software simulation.

4.8 HARDWARE IMPLEMENTATION

The architectures for RCM-RIW algorithm have been implemented using Xilinx System Generator 14.7, in which four possible schemes for hardware implementation are provided. Furthermore, a comparative study has been made between the resource constraint implementation versus the timing constraint implementation. In Figures 4.22–4.25, the system generator implementation for the four different schemes has been shown. Scheme 1 makes use of a multiplier and subtractor to perform the entire operation. Scheme 2 comprises a single subtractor. Scheme 3 makes use of a single adder, and Scheme 4 uses an adder and a subtractor to complete the entire operation. The schemes have been designed such that no scheme makes use of more than one resource of the same kind.

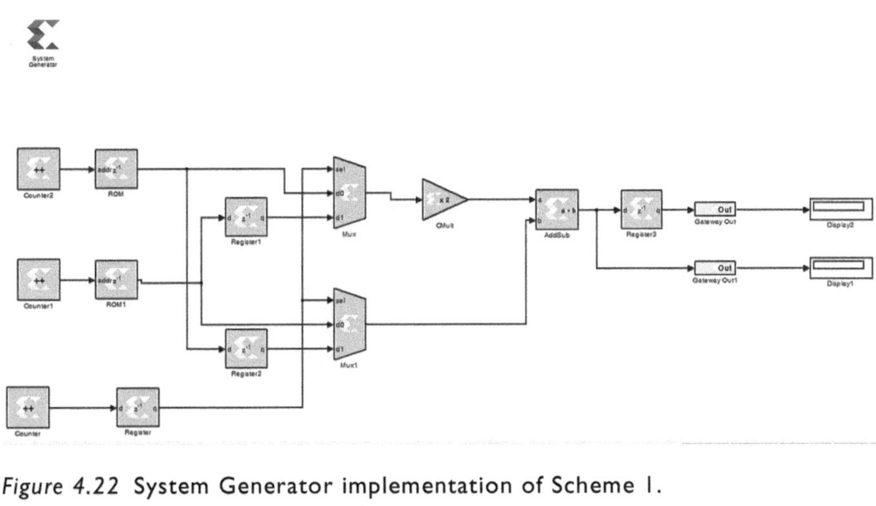

Figure 4.22 System Generator implementation of Scheme 1.

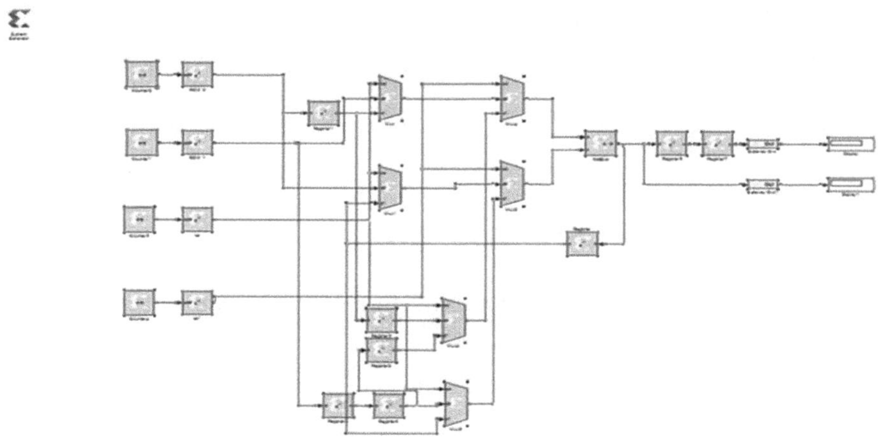

Figure 4.23 System Generator implementation of Scheme 2.

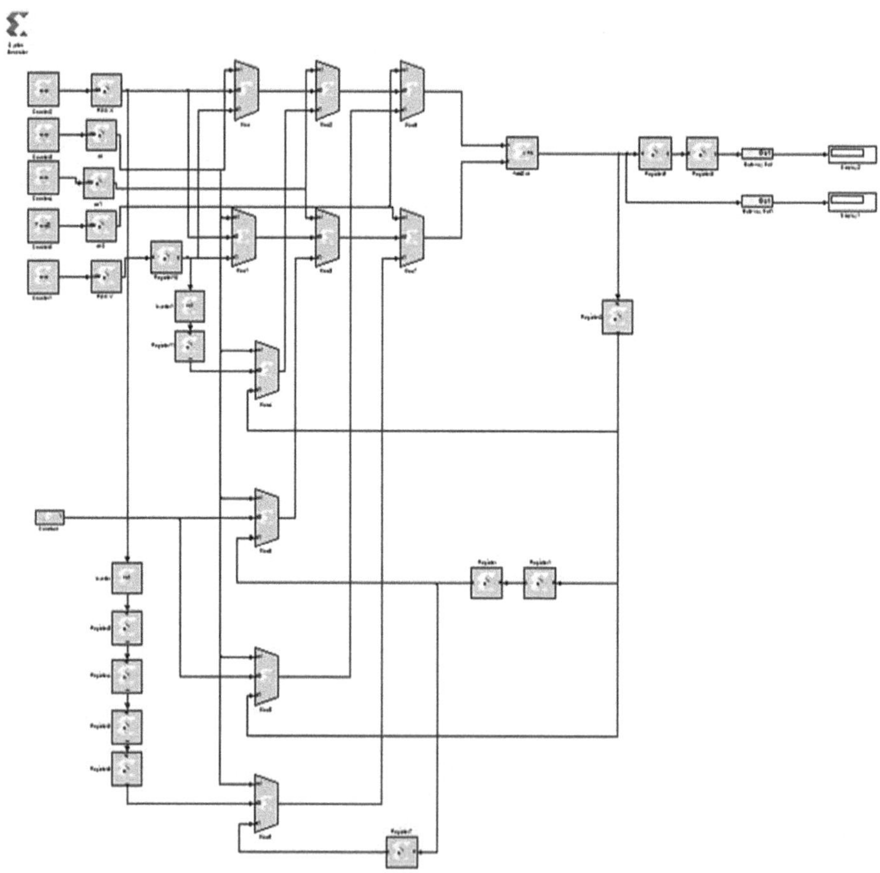

Figure 4.24 System Generator implementation of Scheme 3.

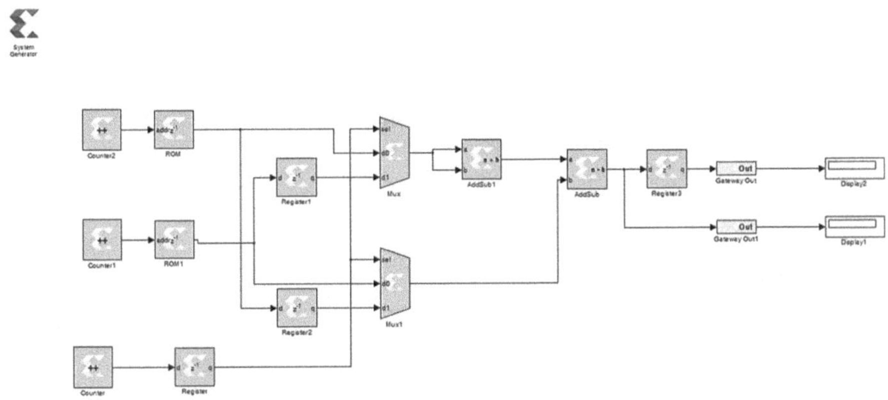

Figure 4.25 System Generator implementation of Scheme 4.

In Tables 4.1–4.12, the synthesis results of the four different schemes on the Spartan-3e FPGA, Zynq-7000 P-SoC and Kintex-7 ultra-scale FPGA are shown. It is seen from Tables 4.1–4.12, Scheme 3 has the maximum device utilization, followed by Scheme 2, Scheme 4, and Scheme 1, respectively.

Table 4.1 Device utilization result for synthesis of Scheme 1 on SPARTAN 3E FPGA

Resources	*Resource-used*	*Total resources*	*Utilized (%)*
#Slices	11	4656	0
#Slice-Flops	19	9312	0
#4-input-LUTs	17	9312	0
#Bonded-IOBs	18	232	7
#GCLKs	1	24	4

Table 4.2 Device utilization result for synthesis of Scheme 1 on ZYNQ-7000 P-SoC

Resources	*Resource-used*	*Total resources*	*Utilized (%)*
#Slice-Registers	19	35200	0
#Slice-LUTs	17	17600	0
#Fully-used LUT-FF-pairs	15	21	71
#Bonded-IOBs	51	100	51
#BUFG/BUFGCTRLs	1	32	3

Table 4.3 Device utilization result for synthesis of Scheme 1 on KINTEX-7 ultrascale FPGA

Resources	*Resource-used*	*Total resources*	*Utilized (%)*
#Slice-Registers	19	407600	0
#Slice-LUTs	17	203800	0
#Fully-used LUT-FF-pairs	15	21	71
#Bonded-IOBs	51	500	10
#BUFG/BUFGCTRLs	1	32	3

Table 4.4 Device utilization result for synthesis of Scheme 2 on SPARTAN 3E FPGA

Resources	*Resource-used*	*Total resources*	*Utilized (%)*
#Slices	52	4656	1
#Slice-Flip-Flops	66	9312	0
#4-input-LUTs	25	9312	0
#Bonded-IOBs	17	232	7
#GCLKs	1	24	4

Table 4.5 Device utilization result for synthesis of Scheme 2 on ZYNQ-7000 P-SoC

Resources	*Resource-used*	*Total resources*	*Utilized (%)*
#Slice-Registers	66	35200	0
#Slice-LUTs	17	17600	0
#Fully-used LUT-FF-pairs	16	67	23
#Bonded-IOBs	33	100	33
#BUFG/BUFGCTRLs	1	32	3

Table 4.6 Device utilization result for synthesis of Scheme 2 on KINTEX-7 ultrascale FPGA

Resources	*Resource-used*	*Total resources*	*Utilized (%)*
#Slice-Registers	66	407600	0
#Slice-LUTs	17	203800	0
#Fully-used LUT-FF-pairs	16	67	23
#Bonded-IOBs	33	400	8
#BUFG/BUFGCTRLs	1	32	3

Table 4.7 Device utilization result for synthesis of Scheme 3 on SPARTAN 3E FPGA

Resources	*Resource-used*	*Total resources*	*Utilized (%)*
#Slices	63	4656	1
#Slice-Flip-Flops	90	9312	0
#4-input-LUTs	59	9312	0
#Bonded-IOBs	17	232	7
#GCLKs	1	24	4

Table 4.8 Device utilization result for synthesis of Scheme 3 on ZYNQ-7000 P-SoC

Resources	*Resource-used*	*Total resources*	*Utilized (%)*
#Slice-Registers	98	35200	0
#Slice-LUTs	50	17600	0
#Fully-used-LUT-FF-pairs	25	123	20
#Bonded-IOBs	33	100	33
#BUFG/BUFGCTRLs	1	32	3

Table 4.9 Device utilization result for synthesis of Scheme 3 on KINTEX-7 ultrascale FPGA

Resources	*Resource-used*	*Total resources*	*Utilized (%)*
#Slice-Registers	98	407600	0
#Slice-LUTs	50	203800	0
#Fully-used-LUT-FF-pairs	25	123	20
#Bonded-IOBs	33	400	8
#BUFG/BUFGCTRLs	1	32	3

Table 4.10 Device utilization result for synthesis of Scheme 4 on SPARTAN 3E FPGA

Resources	*Resource-used*	*Total resources*	*Utilized (%)*
#Slices	11	4656	0
#Slice-Flops	19	9312	0
#4-input-LUTs	17	9312	0
#Bonded-IOBs	21	232	9
#GCLKs	1	24	4

Table 4.11 Device Utilization result for Synthesis of Scheme 4 on ZYNQ-7000 P-SoC

Resources	*Resource-Used*	*Total Resources*	*Utilized (%)*
#Slice-Registers	19	35200	0
#Slice-LUTs	17	17600	0
#Fully-used-LUT-FF-pairs	15	21	71
#Bonded-IOBs	37	100	37
#BUFG/BUFGCTRLs	1	32	3

Table 4.12 Device Utilization result for Synthesis of Scheme 4 on KINTEX-7 ultrascale FPGA

Resources	*Resource-Used*	*Total Resources*	*Utilized (%)*
#Slice-Registers	19	407600	0
#Slice-LUTs	17	203800	0
#Fully-used-LUT-FF-pairs	15	21	71
#Bonded-IOBs	37	400	9
#BUFG/BUFGCTRLs	1	32	3

Figures 4.26 and 4.27 show the timing analysis simulation waveforms of the four different proposed schemes. Scheme-1 and Scheme-4 take 2 clock cycles to give the desired output for a single pair of pixels, making them the most efficient implementations as per timing constraint analysis. Similarly, Scheme-2 takes 4 clock cycles to give the desired output for a single pair of pixels, whereas Scheme-3 appears to be the most timing inefficient implementation, taking a total of 6 clock cycles to give the desired output for a pair of pixels. The clock period for each of the above implementations is 10 ns, as can be seen from Figures 4.26 and 4.27. $x = 200$ and $y = 180$ have been chosen. Schemes 1 and 4 give the desired outputs, i.e., $x' = 220$ and $y' = 160$, after 2 clock cycles. The timing analysis shown here has been performed on a Spartan 3e FPGA.

In Figure 4.28, a resource-based comparative analysis for the 4 different schemes has been shown. Scheme 1 uses a total of 13 different resources consisting of 2 Multiplexers, 1 Multiplier, 1 Subtractor, 4 Registers, 2 ROMs, and 3 Counters. Similarly, Scheme 2 consists of 23 different resources (6 Multiplexers, 1 Subtractor, 8 Registers, 4 ROMs, and 4 Counters). Scheme 3 requires the highest number of resources consisting of 10 Multiplexers, 1 Adder, 12 Registers, 5 ROMs, and 5 Counters, giving a total of 33 different resources. Scheme 4 requires 13 different resources, which are the same as Scheme 1, the only difference lies in the fact that it requires an adder in place of a multiplier. The details of the number of resources required have been obtained from the system generator implementations, as shown in Figures 4.22–4.25.

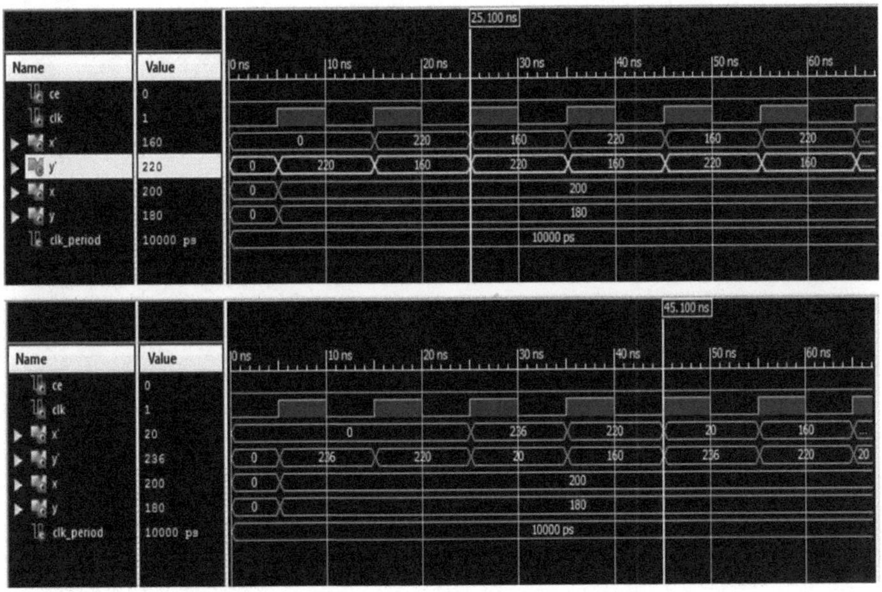

Figure 4.26 Simulation waveform for Scheme 1 and Scheme 2.

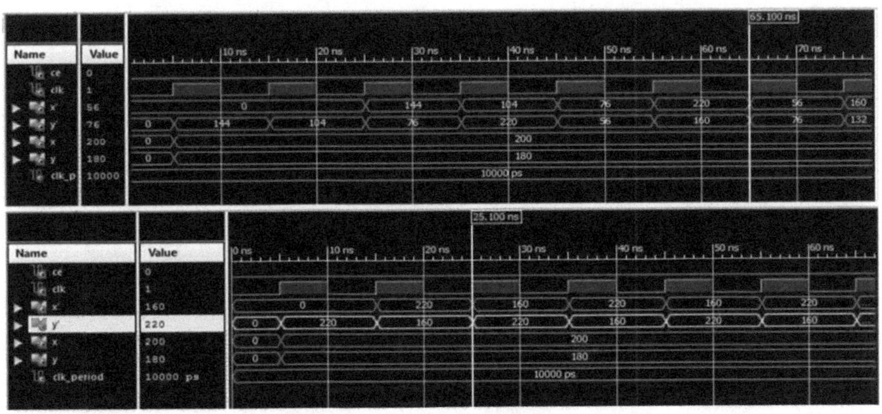

Figure 4.27 Simulation waveform for Scheme 3 and Scheme 4.

Latency is defined as the time taken by a module to give the desired output after the application of the desired input. In Figure 4.29, the latencies of the four different schemes are shown. Scheme 1 and Scheme 4 have the same latency of 2 clock cycles, whereas Scheme 2 and Scheme 3 have latencies of 4 and 6 clock cycles, respectively.

Figure 4.30 gives the latency versus resource trade-off for the four different schemes. Since we desire to have a scheme requiring the least number of resources and having the lowest latency, Schemes 1 and 4 appear to be the most appropriate for the task, as shown in Figure 4.30. Scheme 2 requires

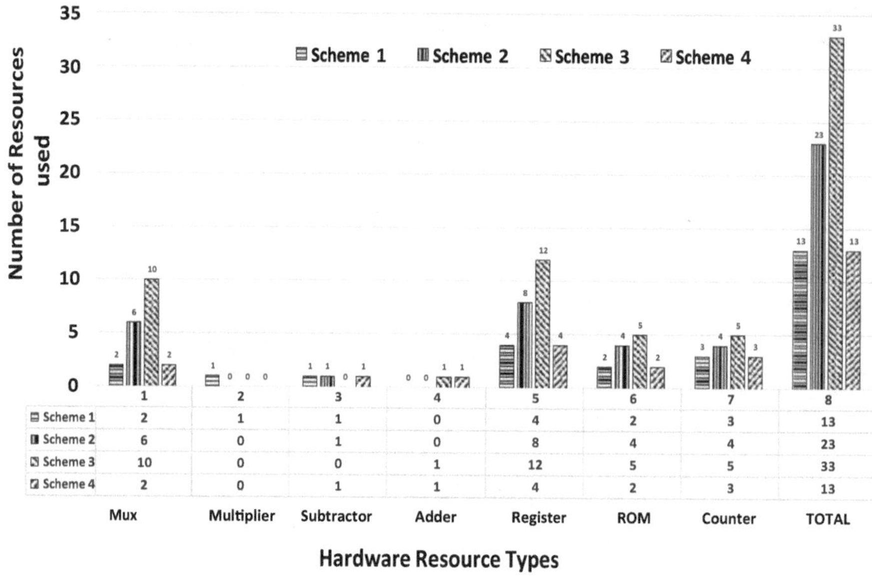

	1	2	3	4	5	6	7	8
Scheme 1	2	1	1	0	4	2	3	13
Scheme 2	6	0	1	0	8	4	4	23
Scheme 3	10	0	0	1	12	5	5	33
Scheme 4	2	0	1	1	4	2	3	13
	Mux	Multiplier	Subtractor	Adder	Register	ROM	Counter	TOTAL

Figure 4.28 Resource utilization-based comparative analysis for Scheme 1 to Scheme 4.

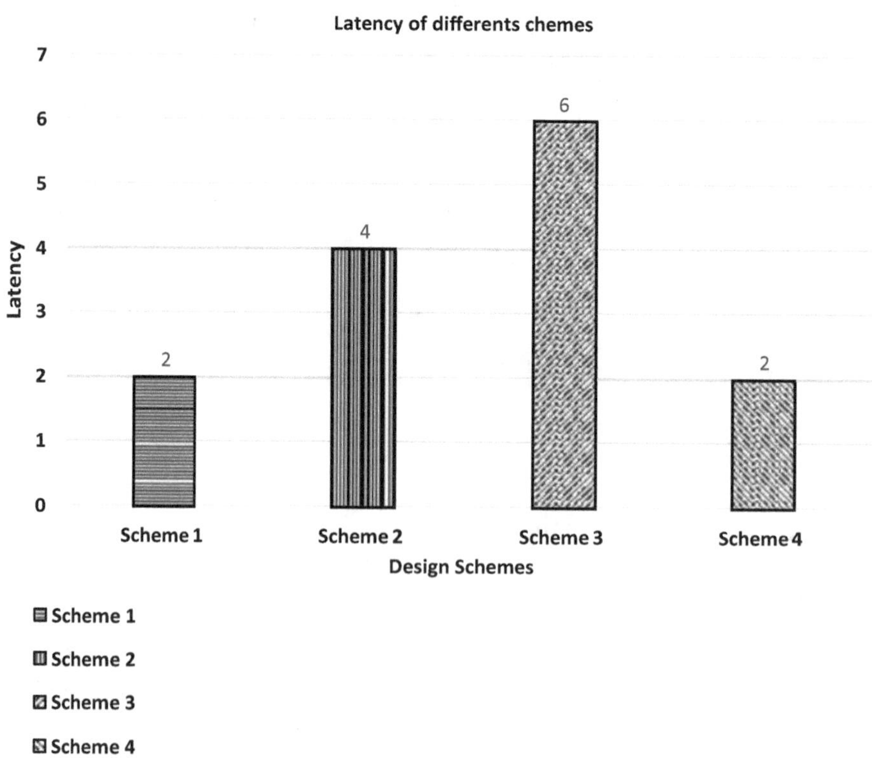

Figure 4.29 Latency for Scheme 1 to Scheme 4.

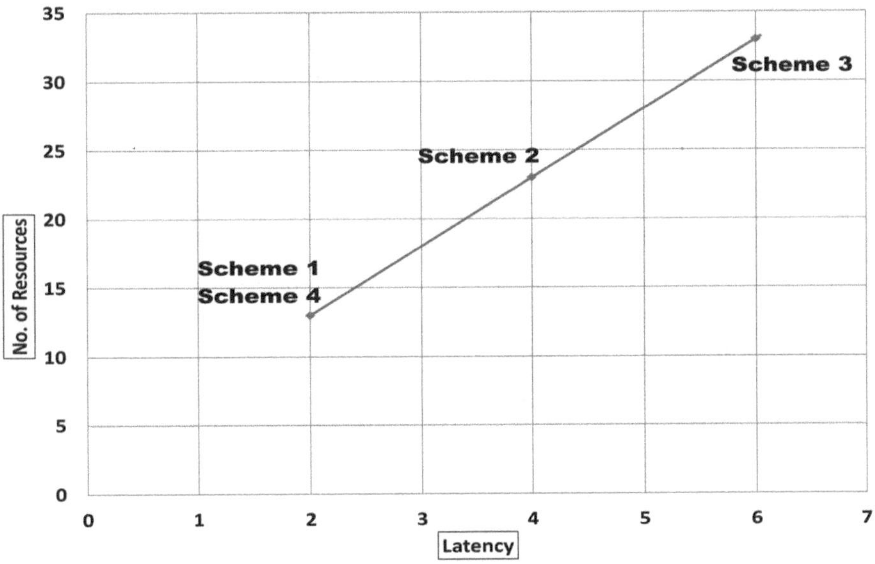

Figure 4.30 Resource and latency trade-off for Scheme 1 to Scheme 4.

more latency and number of resources than Scheme 1 and 4. Scheme 3 requires the most latency and number of resources among all the schemes. Figure 4.30 shows a linear nature of the graph for Scheme 1 to Scheme 4 for latency versus number of resources required. A generalized formula for the total number of clock cycles required is given as:

$$\text{Total no.of clocks required} = \text{Total no.of pixels} + \text{Latency} \quad (4.10)$$

Table 4.13 shows study of total no. of clock cycles of different schemes. Images of different sizes, starting with (32 × 32) to (512 × 512) are chosen for this analysis.

In Figure 4.31, a throughput-based comparative analysis for the various proposed schemes is shown. Throughput is defined as the number of bits

Table 4.13 Clock cycle requirement for various sizes of images for Scheme 1 to Scheme 4

	Scheme 1	*Scheme 2*	*Scheme 3*	*Scheme 4*
(32 × 32)	1026	1028	1030	1026
(64 × 64)	4098	4100	4102	4098
(128 × 128)	16386	16388	16390	16386
(256 × 256)	65538	65540	65542	65538
(512 × 512)	262146	262148	262150	262146

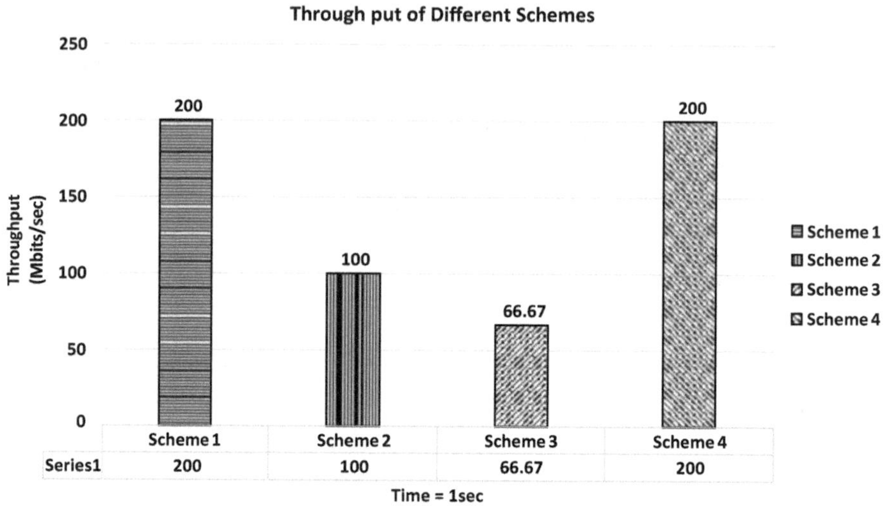

Figure 4.31 Throughput of Scheme 1 to Scheme 4.

being processed per unit time, given in bits/sec. A generalized formula for throughput is given as:

$$\text{Throughput} = \left[8/\left(\text{Latency} * \text{ClockPeriod}\right)\right]\text{Mbits/sec} \tag{4.11}$$

Considering the maximum clock period to be 20 ns (Since the maximum clock frequency supported by Spartan 3E FPGA Starter Kit is 50 MHZ), the throughputs for various schemes are given in Figure 4.31.

Figure 4.32 gives the gate count(area) based comparative studies for the 4 different schemes. Some of the prerequisites that have been taken into account for the above analysis are given as:

1. Every 1-bit D-Register consists of 5 NAND gates.
2. Every 1-bit 2-to-1 Mux consists of 4 NAND gates.
3. Every 1-bit Full Adder consists of 9 NAND gates.
4. Every 1-bit Full Subtractor consists of 9 NAND gates.
5. Every 1-bit Half Adder consists of 4 NAND gates.
6. Every 8-bit Multiplier consists of 49 1-bit Full Adders and 7 1-bit Half Adders. Since each 1-bit Full Adder requires 9 NAND gates and each 1-bit Half Adder requires 4 NAND gates, so total no. of NAND gates in each 8-bit Multiplier is given as:

$$\text{No.of NAND gates} = 49 \times 9 + 7 \times 4 = 469.$$

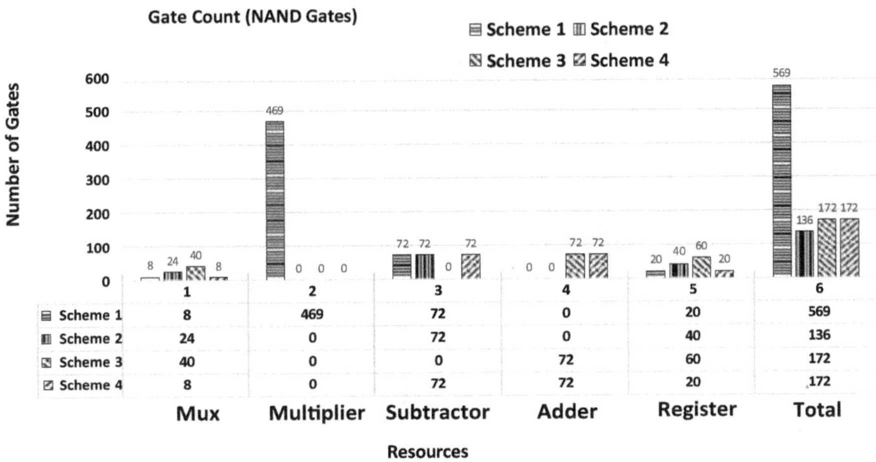

	Mux	Multiplier	Subtractor	Adder	Register	Total
	1	2	3	4	5	6
Scheme 1	8	469	72	0	20	569
Scheme 2	24	0	72	0	40	136
Scheme 3	40	0	0	72	60	172
Scheme 4	8	0	72	72	20	172

Figure 4.32 Gate Count(Area) based comparative analysis of Scheme 1 to Scheme 4.

An in-depth analysis of Figure 4.28 shows that the Scheme 1 requires the maximum number of gates, whereas the Scheme 2 requires the minimum number of gates. But at the same time, the Scheme 2 is not the best implementation, as it has low throughput. The Scheme 4 has higher throughput as well as comparatively lower gate count than the Scheme 1, and can be stated as one of the most optimal solutions, taking a trade-off between the latency and the gate count.

4.9 CONCLUSIONS AND SCOPE OF FUTURE WORKS

The proposed architecture for the Scheme 4 consists of 1 Adder, 1 Subtractor, for watermark embedding, along with 4 Registers and 2 Multiplexers. The decoder also needs only 3 Adders, 1 Subtractor, and 1 Divider, along with 4 Regs and 2 Muxes. The implementation is efficient and fast. To process (512 × 512) the embedding hardware requires 5.24 msec with latency 40 ns for a 50 MHz clock and a throughput of 200 Mbits/sec.

This work can also be extended for hardware design of PEE-based RW on digital images (Table 4.14).

Table 4.14 Comparison of hardware implementation of different watermarking techniques in the literature

Research work	*Watermarking type*	*Hardware platform*	*Data type*	*Watermarking domain*	*Chip statistics*
Maity and Kundu [18]	Invisible Fragile	FPGA Board	Image	Spatial	Xilinx Spartan, XCS05, 85 CLB, 80 MHz
Ghosh et al. [15]	Invisible Semi-Fragile	FPGA Board	Image	RCM-RIW	Xilinx Spartan 3E, decoder clk freq 45 MHz
Maity et al. [16]	Invisible Semi-Fragile	FPGA Board	Image	RCM-RIW	Xilinx Virtex, xcv3200efg1156-6, 67.98 MHz (embedding)
Maity et al. [17]	Invisible Semi-Fragile	FPGA Board	Image	RCM-RIW	Xilinx Spartan-3E, XC3S1600E-4FG320, 98.7 MHz
Ghosh et al. [19]	Invisible Semi-Fragile	FPGA,P-SoC, Ultra-Scale FPGA	Image	(DE-IRI)-RIW	Xilinx, Power (Embedding) = 7.2 Watt (Virtex-7), 3.7 Watt (Zynq-7000), 1.4 Watt (Ultra-Scale FPGA). Frequency (Embedding) = 123.4 MHz (Virtex-7), 168.23 MHz (Zynq-7000), 204.6MHz (Ultrascale FPGA). Power (Decoding) = 9.2 Watt (Virtex-7), 6.7 Watt (Zynq-7000), 3.4 Watt (Ultrascale FPGA). Frequency (Decoding)= 138.1 MHz (Virtex-7), 178.33MHz (Zynq-7000), 210.436 MHz (Ultrascale FPGA).
Ghosh et al. [20]	Invisible Semi-Fragile	FPGA Board	Image	DE-RIW	NA
This work	Invisible Semi-Fragile	FPGA Programmable SoC Ultra-Scale FPGA	Image	RCM-RIW	Xilinx SPARTAN-3E FPGA, Zynq-7000 P-SoC, Ultra-scale Kintex-7 FPGA, 5.24 msec (for 512 × 512), latency = 40 ns, clock = 50 MHz, throughput = 200 Mbps.

REFERENCES

[1] N. A. Loan, N. N. Hurrah, S. A. Parah, J. W. Lee, J. A. Sheikh and G. M. Bhat, Secure and Robust Digital Image Watermarking Using Coefficient Differencing and Chaotic Encryption, *IEEE Access*, vol. 6, pp. 19876–19897 (2018). doi: 10.1109/ACCESS.2018.2808172

[2] S.J. Horng, D. Rosiyadi, T. Li, T. Takao, M. Guo and M. K. Khan, A Blind Image Copyright Protection Scheme for e-government, *Journal of Visual Communication and Image Representation*, vol. 24, no. 7, pp. 1099–1105 (2013).

[3] S.J. Horng, M. E. Farfoura, P. Fan, X. Wang, T. Li and J.M. Guo, A Low Cost Fragile Watermarking Scheme in H.264/AVC Compressed Domain, *Multimedia Tools and Applications*, vol. 72, pp. 2469–2495 (October 2014).

[4] A. H. M. Mohashin, P. Kumar Dhar and T. Shimamura, Blind Image Watermarking Based on Discrete Hilbert Transform and Polar Decomposition, *2019 11th International Conference on Knowledge and Smart Technology (KST)*, Phuket, Thailand, pp. 78–81 (2019). doi: 10.1109/KST.2019.8687512

[5] J. O. Ruanaidh and T. Pun, Rotation Scale and Translation Invariant Spread Spectrum Digital Image Watermarking, *Signal Process*, vol. 66, pp. 303–317 (1998).

[6] O. Kwon, S. Choi and B. Lee, A watermark-based Scheme for Authenticating JPEG Image Integrity, *IEEE Access*, vol. 6, pp. 46194–46205 (2018). doi: 10.1109/ACCESS.2018.2866153

[7] S. Ghosh, A. Biswas, S. P. Maity and H. Rahaman, Field Programmable Gate Array and System-on-Chip Based Implementation of Discrete Fast Walsh-Hadamard Transform Domain Image Watermarking Architecture For Real-Time Applications, *Journal of Low Power Electronics (JOLPE)*, vol. 11, no. 3, pp. 375–386, (September 2015).

[8] Y. Tan, J. Qin, X. Xiang, W. Ma, W. Pan and N. N. Xiong, A Robust Watermarking Scheme in YCbCr Color Space Based on Channel Coding, *IEEE Access*, vol. 7, pp. 25026–25036, (2019). doi: 10.1109/AC-CESS.2019.2896304

[9] J. Tian, Reversible Data Embedding Using a Difference Expansion, *IEEE Transactions on Circuits Systems and Video Technology*, vol. 13, no. 8, pp. 890–896 (2003).

[10] D. Coltuc and J. M. Chassery, Very Fast Watermarking by Reversible Contrast Mapping. *IEEE Signal Processing Letters*, 14, pp. 255–258 (2007).

[11] C. Dragoi and D. Coltuc, Improved Rhombus Interpolation for Reversible Watermarking by Difference Expansion, *IEEE EUSIPCO*, pp. 1688–1692 (2012).

[12] Z. Ni, Y. Q. Shi, N. Ansari and W. Su, Reversible Data Hiding, *IEEE Transactions on Circuits and Systems for Video Technology*, vol. 16, no. 3, pp. 354–362 (2006).

[13] D. M. Thodi and J. J. Rodriguez, Prediction-error based reversible watermarking, in *Proceedings of the ICIP International Conference on Image Processing*, vol. 3, Genova, pp. 1549–1552 (2004).

[14] A. Vashistha and A. M. Joshi, Fingerprint based biometric watermarking architecture using integer DCT, *2016 IEEE Region 10 Conference (TENCON)*, Singapore, pp. 2818–2821 (2016). doi: 10.1109/TENCON.2016.7848556

[15] S. Ghosh, B. Kundu, D. Datta, S. P. Maity and H. Rahaman, Design and Implementation of Fast FPGA based Architecture for Reversible Watermarking, in *IEEE International Conference on Electrical Information and Communication Technology (EICT)*, pp. 1–6, 19–21 December (2013).
[16] H. K. Maity, S. P. Maity and C. Delpha, A Modified RCM for Reversible Watermarking with FPGA Implementation, *4th European Workshop on Visual Information Processing (EUVIP)*, pp. 100–105, 10–12 June (2013).
[17] H. K. Maity and S. P. Maity, FPGA Implementation of Reversible Watermarking in Digital Images Using Reversible Contrast Mapping, *Elsevier Journal of Systems and Software*, vol. 96, pp. 93–104, October (2014).
[18] S. P. Maity and M. K. Kundu, Distortion Free Image-in-image Communication with Implementation in FPGA, *International Journal of Electronics and Communications*, vol. 67, pp. 438—447 (2013).
[19] S. Ghosh, N. Das, S. Das, S. P. Maity and H. Rahaman, FPGA and SoC Based VLSI Architecture of Reversible Watermarking Using Rhombus Interpolation By Difference Expansion, in *11th IEEE India Conference INDICON 2014* at Yashada, Pune, India, pp. 1–6 (December 2014).
[20] S. Ghosh, N. Das, S. Das, S. P. Maity and H. Rahaman, Digital Design and Pipelined Architecture for Reversible Watermarking Based on Difference Expansion using FPGA, in *13th IEEE International Conference on Information Technology (ICIT 2014)* at Bhubaneswar, Orrisa, India, pp. 123–128 (December 2014).
[21] L. D. Strycker, P. Termont, J. Vandewege, J. Haitsma, A. Kalker, M. Maes, and G. Depovere, Implementation of a Real-Time Digital Watermarking Process for Broadcast Monitoring on Trimedia VLIW Processor, *IEE Proceedings on Vision, Image and Signal Processing*, vol. 147, no. 4, pp. 371—376 (2000).
[22] Md. Shohidul Islam, Cheol-Hong Kim, Jong-Myon Kim, A GPU-based (8, 4) Hamming decoder for secure transmission of watermarked medical images. *Springer Journal of Cluster Computing*, vol. 18, no. 1, pp. 333–341 (2015). doi: 10.1007/s10586-014-0392-x
[23] P. Karthigai Kumar, K. Baskaran, An ASIC Implementation of a Low Power Robust Invisible Watermarking Processor, *Elsevier Journal of Systems Architecture*, vol. 57, no. 4, pp. 404–411 (2011).
[24] S. Das, S. Ghosh, N. Das, et al. VLSI-Based Pipeline Architecture for Reversible Image Watermarking by Difference Expansion with High-Level Synthesis Approach, *Springer Circuits Syst Signal Process (CSSP)*, vol. 37, no. 4, pp. 1575–1593 (2018). doi: 10.1007/s00034-018-0979-1

Chapter 5

Framework for UAV-based wireless power harvesting

Tanishk Singhal, Manoj Kumar Shukla, Harpreet Singh Bedi and Yogesh Kumar Verma

5.1 INTRODUCTION

It traces the evolutionary path of Unmanned Aerial Vehicles (UAVs), which illuminates inadequacies inherent in traditional sources of power. Since UAVs are increasingly ubiquitous in a wide range of applications, the need for sustainable and efficient power solutions increases. This chapter introduces a novel framework that integrates solar power, radio frequency harvesting, and energy scavenging to present a comprehensive solution to the specific energy needs of UAVs [1]. Based on synthesis from literature review, methodology, and results, this chapter becomes a backdrop to the study of the theoretical and practical implications of UAV-based wireless power harvesting [2]. In that line, a transformative period would then start to consider where this framework can transform unmanned aerial system capabilities to promise sustainability, efficiency, and autonomy.

5.2 LITERATURE REVIEW

On analyzing this, it requires looking at the current scenario of what options are available with power sources. The usual batteries – the same for all applications with reliability – make terrible choices because energy density as well as their weight hurt endurance and UAV ranges so considerably. More efficient alternatives have been sought and, in turn, brought forward the fuel-based systems. These have their own problems regarding environmental impact and logistics [3, 4]. In UAV applications, which now range from surveillance to delivery services, shortcomings of these traditional power solutions have begun to surface more and more.

Among these challenges, wireless power harvesting technologies are a promising frontier. Solar power is one of the stalwart renewable energy sources that find application in UAVs by integrating solar panels on the surface [5]. It capitalizes on sunlight to convert it into electrical power and extend operational windows, especially in regions with ample sunshine. The other route is RF harvesting. It converts ambient radio frequency signals to

DOI: 10.1201/9781003643173-5

electrical energy. Although continuous power sources can be available, the RF harvesting devices operate at relatively low powers, requiring caution in addressing the trade-offs [6]. Energy scavenging is an emerging field to scavenge ambient environmental energy, such as vibrations or airflow, for UAV systems.

Relevance to UAVs: Application of wireless power harvesting technology to UAVs gives an innovative dimension to the power dynamics of UAVs. Solar-powered UAVs with efficient solar panels are equipped to provide extended flight durations. These UAVs can, therefore, be used in regions where the sun shines more brightly [7]. RF harvesting, though having a continuous but low-power output, is an opportunity for on-the-go recharging that might reduce the dependency on conventional batteries during flight. Energy scavenging, though in its nascent stages, promises to tap and capture untapped ambient energy, thereby augmenting the overall efficiency of energy utilization. As UAVs are developed with increasing complexity and evolving mission profiles, so will the adaptability and versatility of these wireless power harvesting technologies.

5.3 PROPOSED FRAMEWORK

Components of the framework: In this endeavor to overcome the limitations of traditional sources of power for UAVs and to pursue opportunities of wireless power harvesting technologies, our proposed framework provides a holistic solution [8, 9]. The elements woven into the framework are not just simple constituents but rather synergistic building blocks that have been carefully designed to optimize energy acquisition, distribution, and utilization in the dynamic operational context of a UAV.

The basis of our framework is built upon advanced energy harvesting devices chosen based on the specific sources that can be found within a UAV's operational environment. It includes high-efficiency solar panels for sun-rich areas, tuned RF harvesting antennas for continuous on-the-go recharging, and state-of-the-art energy scavenging modules that will scavenge ambient environmental energy when flight takes place. This ensures a diversified and efficient energy acquisition strategy through the synergy of these devices.

Communication systems: Any framework is effective only if communication is perfect. Our proposed framework is rich in robust communication systems that enable real-time data transfer between the energy harvesting devices, the UAV's power management system, and ground

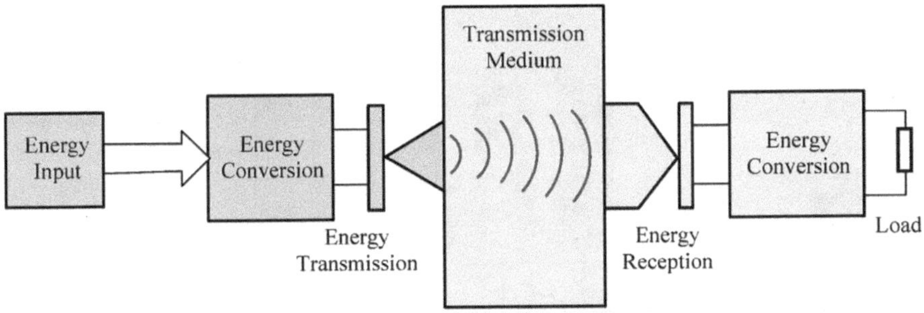

Figure 5.1 Proposed framework for energy harvesting and transmission using UAV technology.

control [10]. This ensures not only the efficient monitoring of energy levels but also dynamic adaptation to changing environmental conditions [11]. The communication capabilities of the framework allow for a symbiotic relationship between the UAV and its energy sources, with optimum performance in all scenarios.

Power management: The operational capabilities of UAVs can be extended efficiently if energy is utilized correctly. In this regard, advanced power management algorithms have been incorporated into the framework. These algorithms work like surgeons, optimizing the way in which harvested energy is distributed to different on-board systems [12]. It encompasses an integral component, where the vital functions are prioritized, with dynamic adjustment of the power distribution in response to the real-time demand and smart management of surplus energy with storage or transmission of power for a judicious utilization of the available power. Figure 5.1 shows a proposed framework for energy harvesting and transmission using UAV technology.

5.4 INTEGRATION WITH UAV SYSTEMS

What will mark the distinctive feature of our proposed framework is the ability to integrate seamlessly into current UAV systems. Knowing that UAVs come in every form and size, it is designed to be adaptive. From compact surveillance drones to large cargo UAVs, components fit together with the aim of enhancing energy efficiency without losing any of the core functionalities of the UAV. This integration is not an overlay but an enhancement that transforms the framework to make it a part of the UAV architecture [13, 14]. It would further increase the overall success and operational sustainability of missions as it turns the UAV into more than a flying device but rather an integrated, sophisticated, self-sustaining system. Figure 5.2 shows energy optimization in UAV-based cellular networks and approaches.

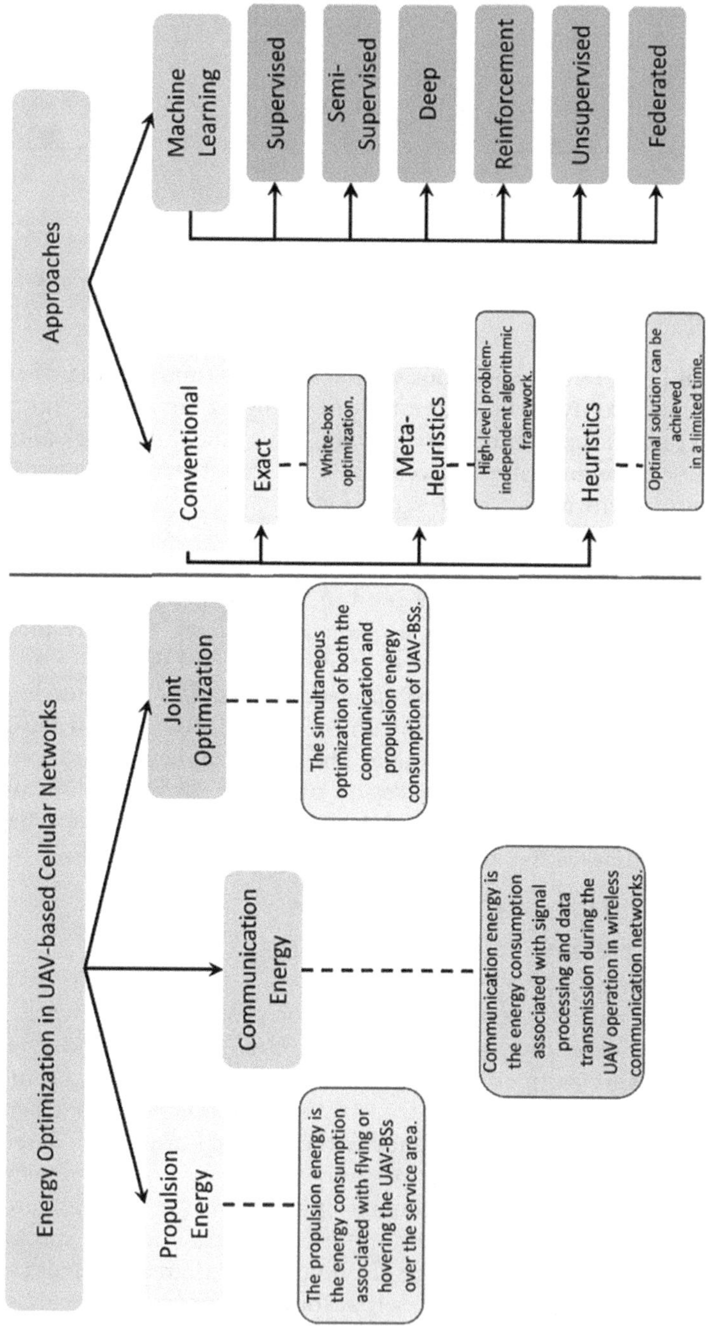

Figure 5.2 Energy optimization in UAV-based cellular networks and approaches.

Adaptability and scalability: Given the continued revolution of UAV technology, our proposed framework is based on adaptability and scalability as its core elements [15]. As technology for new energy harvesting devices emerges, and as the types of missions of UAVs change, this framework can expand to accept newer technologies. Modular in design, its advantage lies not only in the fact that it can be scaled to accommodate different sizes and capabilities of UAVs ranging from micro-drones to high-altitude, long-endurance vehicles [16, 17]. This aspect goes to show how the framework will definitely keep up with the ever-evolving UAV technologies. We shall look at each of the subparts in detail in the following sections and break down their combinations, testing methods, and actual applications.

The proposed framework is better described as a dynamic set of principles and a kind of living blueprint that might dynamically adjust to the pacing of technological change. Guiding UAV technology toward future wireless power harvesting, these unmanned aerial systems will make it possible to transcend limits and redefine what is perceived as sustainable, efficient, and autonomous operation [18]. Join us as we work out the complex machinations of this revolutionary framework to discover how it will likely change the future of UAV energy sustainability.

5.5 METHODOLOGY

Simulation and testing: The journey toward the realization of the proposed framework is a careful intermingling of the conceptual and practical validation of the proposed framework [19]. Methodologies on simulation and testing are the linchpins to this process, making sure the envisioned framework translates smoothly into a robust and adaptive solution for UAVs.

In simulation, we use state-of-the-art computer-aided tools to expose the framework to an immense variety of virtual scenarios. These simulations reproduce in high detail the many varied environmental conditions, mission profiles, and energy harvesting challenges that UAVs will be exposed to [20]. This iterative process enables us to adjust the algorithms that govern the framework so as to fine-tune them for actual applications. The simulations will act as a proving ground where the responses of the framework are analyzed under different parameters to give critical insights for further refinement. Figure 5.3 is an illustration of UAV-enabled wireless-powered MEC.

Real-world testing follows suit by taking the framework from the digital world into the dynamic complexities of the real world [21]. The field tests deploy the UAVs carrying the framework in controlled and dynamic environments that would closely mirror the unpredictable conditions of real missions

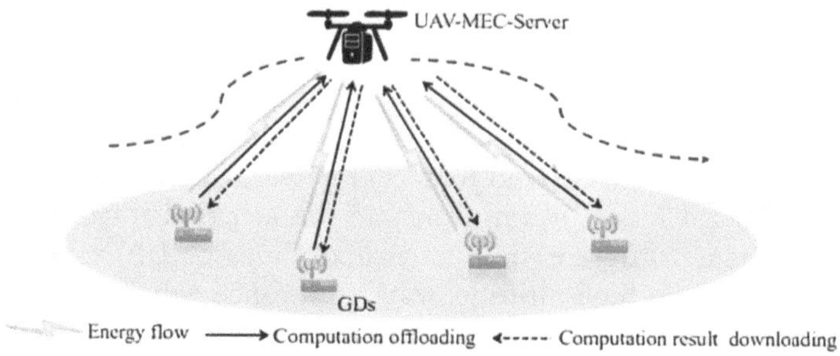

Figure 5.3 Illustration of UAV-enabled wireless-powered MEC.

[22, 23]. These tests are followed by an essential feedback loop: data gathering about energy harvesting rates, power distribution efficiency, and adaptability to unforeseen challenges. The framework developed and refined in this iterative cycle of simulation and testing grows not in isolation but with continuous dialogue with the practical demands of UAV operations.

Data collection: The methodology's robustness lies in its commitment to comprehensive data collection. As the framework undergoes simulation and real-world testing, a wealth of information is gathered, spanning energy consumption patterns, environmental variables, communication system performance, and the overall health of the UAV's power management system [24].

Quantitative measures are found in terms of energy efficiency ratios, power transmission rates, and adaptive response times. The qualitative data is derived from the usage of user feedback and operational observation to understand the real implications of the framework in reality [25]. This brings together both quantitative and qualitative measures that form a comprehensive data set, which not only establishes the theoretical basis of the framework but also forms the basis for iterative optimization.

5.6 RESULTS AND DISCUSSIONS

Performance metrics: The culmination of simulation, testing, and data collection yields a rich tapestry of performance metrics that quantifies the framework's efficacy. Energy efficiency emerges as a pivotal metric, measured in terms of the energy harvested per unit time. This metric serves as a litmus test for the framework's ability to sustain UAV operations over extended periods. Metrics in terms of rates of power transmission and system reliability of communication systems indicate how the framework responds to dynamic conditions of operation characteristically typical of UAVs in real time. Figure 5.4 achieved

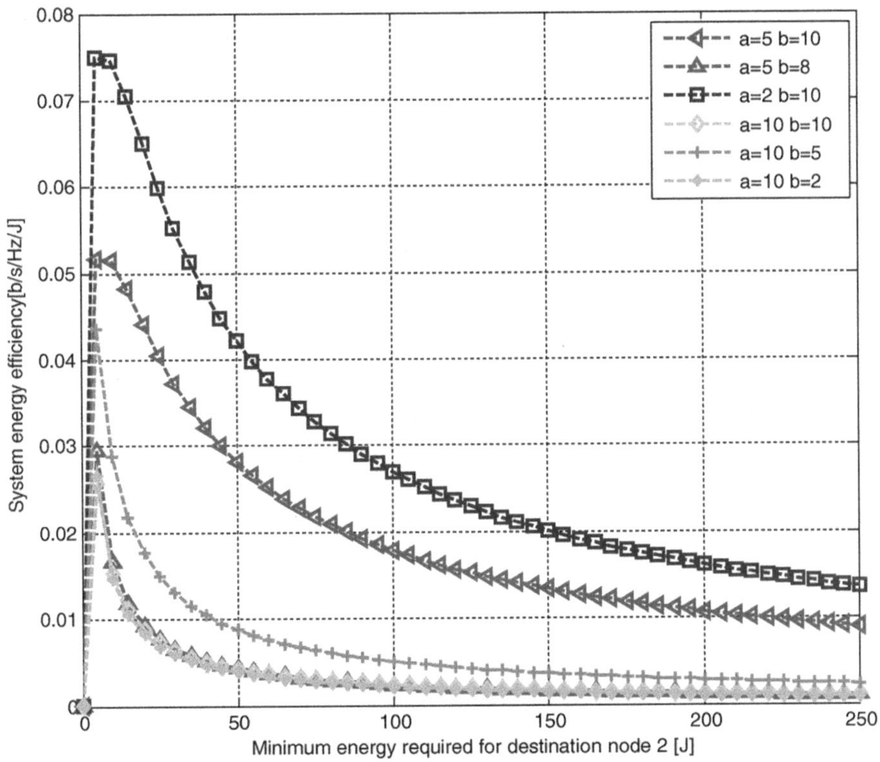

Figure 5.4 Achieved energy efficiency of the whole IoT systems by the proposed scheme at the power splitting factor ρ = 0.5.

energy efficiency of the whole IoT system by the proposed scheme at the power splitting factor ρ = 0.5.

The granularities of such metrics allow fine distinctions between strengths of the framework and areas of improvement. Ratios of energy efficiency will then portray how efficiently the given structure can convert ambient energies into usable power, while transmission rates of power will portray how adaptable the framework will be to environmental changes. Instead of being a mere set of numbers, now these become measures for the performance of the framework, orienting its course toward ideal functionality. Figure 5.5: Achieved energy efficiency of the whole IoT system by the proposed scheme at the power splitting factor ρ = 0.1.

Comparison with existing solutions: To place the significance of the proposed framework in perspective, a detailed comparative analysis with existing UAV power solutions has been carried out. Here, traditional battery-powered UAVs and those that exploit solar, RF, or other wireless power harvesting technologies have been used as references for this

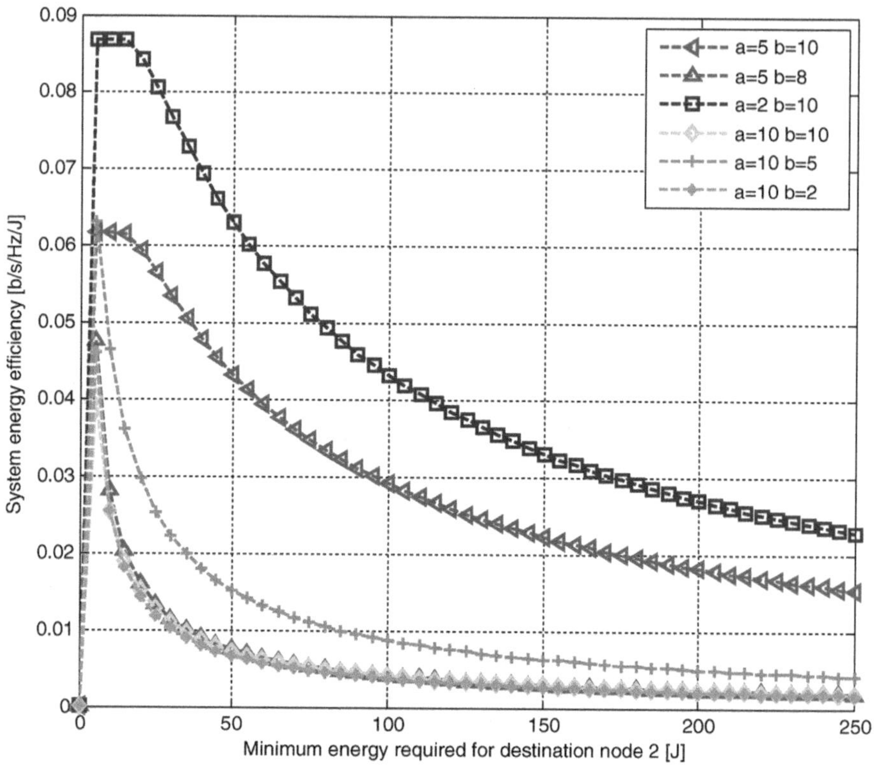

Figure 5.5 Achieved energy efficiency of the whole IoT systems by the proposed scheme at the power splitting factor ρ = 0.1.

comparative analysis [26]. Besides merely pointing out the strengths of the proposed framework, the objective is to situate it within the context of UAV power solutions as a whole. Figure 5.6 achieved energy efficiency of the whole IoT system by the proposed scheme at the power splitting factor ρ = 0.9.

This comparative approach develops the innovations and differentiating advantages that the framework under consideration portrays. We are contributing toward the academic discussion as we are providing practical insights in industries and organizations seeking ways to uplift the capability of their UAV fleets through this rigorous comparison: a critical bridge between theoretical advancement and real-world application that propels the proposed framework into a transformative force for UAV power solutions. Figure 5.7 achieved energy efficiency of the whole IoT system by the proposed scheme at different power splitting factors ρ with a = 5, b = 10.

We progress from the groundwork of methodology into a deeper examination of results. With rigorous interpretation and discussion, we open the implications of our findings. This will guide the way forward with the

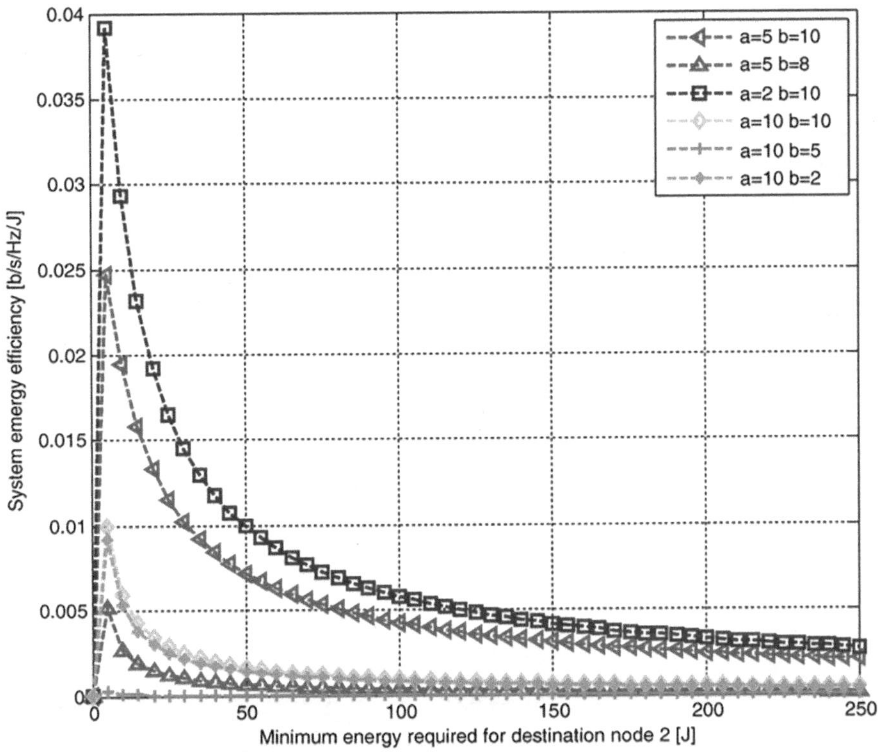

Figure 5.6 Achieved energy efficiency of the whole IoT systems by the proposed scheme at the power splitting factor ρ = 0.9.

paradigm shift in power solutions for UAVs, as the proposed framework promises. With that, we begin our expedition through the nuances of the methodology and the impressive results that set up a transformational era for UAV-based wireless power harvesting. Figure 5.8 shows the comparison between the asymptotically optimal result and the globally optimal result.

Interpretation of data: Data interpretation is not only a statistical analysis but also a very fine exploration of the underlying stories in the dataset. As we go through the large amount of information collected from simulations and real-world tests, subtle patterns and some unexpected correlations start to emerge [27]. The interpretation goes beyond numerical values, seeking to understand the context behind the data points and uncovering the intricate interplay of variables. It's not just about what the numbers convey but how they tell the story of the framework's performance under diverse conditions.

The metrics, such as energy efficiency ratios or power transmission rates, are statistically analyzed. We investigate how the variability of environmental conditions, mission profiles, and UAV configurations impacts the performance

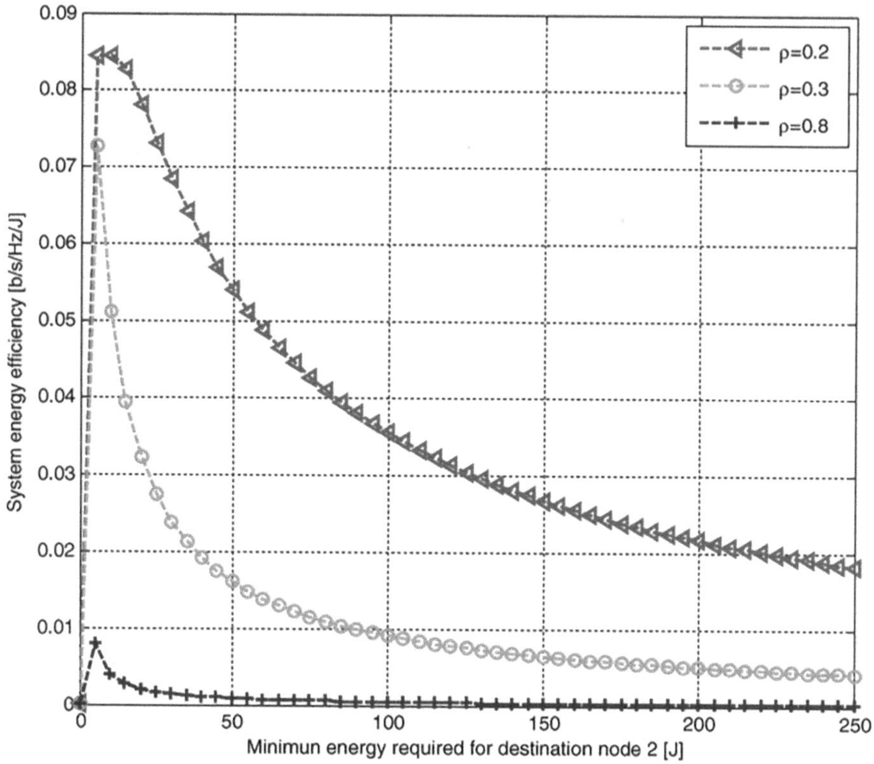

Figure 5.7 Achieved energy efficiency of the whole IoT systems by the proposed scheme at the different power splitting factor ρ with a = 5, b = 10.

of these frameworks. Synthesis of the results enables us to clearly define the strengths and advantages – and challenges in some operational scenarios.

5.7 CONCLUSION

All of the above aspects have been investigated through this study, and so, a basis has been established for further research, which is needed to uncover the reasons for the use of UAVs for wireless power harvesting. The part includes UAV transition, key problems of traditional power sources, the major advantages of the newly emerging wireless power harvesting technologies, and other aspects needed for the development of an efficient nationwide power system.

We have observed simulations, testing, and data interpretation in the course of this journey, and now we know what it has for the UAV operators of the future. Key areas of this chapter – Positive Ethical Impacts, Related Regulatory Scene, and Recommendations for Future Articles – are explored to zero in on the overall essence of the proposed framework for the UAV industry.

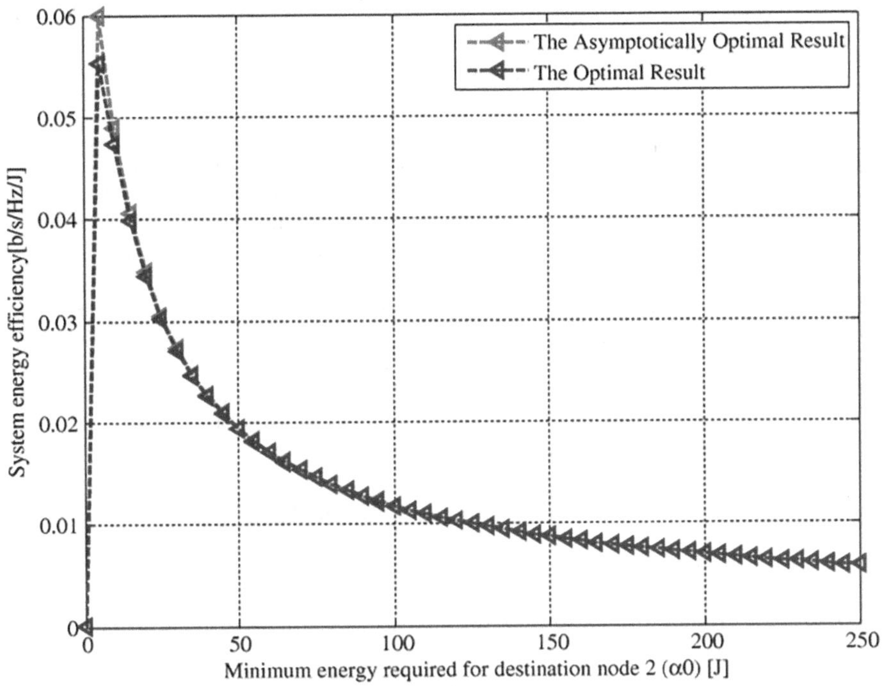

Figure 5.8 The comparison between the asymptotically optimal result and the globally optimal result.

DECLARATION OF COMPETING INTEREST

The authors declare that they have no known competing financial interests or personal relationships that could have appeared to influence the work reported in this paper.

CONSENT FOR PUBLICATION

Not applicable.

CONFLICT OF INTEREST

The authors declare no conflict of interest, financial or otherwise.

ACKNOWLEDGMENT

Declared none.

REFERENCES

[1] H. Wang et al. "Power control in UAV-supported ultra dense networks: communications, caching, and energy transfer," *IEEE Commun. Mag.*, vol. 56, no. 6, pp. 28–34, 2018.

[2] A. Orsino et al. "Energy efficient IoT data collection in smart cities exploiting D2D communications," *Sensors*, vol. 16, no. 6, p. 836, 2016.

[3] Y. Peng et al. "Prolonging sensor network lifetime through wireless charging," in *Proceedings of the 31st IEEE Real-Time Systems Symposium (RTSS)*, San Diego, CA, USA, Dec. 2010, pp. 129–138.

[4] Z. Sheng et al. "Surfing the internet-of-things: lightweight access and control of wireless sensor networks using industrial low power protocols," *EAI Endorsed Trans. Ind. Netw. Intell. Syst.*, vol. 1, no. 1, p. e2, 2014.

[5] K.-W. Fan et al. "Steady and fair rate allocation for rechargeable sensors in perpetual sensor networks," in *Proceedings of the 6th ACM Conference on Embedded Networked Sensor Systems (SenSys)*, Raleigh, NC, USA, Nov. 2008, pp. 239–252.

[6] M. Mozaffari et al. "A tutorial on UAVs for wireless networks: applications, challenges, and open problems," *IEEE Commun. Surv. Tutor.*, vol. 21, no. 3, pp. 2334–2360, 2019.

[7] M. Mozaffari et al. "Efficient deployment of multiple unmanned aerial vehicles for optimal wireless coverage," *IEEE Commun. Lett.*, vol. 20, no. 8, pp. 1647–1650, 2016.

[8] A. Mairaj et al. "Application specific drone simulators: recent advances and challenges," *Simul. Modell. Pract. Theory*, vol. 94, pp. 100–117, 2019.

[9] L. Gupta et al. "Survey of important issues in UAV communication networks," *IEEE Commun Surv. Tutor.*, vol. 18, no. 2, pp. 1123–1152, 2015.

[10] X. Lu et al. "Wireless networks with RF energy harvesting: acontemporary survey," *IEEE Commun. Surv. Tutor.*, vol. 17, no. 2, pp. 757–789, 2014.

[11] M. Simic et al. "Investigation in wireless power transmission for UAV charging," *Proc. Comput. Sci.*, vol. 60, pp. 1846–1855, 2015.

[12] R. Want et al. "Enabling the internet of things," *Computer*, vol. 48, no. 1, pp. 28–35, 2015.

[13] S. Pattar et al. "Searching for the IoT resources: fundamentals, requirements, comprehensive review, and future directions," *IEEE Commun. Surv. Tutor.*, vol. 20, no. 3, pp. 2101–2132, 2018.

[14] S. Zhang et al. "Fundamental green tradeoffs: progresses, challenges, and impacts on 5G networks," *IEEE Commun. Surv. Tutor.*, vol. 19, no. 1, pp. 33–56, 2016.

[15] Q. Wu et al. "An overview of sustainable green 5G networks," *IEEE Wireless Commun.*, vol. 24, no. 4, pp. 72–80, 2017.

[16] C. Zhong, J. Yao, and J. Xu, "Secure UAV communication with cooperative jamming and trajectory control," *IEEE Commun. Lett.*, vol. 23, no. 2, pp. 286–289, 2019.

[17] H. Wang, J. Wang, G. Ding, J. Chen, Y. Li, and Z. Han, "Spectrum sharing planning for full-duplex UAV relaying systems with underlaid D2D communications," *IEEE J. Sel. Areas Commun.*, vol. 36, no. 9, pp. 1986–1999, 2018.

[18] M. Mozaffari, W. Saad, M. Bennis, and M. Debbah, "Unmanned aerial vehicle with underlaid device-to-device communications: performance and tradeoffs," *IEEE Trans. Wireless Commun.*, vol. 15, no. 6, pp. 3949–3963, 2016.

[19] A. Asheralieva and D. Niyato, "Game theory and lyapunov optimization for cloud-based content delivery networks with device-to-device and UAV-enabled caching," *IEEE Trans. Veh. Technol.*, vol. 68, no. 10, pp. 10094–10110, 2019.
[20] A. Al-Hourani and K. Gomez, "Modeling cellular-to-UAV path-loss for suburban environments," *IEEE Wireless Commun. Lett.*, vol. 7, no. 1, pp. 82–85, 2018.
[21] Y. Wang, W. Yang, X. Shang, and Y. Cai, "Energy-efficient secure transmission for UAV-enabled wireless powered communication," in *Proc. IEEE International Conference on Wireless Communications and Signal Processing (WCSP)*, Hangzhou, China, 2018, pp. 1–5.
[22] J. Wang, C. Jin, Q. Tang, N. Xiong, and G. Srivastava, "Intelligent ubiquitous network accessibility for wireless-powered MEC in UAV-assisted B5G," *IEEE Trans. Netw. Sci. Eng.*, vol. 8, no. 4, pp. 2801–2813, 2020.
[23] X. Mo, Y. Huang, and J. Xu, "Radio-map-based robust positioning optimization for UAV-enabled wireless power transfer," *IEEE Wireless Commun. Lett.*, vol. 9, no. 2, pp. 179–183, 2020.
[24] L. Xie, J. Xu, and Y. Zeng, "Common throughput maximization for UAV-enabled interference channel with wireless powered communications," *IEEE Trans. Commun.*, vol. 68, no. 5, pp. 3197–3212, 2020
[25] Y. Zeng and R. Zhang, "Energy-efficient UAV communication with trajectory optimization," *IEEE Trans. Wireless Commun.*, vol. 16, no. 6, pp. 3747–3760, 2017.
[26] C. E. Miller, A. W. Tucker, and R. A. Zemlin, "Integer programming formulation of traveling salesman problems," *J. ACM*, vol. 7, no. 4, pp. 326–329, 1960. [Online]. Available: https://doi.org/10.1145/321043.321046
[27] F. Wu, D. Yang, L. Xiao, and L. Cuthbert, "Energy consumption and completion time tradeoff in rotary-wing UAV enabled WPCN," *IEEE Access*, vol. 7, pp. 79617–79635, 2019.

Chapter 6

Smart healthcare devices

Chandrappa S, Guru Prasad M S, Naveen Kumar H N, Praveen Gujjar J, Amith K Jain and Aditya Pai H

6.1 EMERGING SMART HEALTHCARE SYSTEMS

Emerging smart healthcare systems present a major shift in healthcare services, such as delivered, managed, and accessed. It is based on cutting-edge technologies such as artificial intelligence (AI), the Internet of Things (IoT), Big Data Analytics, and Wearable Devices to revolutionize patient care and streamline healthcare operations (Prasad et al., 2019). The technologies associated with smart healthcare are shown in Figure 6.1. The primary feature of this system is to monitor the patient remotely, where wearable sensors collect real-time health data, and the healthcare analyst will analyze the patient data remotely in patients' vital signs and health parameters. The advantages of this are early detection of health-related issues, personalized care, and proper management of chronic conditions.

Due to smart healthcare systems, there is an improvement in healthcare operations through effective resource utilization and less administrative burdens. The algorithms based on AI analyze large amounts of healthcare data to identify meaningful patterns, trends, and correlations between the features that can help clinical decision-making and improve treatment practices. Healthcare providers make use of predictive analytics to fulfill patient needs and resource allocations. These systems additionally provide easy communication and collaboration with healthcare professionals, allowing them to share patient information and collaborate on treatment plans (Kirubasri et al., 2024).

The emerging smart healthcare systems help patients to take a more active role in managing their health and wellness. Smart health applications and wearable devices allow individuals to track their health metrics and access personalized health recommendations in real time. Patients can make use of telemedicine consultations with healthcare providers, virtual coaching assistance, and participation in health-related virtual programs. By promoting patient engagement and self-management, smart healthcare systems allow individuals to make informed decisions about their health and well-being, leading to better health outcomes and improved quality of life. Emerging

 DOI: 10.1201/9781003643173-6

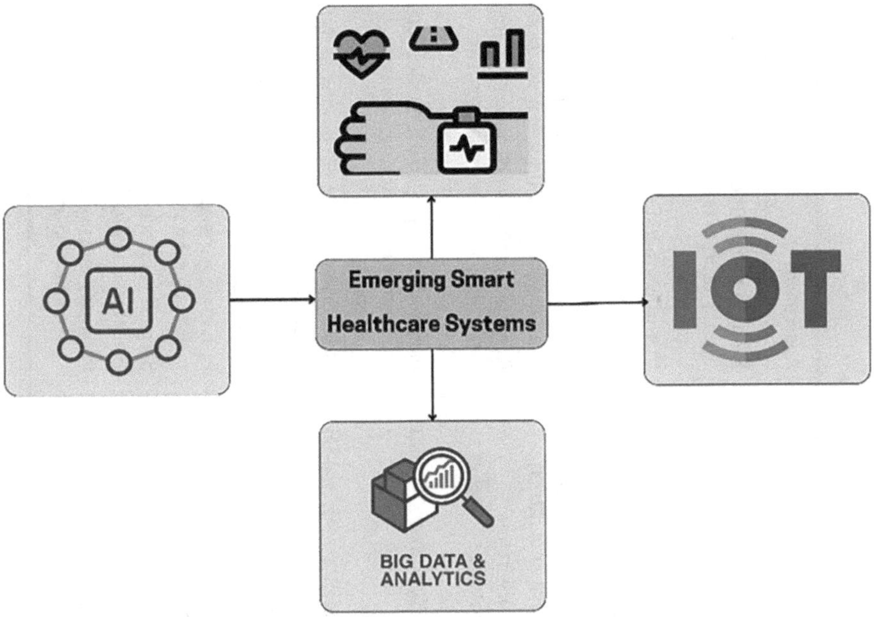

Figure 6.1 Technology associated with smart healthcare.

smart healthcare systems hold tremendous potential to revolutionize the healthcare industry by making use of advanced technologies to enhance patient care, improve operational efficiency, and empower individuals to take control of their health. By embracing innovation and adopting a patient-centered approach, healthcare providers can harness the benefits of these systems to deliver more personalized, efficient, and effective care to patients around the globe. Table 6.1 shows the tools, techniques, and applications of emerging smart healthcare systems.

6.2 INTERNET OF THINGS FOR SMART HEALTHCARE

Healthcare delivery and management are modernized with the help of IoT technology. Because of its working principles, it interconnects with various devices and sensors; IoT enables the continuous exchange of health data and supports remote monitoring (Gujjar et al., 2023). The main aim of IoT in smart healthcare is remote patient monitoring. The real-time tracking of patients' vital signs and health parameters is possible through wearable sensors and medical devices connected through IoT technology (Guru Prasad et al., 2023). This feature helps early detection of health issues, timely interventions, and personalized treatment plans, leading to improved patient outcomes.

Table 6.1 The tools, techniques, and applications of emerging smart healthcare systems

Category	*Tools*	*Techniques*	*Applications*
Wearable devices	Smartwatches, fitness trackers, health monitoring patches	Machine learning algorithms, sensors, Bluetooth connectivity	Remote patient monitoring, fitness tracking, real-time health data collection
Telemedicine	Telemedicine platforms, video conferencing tools	AI-driven chatbots, real-time video streaming, secure data transmission (Anand Kumar et al., 2023)	Virtual consultations, remote diagnosis, follow-up appointments
AI in healthcare	AI algorithms, deep learning models, NLP	Predictive analytics, pattern recognition, voice recognition (Pai et al., 2023)	Clinical decision support, personalized treatment plans, disease prediction
IoT in healthcare	IoT sensors, medical devices, smart home devices	Edge computing, RFID, wireless communication protocols (Patel et al., 2023)	Remote monitoring of patients, asset tracking, inventory management
Mobile health apps	Health tracking apps, symptom checkers, medication reminders	Cloud computing, GPS, push notifications	Health data management, medication adherence monitoring, wellness coaching
Remote monitoring	Remote monitoring devices, IoT sensors	Data analytics, machine learning, real-time alerts (Nagesh and Prabhu, 2017)	Monitoring chronic conditions, early detection of health issues
Blockchain in healthcare	Blockchain platforms distributed ledgers	Cryptography, decentralized data storage	Secure health data exchange, patient records management
Virtual reality (VR)	VR headsets, immersive simulations	3D modeling, haptic feedback (Anand Kumar et al., 2022)	Pain management, medical training, therapy sessions
Augmented reality (AR)	AR glasses, heads-up displays	Computer vision, spatial mapping	Surgical guidance, medical education, anatomical visualization
Robotics in healthcare	Surgical robots, exoskeletons	Computer vision, motion planning (Avinash et al., 2023)	Minimally invasive surgery, rehabilitation assistance, elder care

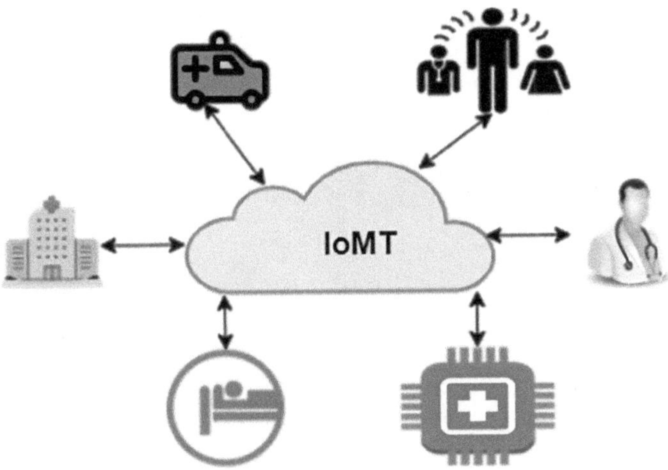

Figure 6.2 Internet of medical things.

IoT improves the easy management of chronic disease by empowering patients to monitor their health status and follow treatment plans. Connected devices allow individuals with chronic conditions such as diabetes, hypertension, or heart disease to track relevant metrics, receive medication reminders, and share data with their healthcare providers remotely (Chandrappa, 2015, 2017a, 2017b, 2019, 2021, 2023a, b), as shown in Figure 6.2. This continuous monitoring through the feedback process allows healthcare providers to intervene promptly in case of any deviations from the treatment plan, thereby preventing complications and improving disease management.

Improving age-old person care is also done by the IoT. Smart home devices attached to IoT sensors can monitor daily activities, detect falls, etc. The integration of this device with emergency response systems provides immediate assistance in case of emergencies. Additionally, IoT-enabled telemedicine solutions facilitate virtual consultations and remote healthcare services. The process of using the IoT for smart healthcare is shown in Figure 6.3.

Table 6.2 Sensors and communication standards for smart health care system

Sl. No	*Sensors*	*Communication standards*
1	Biometric sensors	Bluetooth
2	Activity trackers	Wi-Fi
3	Blood glucose monitors	Cellular networks
4	Pulse oximeters	Zigbee
5	Respiratory sensors	RFID
6	Temperature sensors	Near field communication
7	Weight scales	Long-range, low-power wireless protocol
8	Medication adherence sensors	Constrained application protocol
9	Environmental sensors	Message queuing telemetry transport

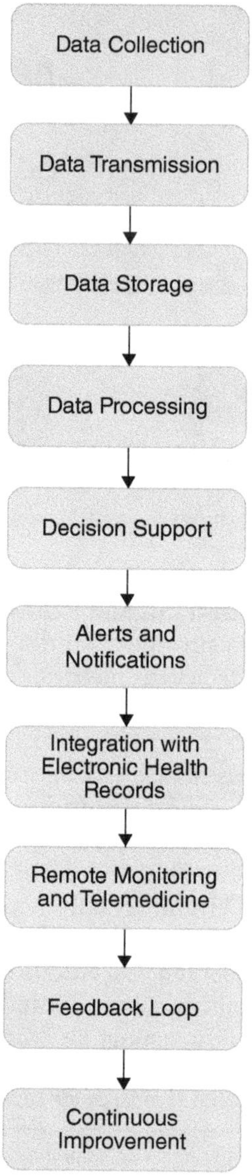

Figure 6.3 Flowchart Internet of Things for smart healthcare system.

The data is collected through sensors and IoT devices regarding patient vitals, environmental conditions, etc. The collected data is transmitted securely to a data repository through network protocols such as Wi-Fi, Bluetooth, or cellular networks to a central data repository. The data is stored in a secure database or cloud platform to ensure data integrity and accessibility. Data is processed by making use of analytics tools and algorithms to extract meaningful insights, detect anomalies, and predict trends.

6.2.1 Sensors and communication standards

The sensors and communication standards play a crucial role in enabling seamless data collection, transmission, and analysis in smart healthcare systems.

6.3 MACHINE LEARNING TECHNOLOGIES, CHALLENGES, AND OPPORTUNITIES IN SMART HEALTHCARE DEVICES

SHD allows enhanced investigative competencies using ML technologies. Nowadays, these ML technologies are progressively essential to the growth and functionality of SHD. This section illustrates the vital technologies, along with opportunities and related challenges, that ML brings to SHD. Deep learning (DL) is a sub-branch of ML, which includes innovative and advanced capabilities. DL is more powerful in handling large amounts of data and identifying patterns in big data, which are normally not noticeable by humans. In the healthcare industry, DL technologies are widely used because of their ability to learn from large datasets and make intelligent decisions.

6.3.1 Predictive analytics

In healthcare, predictive analytics uses several ML algorithms to analyze past data and predict upcoming health events, such as predicting the probability of a patient experiencing a chronic disorder. This consists of generating methods that can estimate the probability of conditions such as heart diseases, diabetes, brain tumors, breast cancer, or other diseases based on patients' lifestyles and other significant factors.

Table 6.3 Commonly used ML algorithms for predictive analytics

Sl. No	*ML algorithm*	*Key characteristics*
1	Logistic regression	• Simple and Interpretable • Binary Classification Tasks
2	Decision trees and random forests	• Powerful for handling complex datasets with multiple variables • Effective for classification and regression tasks
3	Gradient boosting machines (GBMs)	• Includes XGBoost • Suitable for error minimization and improved prediction accuracy
4	Support vector machines (SVM)	• Effective for high-dimensional spaces
5	Neural networks	• Effective for larger datasets or more complex relationships

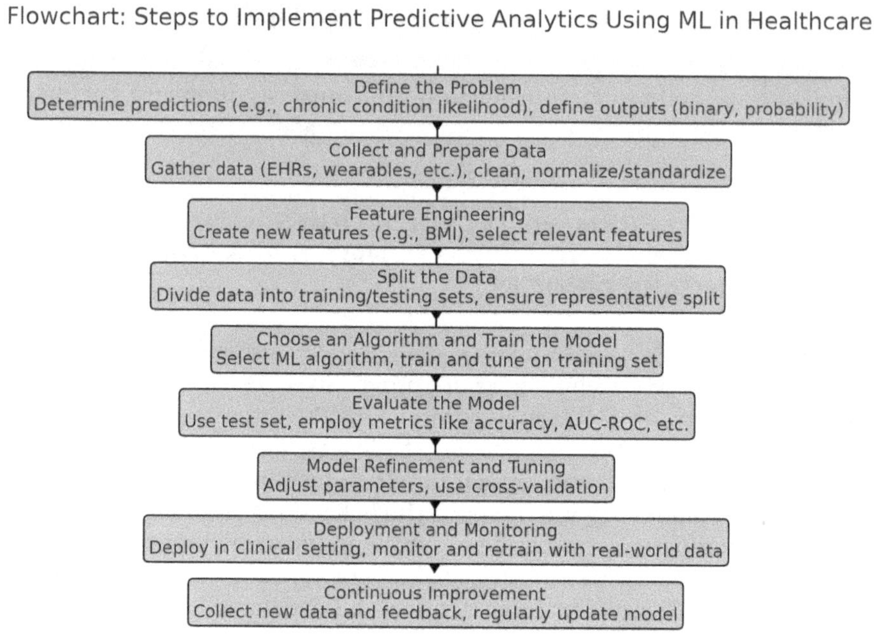

Figure 6.4 Steps to implement the predictive analysis.

General steps to implement predictive analytics using available ML algorithms are illustrated in the Figure 6.4

6.3.2 Diagnostic algorithms

Disease detection in the early stages plays a major role in the healthcare system. To predict and treat the disease at the early stage, many ML models are used, as shown in Table 6.4. These ML models help to diagnose diseases by identifying patterns, which are not readily visible to the human eye, based on trained datasets of clinical images like MRIs and X-rays. All diagnostic algorithms motorized by ML have expressively enhanced the area of medical imaging, providing tools to diagnose diseases by detecting patterns in clinical images. These ML models also detect sensitivities in images, which a human eye will not easily discern. This leads to earlier and more accurate diagnoses.

6.3.3 Wearable health monitors

Wearable health monitors are used to track important signs related to human organs like body temperature, heart rate, and blood pressure, and this system is helpful for disease diagnosis and patient monitoring. This system is also useful for sleep tracking and physical activity detection in real-time. Wearable health monitors have progressively extended their usability into

Table 6.4 Commonly used ML algorithms for diagnostic, wearable health monitors, natural language processing, and personalized treatment plans algorithms

Sl. No	*Diagnostic algorithms*	*Wearable health monitors algorithms*	*Natural Language Processing (NLP) algorithms*	*Personalized treatment plans algorithms*
1	Convolutional neural networks	**Time series analysis models**: Recurrent neural networks (RNNs), long short-term memory (LSTM) networks	**Rule-based systems**: Regular expressions (Regex)	**Supervised learning algorithms**: Regression models, support vector machines (SVM)
2	Transfer learning models	**Classification and regression models**: Support vector machines (SVM), random forests	**Classical machine learning algorithms**: Support vector machines (SVM), Naive Bayes classifiers	**Ensemble methods**: Random forests and gradient boosting machines (GBMs),
3	Autoencoders	**Signal processing models**: Fourier transforms and wavelet transforms **Deep learning models**: Convolutional neural networks (CNNs), deep belief networks (DBNs)	**Deep learning algorithms**: Recurrent neural networks (RNNs), long short-term memory (LSTM) networks, gated recurrent units (GRUs), convolutional neural networks (CNNs)	**Reinforcement learning**: Q-learning and deep Q-networks (DQN)
4	U-Net	**Anomaly detection models**: Autoencoders	**Hybrid models**: Conditional random fields (CRFs)	**Bayesian methods**: Bayesian networks

an extensive range of applications. These are works on sensors that typically assess and quantify the patient or person's physiology. Systems furnished with wearable sensors and ML abilities will monitor vital signs and other health parameters in real time by providing continuous health monitoring. These devices are used to analyze the data collected from the sensors, giving brief insights into a person's health condition and detecting potential health problems at an early stage before they become more severe. Table 6.4 illustrates the ML algorithms used for Wearable Health Monitors.

6.3.4 Natural Language Processing (NLP)

It facilitates technologies to recognize and reply to text or voice data. Most products and services in everyday use NLP, such as voice-based digital assistants on mobiles, translation applications that decipher countries' languages, and spam identification by scanning emails. These technologies are used to extract meaningful patterns from structured, semi-structured, and unstructured data. Extracted meaningful patterns from data such as doctors' descriptions, patient test results, and clinical publications, enabling patient supervision and supporting medical decision-making. The most important benefits of NLP are the ability to analyze both structured and unstructured data, refine patient satisfaction and understanding by detecting insights using sentiment analysis, reduce overheads by engaging NLP-enabled AI to accomplish definite tasks, and improve understanding by accompanying NLP analysis on appropriate data. The following Figure 6.5 illustrates the steps to implementing NLP in the healthcare system.

6.3.5 Personalized treatment plans

ML algorithms enable the analysis of patient data to enhance and personalize treatment plans. ML algorithms also support refining the usefulness of treatments and dealing with dosages to reduce side effects. In the healthcare system, personalized treatment plans are critical applications of ML, and training to adapt medical treatment based on individual patient data. The most commonly used ML methods used to develop personalized treatment plans are illustrated in Table 6.4.

6.3.6 Challenges in smart healthcare devices

1. **Data privacy and security:** Sensitive health data in healthcare systems generated by SHD raises substantial concerns. Ensuring data privacy and security with government regulations is critical and challenging.
2. **Data quality and integration:** Handling large amounts of data collected from various healthcare devices and combing data from various sources is more challenging to ensure it is usable for ML, posing additional challenges.

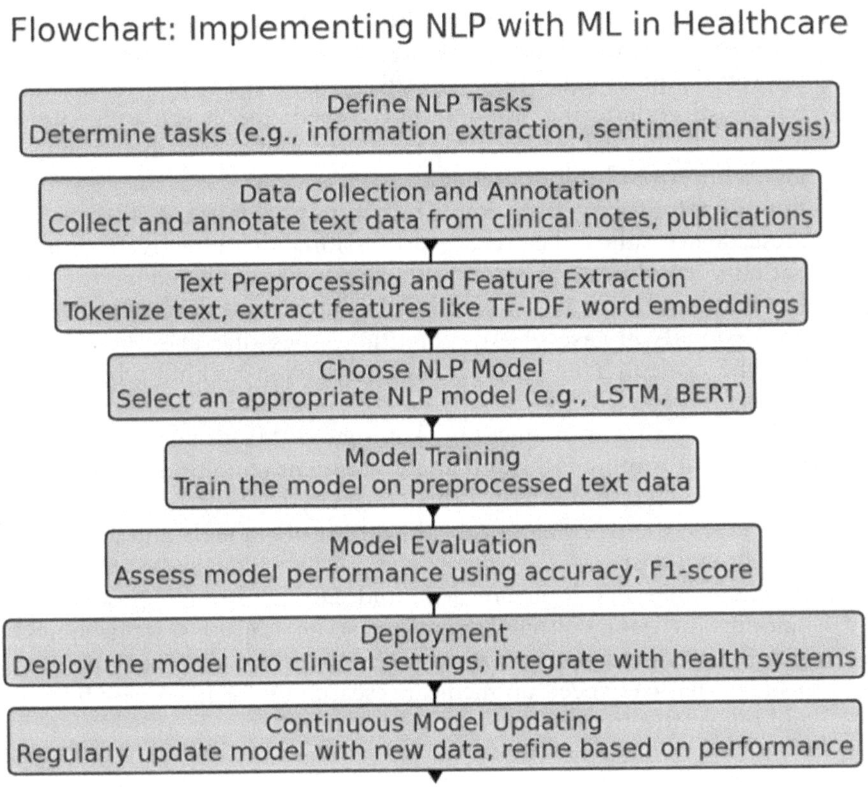

Figure 6.5 Steps to implement the NLP with ML in healthcare.

3. **Algorithmic bias:** ML methods trained on non-representative datasets develop biases. This in turn leads to unequal treatment.
4. **Regulatory compliance:** All ML-based SHD available in the market strictly adhere to regulatory standards. These regulatory standards are varying from country to country and need extensive authentication and testing before providing approval to use in the public domain.
5. **Adoption and Trust:** Because of potential fears about the accuracy and reliability of ML-based SHD, it is difficult to advise and convince all healthcare professionals and patients to trust and adopt these ML-driven technologies.

6.3.7 Opportunities in smart healthcare devices leveraged by machine learning

1. **Enhanced diagnostic capabilities:** The development of ML-driven technologies for smart health monitoring is playing a major role in the healthcare industry and creating more opportunities. ML-driven smart healthcare technologies expressively enrich diagnostic precision

and efficiency. This is helpful in the early detection and treatment of diseases like cancer, glaucoma, etc.

2. **Improved patient outcomes:** ML-driven technologies provide continuous health monitoring, reminders, and medicine. It is also helpful to generate individual personalized treatment plans, hypothetically leading to improved health outcomes.
3. **Remote monitoring and care:** ML-driven devices are self-driven, and these devices allow healthcare professionals to effectively monitor patients' conditions in remote locations. This will reduce the need for frequent hospital visits and also provide an alert to healthcare professionals in case of any potential issues that arise before they become critical. This leads to earlier decision-making when health issues are detected.
4. **Efficiency and cost reduction:** Automated ML-driven technologies perform all routine tasks and analyze patient conditions in real-time. By optimizing resource allocation, automated ML-driven technologies will reduce the workload on healthcare professionals and also lower operational costs.
5. **Research and development:** ML and DL algorithms process vast amounts of data collected from smart healthcare systems quickly at a scale impossible for human researchers and discover important insights that can speed up medical research and lead to new findings and innovative treatments.

6.4 MEDICAL APPLICATIONS

Patient adherence to treatment programs, vital signs, and symptoms can all be continuously and remotely monitored with the use of SHD. Telemedicine platforms that are integrated with SHD allow for remote consultations, monitoring, and follow-ups, which play a vital role for patients from rural areas. Personalized medicine is on the rise thanks to SHDs, which make it easier to gather vast amounts of health data from individual patients. This data can then be analyzed using ML and AI algorithms to find patterns, forecast health outcomes, and customize treatments to meet the specific needs of each patient. The generalized architecture of SHDs for the analysis of medical data is depicted in Figure 6.6. The applications of SHDs in the detection of tumors are discussed in the following section.

Optimized modified ResNet 18, a modified version of ResNet18 that uses CNN to detect tumors. The CNN parameters, such as optimization algorithm, learning rate, mini-batch size, and dropout rate, are varied, and their impact on model performance is analyzed (El-Feshawy et al., 2023). The novel approach integrates the spatially constrained fish school optimization

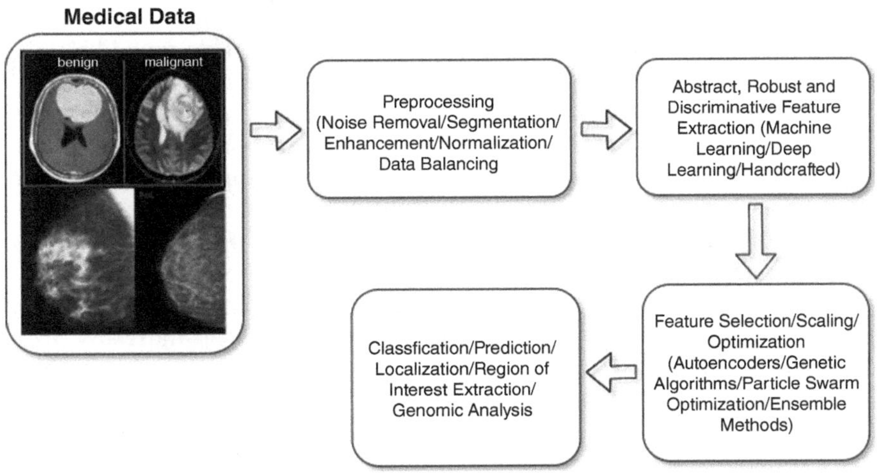

Figure 6.6 The generalized architecture of the SHDs for the medical data analysis using machine learning and deep learning approaches.

algorithm with interval type-II fuzzy logic system techniques to correct erroneous predictions of anomalies in brain subjects using MRI (Alagarsamy et al., 2020). The reported work yields accurate segmentation results for diverse MRI sequences and can detect tumors/anomalies of varying sizes, whether single or multiple occurrences.

Extracting abstract and discriminative features from X-ray images of the lung can aid in tumor categorization. K-Nearest Neighborhood Classifier and Feed-Forward Neural Network with Back Propagation are used for training and classification (Avinash et al., 2023). The DL model's predicting accuracy is enhanced using transfer learning and data augmentation methodologies (Khan et al., 2023). The reported work contributes significantly to the healthcare industry 5.0 by developing an IoMT-enabled intelligent system for predicting lung cancer. In the smart healthcare sector 5.0, precise disease prediction is achieved through the utilization of the Google Net Deep Machine-Learning Model.

The human body generates several biological signals, such as electrophysiological signals, bodily symptoms, and habitual physical activity (Wang et al., 2019). It may signal the presence of deadly diseases. Wearable health monitoring technologies can detect biological signals, leading to early disease detection. A biocompatible gas sensor based on butterfly wings detects acetone gas in breath for early diabetes diagnosis. Many flexible biochemical sensors have been used recently to detect human cancer molecules and protein markers in real time. Using chemical treatment and antibody immobilization on rGO@SFP-based biosensors, pathogenic prostate-specific antigen (PSA) cancer markers can be detected in real-time with high sensitivity. Analyzing body movements regularly can identify irregular gait patterns and hand tremors, which are precursors to fatal diseases including Parkinson's, Alzheimer's, and diabetes. This can aid in early detection and treatment. Bio-multifunctional wearable sensors are now often utilized for motion detection in various applications, such as assisted living, rehabilitation, and monitoring. Mobile sensing can identify behavioral variations in cancer-affected persons that may indicate significant differences in their functional status, symptom load, quality of life, likelihood of readmission, and other unfavorable outcomes (Low, 2020). A study of 62 cancer patients found that smartphone accelerometer data could distinguish between those who experienced postoperative complications, emergency department visits, readmissions, reoperations, or mortality from those who recovered well.

A novel, cozy wearable technology that uses microwave imaging to securely assist women in routinely undergoing breast cancer screening, especially young, densely breasted women (Elsheakh et al., 2023). This method detects changes in electrical characteristics between normal and tumor-affected breast tissue using electromagnetic microwave technology. The reported SHD is a conductive fabric with a low SAR value, making it safe, portable, and easily worn as a "Smart Bra". Researchers at MIT developed an SHD based on ultrasound technology for the early detection of breast tumors. A flexible 3D-printed patch with openings resembling a honeycomb is used to make a wearable SHD. The patch fastens to a bra that has holes in it that let it come into contact with skin so that it may examine breast tissue. SHD uses sound waves to penetrate breast tissue, and as it passes over the breast, it creates detailed pictures that show any cysts that would require further examination by a breast cancer expert.

Probability maps of the tumor's location and dimensions would be an important screening tool for breast cancer. A neural network model is used in (Borghouts et al., 2023) to generate this kind of probability map straight from the scattering matrix. The framework used in this study is based on a U-Net design with a single dense layer. Using a system of 24 antennas, it was possible to identify the precise size and position of the tumor in addition to determining its existence inside the breast (Moussa et al., 2022). The array antennas detect the presence of a tumor by varying the scattering signals they receive. The electric field passing through the breast phantom layers illuminated by the antenna array is used to detect the size and position of the tumor. The work in (Lu et al., 2022) employs a CNN-LSTM framework for the location and identification of breast tumors based on ultrawideband backscatter data. The UWB microwave signal is a small pulse with high resolution and penetration depth, making it suitable for medical detection. The convolution layer utilizes CNN's excellent feature extraction capabilities to process raw backscatter signals. The collected features are molded in a flattened layer before being sent into an LSTM network for time series processing.

6.5 CLOUD- AND FOG-BASED SMART HEALTHCARE SYSTEMS

A Cloud-based architecture typically has two layers: the Cloud layer, which includes cloud servers with high-performance processing and storage capabilities. The core optical network provides high-bandwidth communication, whereas the end-user layer consists of individual devices. The generalized architecture of cloud- and fog-based SHDs is depicted in Figure 6.7. Fog computing is a viable solution for Internet of Healthcare Things (IoHT) applications, including emergency rescue and real-time patient monitoring for patients with heart disease, blood pressure, and stroke, due to its fast response time. Fog-based healthcare IoT solutions played a significant role during the COVID-19 epidemic and are expected to continue to grow in the future.

Fog computing is an innovative standard for healthcare monitoring systems (Elhadad et al., 2022). Fog computing allows for faster data processing while lowering network congestion. The remote healthcare system employs body sensors that are incorporated in or ideally positioned on the patient's body to monitor their health state by detecting specific signals and assisting them in receiving precise remedies. Due to the tremendous scalability of IoT systems in fog computing, new micro-fog centers could be installed to offset an increase in end-user traffic.

The fog computing platform has been used to propose a fine-grained searchable data exchange method (Mamta, 2022). The fog node aids end users by doing computationally demanding activities on their behalf, making the resulting scheme lightweight and efficient. The suggested design, which represents the resource-constrained IoT devices, has significantly reduced the storage and computational costs at the data owner's end. The system uses fog assistance to generate keyword ciphertext, resulting in constant storage and processing costs at the patient's end. Data is collected via resource-constrained IoT devices. The semantic security is demonstrated by using a generic bilinear group model. IoT-based smart ICU systems will monitor critical ICU patients in real-time, collecting parameters such as temperature, blood pressure, ECG,

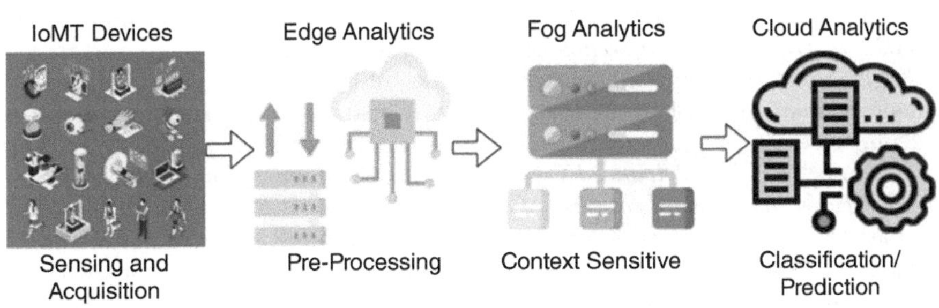

Figure 6.7 The generalized architecture of cloud and fog-based SHDs.

SpO2, and fluid measurements from their bodies via IoT devices (Al Mudawi, 2022). The gathered information will be stored in fog nodes, where it will be processed, and reports forwarded to the cloud for storage. Additionally, following real-time data processing, the system will alert the doctor if any aberrant parameters are found.

6.6 CONCLUSION

The incorporation of evolving technologies in smart healthcare systems provides more reliable, secure, and efficient patient-centered healthcare services. These technologies provide real-time monitoring, predictive analytics, and diagnostic and personalized interventions. The IoT provides secure connectivity among medical devices. Standard communication protocols over a network consisting of sensors and storage devices ensure secure and reliable data transfer. It also supports care coordination and decision-making. Predictive analytics and diagnostic ML technologies take control of vast amounts of patient data for proactive health management and early disease detection. Wearable health monitors allow individuals to take control of their health. Where NLP, extracts the meaningful insights present in unstructured or structured healthcare data. Personalized treatment plans leverage patient-specific information to support healthcare professionals in decision-making and optimizing outcomes. Cloud and fog computing technologies afford expandable and secure structures for data storage and processing.

REFERENCES

Alagarsamy, Saravanan, Yu-Dong Zhang, Vishnuvarthanan Govindaraj, Murugan Pallikonda Rajasekaran, and Sakthivel Sankaran. "Smart identification of topographically variant anomalies in brain magnetic resonance imaging using a fish school-based fuzzy clustering approach." *IEEE Transactions on Fuzzy Systems* 29, no. 10 (2020): 3165–3177.

Anand Kumar, M., N. Abirami, M. S. Guru Prasad, and M. Mohankumar. "Stroke disease prediction based on ECG signals using deep learning techniques." In *2022 International Conference on Computational Intelligence and Sustainable Engineering Solutions (CISES)*, pp. 453–458. IEEE, 2022.

Anand Kumar, M., A. H. Pai, J. Agarwal, S. Christa, M. S. Guru Prasad, and S. Saifi. "Deep learning model to defend against covert channel attacks in the SDN networks." In *2023 Advanced Computing and Communication Technologies for High Performance Applications (ACCTHPA)*, pp. 1–5. IEEE, 2023.

Avinash, S., H. N. Naveen Kumar, M. S. Guru Prasad, R. Mohan Naik, and G. Parveen. "Early detection of malignant tumor in lungs using feed-forward neural network and k-nearest neighbor classifier." *SN Computer Science* 4, no. 2 (2023): 195.

Borghouts, Marijn, Michele Ambrosanio, Stefano Franceschini, Maria Maddalena Autorino, Vito Pascazio, and Fabio Baselice. "Microwave breast sensing via deep learning for tumor spatial localization by probability maps." *Bioengineering* 10, no. 10 (2023): 1153.

Chandrappa, S., P. Chandra Shekar, P. Chaya, L. Dharmanna, and M. S. Guruprasad. "Machine learning algorithms for identifying fake currencies." *SN Computer Science* 4, no. 4 (2023a): 368.

Chandrappa, S., L. Dharmanna, B. Anami, C. Nandini, and M. Rafi. "A novel approach to glaucoma screening using optic nerve head through image fusion and fractal geometry." *International Journal of Mathematics and Computer Science* 11, no. 4 (2021): 4284–4300.

Chandrappa, S., L. Dharmanna, and K. I. R. Neetha. "Automatic elimination of noises and enhancement of medical eye images through image processing techniques for better glaucoma diagnosis." In *2019 1st International Conference on Advances in Information Technology (ICAIT)*, pp. 551–557. IEEE, 2019.

Chandrappa, S., L. Dharmanna, U. V. Shyama Srivatsa Bhatta, M. S. Chiploonkar, M. N. Suraksha, and S. Thrupthi. "Design and development of IoT device to measure quality of water." *International Journal of Modern Education and Computer Science (IJMECS)* 9, no. 4 (2017a): 50–56.

Chandrappa, S., M. S. Guruprasad, H. N. Naveen Kumar, K. Raju, and D. K. Santhosh Kumar. "An IOT-based automotive and intelligent toll gate using RFID." *SN Computer Science* 4, no. 2 (2023b): 154.

Chandrappa, S., D. Lamani, S. Jagadeesha, and R. Kumar. "Segmentation of retinal nerve fiber layer in optical coherence tomography (OCT) images using statistical region merging technique for glaucoma screening." *International Journal of Computer Applications* 128, no. 10 (2015): 32–35.

Chandrappa, S., D. Lamani, S. V. Poojary, and N. U. Meghana. "Automatic control of railway gates and destination notification system using internet of things (IoT)." *International Journal of Education and Management Engineering* 7, no. 5 (2017b): 45.

El-Feshawy, S. A., W. Saad, M. Shokair, and M. Dessouky. "IoT framework for brain tumor detection based on optimized modified ResNet 18 (OMRES)." *The Journal of Supercomputing* 79, no. 1 (2023): 1081–1110.

Elhadad, Ahmed, Fulayjan Alanazi, Ahmed I. Taloba, and Amr Abozeid. "Fog computing service in the healthcare monitoring system for managing the real-time notification." *Journal of Healthcare Engineering* 2022 (2022): 1–11.

Elsheakh, Dalia N., Rawda A. Mohamed, Omar M. Fahmy, Khaled Ezzat, and Angie R. Eldamak. "Complete breast cancer detection and monitoring system by using microwave textile based antenna sensors." *Biosensors* 13, no. 1 (2023): 87.

Gujjar, J. P., H. R. Prasanna Kumar, and M. S. Guru Prasad. "Advanced NLP framework for text processing." In *2023 6th International Conference on Information Systems and Computer Networks (ISCON)*, pp. 1–3. IEEE, 2023.

Khan, Tahir Abbas, Areej Fatima, Tariq Shahzad, Khalid Alissa, Taher M. Ghazal, Mahmoud M. Al-Sakhnini, Sagheer Abbas, Muhammad Adnan Khan, and Arfan Ahmed. "Secure IoMT for disease prediction empowered with transfer learning in healthcare 5.0, the concept and case study." *IEEE Access* 11 (2023): 39418–39430.

Kirubasri, G., S. Sankar, M. S. Guru Prasad, G. Naga Chandrika, and S. Ramasubbareddy. "LQETA-RP: link quality based energy and trust aware routing protocol for wireless multimedia sensor networks." *International Journal of System Assurance Engineering and Management* 15, no. 1 (2024): 564–576.

Low, Carissa A. "Harnessing consumer smartphone and wearable sensors for clinical cancer research." *NPJ Digital Medicine* 3, no. 1 (2020): 140.

Lu, Min, Xia Xiao, Yanwei Pang, Guancong Liu, and Hong Lu. "Detection and localization of breast cancer using UWB microwave technology and CNN-LSTM framework." *IEEE Transactions on Microwave Theory and Techniques* 70, no. 11 (2022): 5085–5094.

Moussa, Maria N., Mervat A. Madi, and Karim Y. Kabalan. "Breast tumor detection, sizing and localization using a 24-element antenna array." *IEEE Journal of Biomedical and Health Informatics* 26, no. 10 (2022): 5109–5121.

Nagesh, H. R., and S. Prabhu. "High performance computation of big data: performance optimization approach towards a parallel frequent item set mining algorithm for transaction data based on hadoop MapReduce framework." *International Journal of Intelligent Systems and Applications* 9, no. 1 (2017): 75.

Pai, A., M. Anandkumar, M. S. Guru Prasad, J. Agarwal, and S. Christa. "Designing a secure audio/text based captcha using neural network." In *2023 13th International Conference on Cloud Computing, Data Science & Engineering (Confluence)*, pp. 510–514. IEEE, 2023.

Patel, V., M. S. Guru Prasad, H. Aditya Pai, A. Suresh Kumar, J. Praveen Gujjar, and H. N. Naveen Kumar. "Real-time face mask detector." In *2023 IEEE 3rd International Conference on Technology, Engineering, Management for Societal impact using Marketing, Entrepreneurship and Talent (TEMSMET)*, pp. 1–5. IEEE, 2023.

Prasad, G., A. K. Jain, P. Jain, and H. R. Nagesh. "A novel approach to optimize the performance of hadoop frameworks for sentiment analysis." *International Journal of Open Source Software and Processes (IJOSSP)* 10, no. 4 (2019): 44–59.

Prasad, G., H. N. Naveen Kumar, M. Anand Kumar, and A. Kukreti. "Detection of transmission control protocol XMAS attack using pattern analysis with MONOSEK." *Journal of Nano-& Electronic Physics* 15, no. 4 (2023): 1–4.

Wang, Lili, Zheng Lou, Kai Jiang, and Guozhen Shen. "Bio-multifunctional smart wearable sensors for medical devices." *Advanced Intelligent Systems* 1, no. 5 (2019): 1900040.

Chapter 7

Envisioning tomorrow

Exploring the future of smart electronic devices: AI, ML

Suman Rani and Jaibir Singh

7.1 INTRODUCTION

In the steadily developing scene of innovation means (continuous increasing technology), brilliant gadgets have become necessary to our regular routines, molding the manner in which we convey, work, and explore our general surroundings. This presentation makes way for a complete investigation, representing things to come of shrewd gadgets, diving into the expected developments and groundbreaking changes that lie ahead. Figure 7.1 shows the AI technology landscape.

Artificial Intelligence (AI) denotes the capability of machines to imitate human intelligence, executing tasks typically necessitating human cognitive abilities, such as learning, problem-solving, decision-making, and comprehending natural language. AI encompasses various technologies, as illustrated in Figure 7.1, including machine learning (ML), natural language processing, robotics, and computer vision. Machine learning, a subset of AI, entails training computer algorithms to recognize patterns within data and make informed predictions or decisions based on such patterns [1–5]. Deep learning, a specialized form of ML, employs neural networks comprising multiple layers to process intricate data sets, such as images or speech. Natural language processing enables computers to comprehend, interpret, and even generate human language, encompassing both spoken and written forms. Computer vision equips computers with the capacity to scrutinize and decipher visual information, encompassing images and videos. Figure 7.2 shows ML approaches.

Artificial Intelligence (AI) algorithms necessitate data to discern patterns and make predictions or decisions based on that data. Machine learning techniques within AI enable machines to learn patterns and make predictions from data without explicit programming [6]. These methods find broad application across various domains, including natural language processing, image and speech recognition, and recommendation systems. Generally, the availability of more data for an AI algorithm to learn from enhances the accuracy of its predictions or decisions. Building AI systems involves employing several data-learning approaches [6, 7], as outlined below for the sake of completeness:

DOI: 10.1201/9781003643173-7

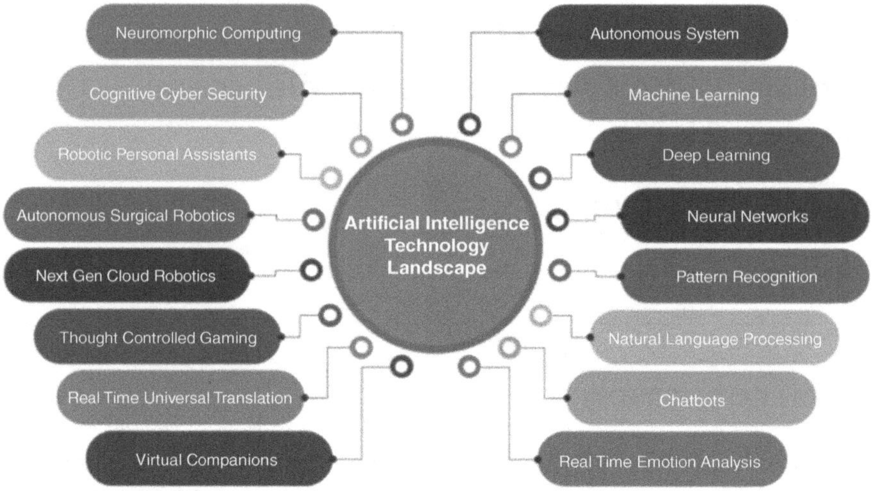

Figure 7.1 AI technology landscape.

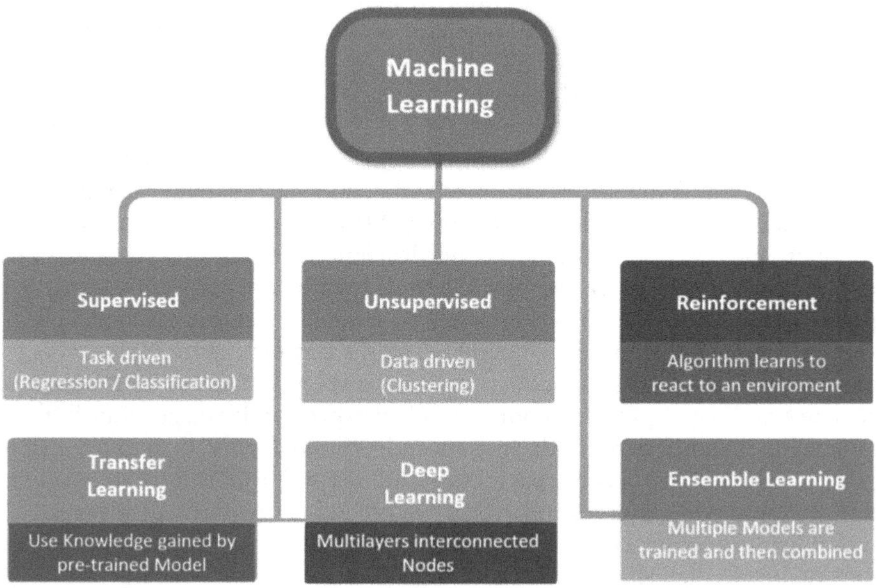

Figure 7.2 Machine learning approaches.

Supervised learning: This approach involves training an AI system on a labeled dataset, where each data point is associated with a label or target variable. The objective is to develop a model capable of accurately predicting the label or target variable for new data points. Supervised learning is commonly used in tasks such as image classification, speech recognition, and natural language processing [8].

Unsupervised learning: In unsupervised learning, an AI system is trained on an unlabeled dataset where no target variable is present. The goal of the AI system is to find patterns, relationships, and structures within the data. Unsupervised learning is used in the applications of clustering, anomaly detection, and dimensionality reduction [9].

Reinforcement learning: The AI system learns to take decisions based on the feedback from the environment. In this process, the system receives rewards or penalties depending on its actions and changes its behavior. Reinforcement learning is used commonly in gaming, robotics, and autonomous driving [10].

Transfer learning: Transfer learning makes use of the knowledge gained from a task to enhance performance on a similar related task. The system is first pre-trained with large data and then fine-tuned with small data specifically to a particular task. It is thus a method that reduces the number of data that an AI model needs to be trained on and also improves the model's accuracy and performance [11].

Deep learningreferstotheuse of deep-learning-based MLtechniques specifically for complex tasks withvery large data and relationships. Deep learning models comprise multiple layers of interconnected nodes capable of learning increasingly complex representations of data. This approach is commonly applied in tasks such as image and speech recognition, natural language processing, and computer vision [12].

7.1.1 Background and importance

The fast headways in man-made consciousness, network, and sensor advancements have moved savvy gadgets beyond simple contraptions, transforming them into basic colleagues. From cell phones that consistently coordinate into our social texture to wearable gadgets that screen our well-being, the development of brilliant gadgets has been downright progressive. As we stand at the cusp of another time, it is vital to comprehend the foundation and meaning of savvy gadgets to anticipate the direction of their future turn of events. The incorporation of the Web of Things (IoT), increased reality, and AI has made a prolific ground for phenomenal conceivable outcomes, promising to rethink the manner in which we communicate with innovation and one another [13]. This investigation will not only explore the innovative aspects but also consider the cultural impact of smart devices. Issues such as security concerns, ethical considerations, and the potential digital divide will be examined to provide a comprehensive understanding of the future landscape shaped by smart technology.

7.1.2 Scope and goals

As we leave on the excursion of investigating the eventual fate of brilliant gadgets, it is fundamental to characterize the degree and targets that will direct our investigation into this dynamic and quickly developing domain.

7.1.2.1 Scope

The extent of this investigation incorporates a complex examination of shrewd gadgets, considering innovative, cultural, and moral perspectives. We will dive into different classifications of shrewd gadgets, including, however not restricted to, cell phones, wearables, smarthome gadgets, and emerging technologies like brilliant urban areas and associated vehicles. The investigation will traverse both existing developments and those not too far-off, guaranteeing an exhaustive comprehension of the advancing scene [14].

7.1.2.2 Goals

Look at Innovative Progressions: Investigate state-of-the-art advancements like man-made consciousness, increased reality, Internet of Things (IoT), and AI prepared to shape the eventual fate of intelligent gadgets.

Comprehend Client Experience: Dig into the developing client experience of brilliant gadgets, taking into account UI plan, client expectations, and the integration of gadgets into daily life. By laying out an unmistakable extension and targets, we plan to explore the intricacies representing things to come of savvy gadgets with accuracy and profundity, offering important insights into the developing scene of innovation and its effect on society.

7.1.3 Structure of the section

To guarantee a precise and complete investigation representing things to come of shrewd gadgets, the part will be organized firmly, guiding the reader through key parts of innovative progressions, cultural ramifications, and moral contemplations. The accompanying layouts show the design of the part.

7.1.3.1 Presentation

A concise recap of the presentation, summing up the meaning of brilliant gadgets in our day-to-day routines and the need to expect their future turns of events.

7.1.3.2 Development of shrewd gadgets

This part will give a verifiable outline of the development of brilliant gadgets, featuring key achievements and innovative progressions that have molded their present status.

7.1.3.3 Technological foundations

Reasonably alludes to the major innovations that support the turn of events and working of smart gadgets. This part would almost certainly cover different angles, such as:

Remote availability: Investigating headways in remote correspondence advancements like Bluetooth, Wi-Fi, and cell networks that empower shrewd gadgets to associate and speak with one another and with the web [15].

Sensors and IoT: Examining the job of sensors and the Web of Things (IoT) in shrewd gadgets, including how sensors gather information from the climate and how IoT empowers gadgets to constantly communicate and share data.

Implanted frameworks: Understanding the significance of inserted frameworks in brilliant gadgets, which incorporate microcontrollers, firmware, and programming that control the gadget's capabilities and activities.

Computerized reasoning and AI: Looking at the reconciliation of AI and AI calculations into brilliant gadgets to empower highlights like voice recognition, prescient analysis, and customized client experiences.

Energy productivity and power the board: Tending to the difficulties of force utilization in brilliant gadgets and investigating advances and procedures to enhance energy effectiveness and expand battery duration.

Security: Featuring the meaning of safety efforts and protection assurances in savvy gadgets, including encryption, verification, and information anonymization to shield client information and prevent unapproved access.

UIs and experience: Talk about headways in UI configuration, including touchscreen interfaces, voice orders, and signal acknowledgment, to improve client experience and ease of use of smart gadgets [16].

Distributed computing and edge registering: Investigating the job of distributed computing and edge computing in intelligent devices, including offloading calculation to far-off servers for handling and storage, as well as utilizing edge processing for continuous data analysis and low-idleness applications. Generally speaking, this part will give an exhaustive outline of the innovative establishments that empower the turn of events and evolution f smart gadgets, molding the eventual fate of innovation and human interaction with the digital world.

7.1.3.4 Classifications of brilliant

Gadgets probably covers the different sorts of savvy gadgets that are imagined for what's in store. These classifications might include:

Wearable gadgets: This class includes gadgets that are worn on the body, for example, smartwatches, wellness trackers, augmented reality glasses, and well-being checking gadgets. These gadgets frequently gather biometric information, track actual work, and give notices and data directly to the client.

Home robotization gadgets: Home mechanization gadgets incorporate shrewd indoor regulators, brilliant lighting frameworks, savvy locks, and smart appliances. These gadgets are intended to make homes more effective, advantageous, and secure by permitting control and computerization of different household functions.

Cell phones and tablets: While currently pervasive, cell phones and tablets keep on advancing with new features and abilities. These gadgets act as the essential connection point for getting to and controlling other brilliant gadgets, as well with respect to correspondence, amusement, and efficiency errands.

Brilliant home collaborators: Shrewd home partners, for example, Amazon Reverberation and Google Home, are voice-activated gadgets that utilize man-made reasoning to perform assignments, answer questions, and control other savvy gadgets in the home. They act as focal center points for overseeing shrewd home biological systems [17].

Associated vehicles: The classification of associated vehicles incorporates vehicles outfitted with web networks and coordinated intelligent elements. These highlights might incorporate route frameworks, infotainment frameworks, driver help advances, and vehicle-to-vehicle correspondence for further developed well-being and efficiency. Modern and IoT Gadgets: Modern and IoT gadgets are utilized in different ventures and applications, like assembling, agribusiness, healthcare, and shrewd urban communities. These gadgets are furnished with sensors and networks to gather information, screen processes, and streamline tasks.

Savvy diversion gadgets: Brilliant amusement gadgets incorporate shrewd televisions, streaming media players, and gaming consoles that offer web networks, content streaming, voice control, and intuitive highlights for upgraded diversion encounters.

Medical services and clinical gadgets: This class incorporates savvy clinical gadgets, for example, far-off quiet checking frameworks, intelligent insulin siphons, and wearable well-being trackers. These gadgets assist people with dealing with their ailments, track imperative signs, and speak with medical services providers. These classes address the assorted scope of brilliant gadgets that are molding the eventual fate of innovation and changing different parts of day-to-day existence, from individual well-being and home computerization to transportation and entertainment [18].

7.1.3.5 Societal impact and implications

It likely delves into the broader effects that the proliferation and advancement of smart devices will have on society. This section may include

Computerized separation: Looking at how the rising dependence on brilliant gadgets might fuel existing disparities in access to innovation, as well as the expected ramifications for people and networks without satisfactory access.

Security and observation: Examining the ramifications of omnipresent brilliant gadgets for protection, freedoms, and reconnaissance, including worries about information assortment, following, and the potential for abuse by state-run administrations, partnerships, or vindictive entertainers [19, 20].

Moral contemplations: Tending to moral difficulties raised by shrewd gadgets, for example, issues connected with information proprietorship, assent, algorithmic predisposition, and the dependable turn of events and sending of AI and AI innovations. Network protection Dangers: Breaking down the network protection dangers presented by interconnected smart gadgets, remembering weaknesses in equipment, programming, and communication protocols, as well as the potential for cyberattacks to disrupt the basic framework or compromise individual information.

Influence on positions and work: Investigating how the reception of brilliant gadgets might prompt changes in the work market, including computerization of routine assignments, the creation of new position opportunities in innovation-related fields, and the requirement for reskilling and upskilling of laborers.

Well-being and prosperity: Taking into account the effect of brilliant gadgets on physical and psychological well-being, including issues connected with screen time, computerized dependence, rest disturbance, and the expected advantages of well-being monitoring, and telemedicine advancements.

Natural supportability: Evaluating the ecological ramifications of assembling, utilizing, and discarding savvy gadgets, includes worries about asset exhaustion, electronic waste, energy utilization, and fossil fuel byproducts [21].

7.2 CURRENT SCENE OF SAVVY GADGETS

To understand the future direction of shrewd gadgets, it is fundamental to first inspect the current scene. This part dives into the present status of brilliant gadgets, offering a detailed investigation of their pervasiveness, innovative highlights, and cultural effects.

7.2.1 Overview of existing savvy gadgets

The outline of existing savvy gadgets probably gives a compact rundown of common innovations forming the ongoing scene. It might include:

Cell phones: These multifunctional gadgets act as the center of individual networks, offering highlights, for example, web perusing, correspondence applications, GPS route, and plenty of other applications [22]. The Figure7.3 shows examples.

Figure 7.3 Savvy gadgets.

Wearable innovation: From smartwatches to wellness trackers, wearable gadgets have become progressively famous for well-being checking, movement following, and getting warnings in a hurry.

Savvy home gadgets: Including brilliant indoor regulators, lighting frameworks, surveillance cameras, and voice-enacted partners like Amazon Reverberation and Google Home, these gadgets offer clients command over their home climate and upgrade accommodation and security [23].

Associated apparatuses: Apparatuses like fridges, clothes washers, and stoves furnished with web availability and shrewd elements empower remote checking and control, energy proficiency, and mechanized schedules.

Savvy theater setups: Shrewd televisions, streaming gadgets, and gaming consoles give admittance to an immense range of computerized content and intuitive encounters, frequently with voice control and customized proposals.

Well-being and well-being contraptions: Going from savvy scales to pulse screens and rest trackers, these gadgets engage clients to screen their well-being measurements, track wellness objectives, and pursue informed ways of life decisions.

Associated vehicles: Vehicles furnished with route frameworks, infotainment reassures, and high-level driver help highlights like versatile journey control and path keeping help are turning out to be progressively normal, upgrading security and comfort out and about.

Modern IoT (Web of Things) Gadgets: In modern settings, IoT gadgets like sensors and actuators are used for observing gear execution, advancing cycles, and prescient support, prompting expanded proficiency and cost savings. This outline features the different clusters of shrewd gadgets previously coordinated into day-to-day existence, laying the basis for imagining the future direction of brilliant innovation.

7.2.2 Impact on daily life

The conversation probably spins around how shrewd gadgets are reshaping and improving different parts of regular day-to-day existence. This effect incorporates:

Comfort and effectiveness: Brilliant gadgets smooth out everyday errands and schedules, making them more advantageous and productive. For example, brilliant home gadgets mechanize lighting, temperature control, and security, saving time and effort for clients.

Connectedness and correspondence: Cell phones and other associated gadgets empower moment correspondence and networking with companions, family, and partners, paying little heed of location. This cultivates closer connections and works with coordinated effort in both individual and expert circles.

Generally, the effect of savvy gadgets on day-to-day existence is diverse, upgrading comfort, network, personalization, and efficiency while additionally raising concerns connected with protection, security, and advanced education [6].

7.2.3 Trends in purchaser reception

The attention is reasonable on the advancing examples of shopper conduct and the reception of savvy gadgets. This outline might include:

Fast development: Featuring the quick extension of the brilliant gadget market, driven by expanding shopper interest for network, comfort, and advancement across different item classes.

The multiplication of associated gadgets: Talking about the expansion of associated gadgets in homes, working environments, and individual spaces, mirroring the developing joining of innovation into regular day-to-day existence.

Shift toward Shrewd home environments: Taking note of the pattern toward the reception of brilliant home biological systems, where interconnected gadgets cooperate to robotize undertakings, upgrade security, and further develop energy effectiveness. Far-reaching Reception of Wearable Innovation: Perceiving the inescapable reception of

Figure 7.4 IoT.

wearable innovation, for example, smartwatches and wellness trackers, for well-being checking, movement following, and getting warnings in a hurry shown in Figure 7.4.

Expanding availability: Stressing the rising availability of gadgets through headways in remote advances like Wi-Fi, Bluetooth, and cell organizations, empowers consistent correspondence and information sharing [6].

Developing interest in IoT gadgets: Taking note of the developing interest in Web of Things (IoT) gadgets for different applications, including modern computerization, medical care, and brilliant city drives, driven by headways in sensor innovation and data analysis.

Interest in personalization and customization: Featuring customer interest in customized and adjustable encounters presented by savvy gadgets, including customized suggestions, versatile settings, and custom-made UIs [24].

Worries about protection and security: Recognizing purchaser worries about protection and security gambles related to shrewd gadgets, for example, information breaks, unapproved access, and reconnaissance, impacting buying choices and reception rates.

Significance of client experience: Perceiving the significance of client experience and UI configuration in driving customer reception of brilliant gadgets, with natural connection points, consistent reconciliation, and solid execution being key variables. Generally, the outline of patterns in purchaser reception gives understanding into the advancing inclinations, ways of behaving, and assumptions of shoppers with respect to shrewd gadgets, molding the future course of mechanical advancement and market development [25].

7.3 INNOVATIVE ESTABLISHMENTS

7.3.1 Man-made brainpower and AI combination

The emphasis is possible on the critical job of AI and AI in molding the mechanical scene of savvy gadgets shown in Figure7.5. This outline might include:

Improved client experience: Feature how computer-based intelligence and AI calculations are being incorporated into savvy gadgets to upgrade client experience by empowering elements, for example, voice acknowledgment, regular language handling, and customized proposals.

Prescient examination: Talk about how simulated intelligence and AI calculations break down tremendous measures of information gathered by savvy gadgets to give prescient bits of knowledge and expect client needs, for example, anticipating traffic designs, suggesting content, or changing energy utilization in smart homes.

Brilliant mechanization: Investigating how simulated intelligence-fueled robotization is changing shrewd gadgets into proactive aides that can computerize routine assignments, expect client inclinations, and

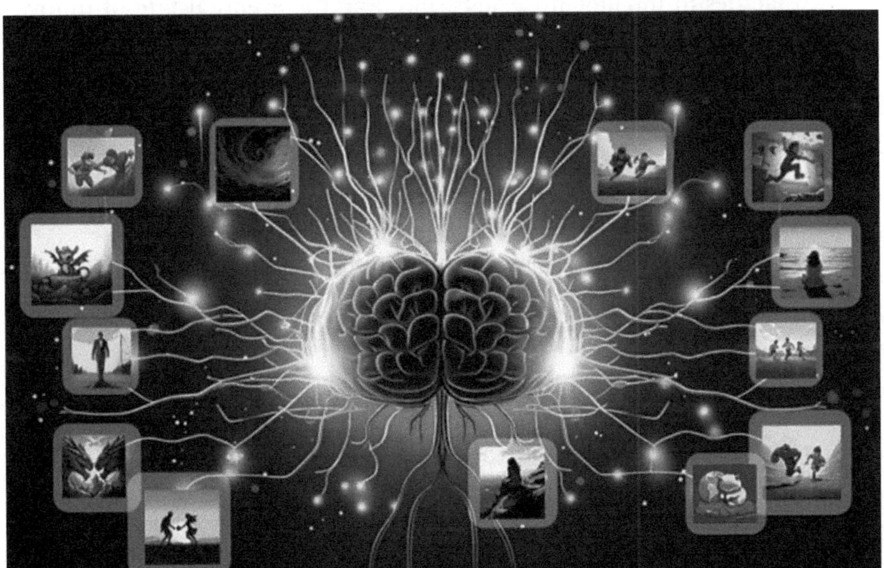

Figure 7.5 Man-made brainpower and AI combination.

improve gadget settings in view of verifiable information and constant data sources [26].

Setting mindfulness: Featuring how computer-based intelligence calculations empower savvy gadgets to comprehend and answer context-oriented prompts, like client area, action, and climate, to convey more customized and significant collaborations and administrations.

Versatile getting the hang of: Examining the idea of versatile learning in shrewd gadgets, where simulated intelligence calculations ceaselessly learn and further develop after some time in light of client criticism and information inputs, prompting more precise expectations and customized encounters [16].

Edge registering: Investigating the joining of AI and AI calculations into edge-figuring gadgets, empowering continuous data processing and examination at the edge of the organization, which diminishes inactivity and improves responsiveness for time-sensitive applications.

Security and oddity recognition: Tending to how simulated intelligence and AI strategies are utilized for online protection purposes, including anomaly detection, danger ID, and conduct analysis to improve the security of smart gadgets and organizations [27].

Moral contemplations: Recognizing the moral ramifications of AI and AI mixed in shrewd gadgets, for example, issues connected with information protection, algorithmic predisposition, and straightforwardness, and examining the significance of mindful AI improvement and sending. Generally speaking, the outline gives understanding of how AI and AI advancements are driving development and empowering new capacities in intelligent gadgets, molding the eventual fate of innovation and human cooperation [22, 28–31].

7.3.2 Internet of Things (IoT) advancements

The focus is likely on the significant advancements and innovations driving the evolution of IoT technologies. This overview may include:

Interconnectivity and integration: Highlighting the increasing interconnectivity and integration of IoT devices across various domains, enabling seamless communication and collaboration between devices and systems [7].

Sensor technology: Discuss advancements in sensor technology, including miniaturization, increased sensitivity, and lower power consumption, which enable IoT devices to gather more accurate and diverse data from the environment.

Edge computing: Explores the integration of edge computing capabilities into IoT devices and enables real-time data processing and analysis at the network edge, with decreased latency and higher responsiveness for applications requiring strict timeliness [32].

Wireless communication: This technology refers to the advancement of wireless communication technologies, such as 5G, Wi-Fi 6, and LPWANs, allowing IoT devices to communicate over longer distances with higher efficiency and reliability.

Energy efficiency: Discuss innovation in energy-efficient design and new power management techniques that significantly extend the life of IoT batteries, thereby making long-term, remote, or resource-constrained deployments feasible [17].

7.3.3 Edge computing and its role in smart devices

This section highlights the emerging importance of edge computing in smart device ecosystems. Key considerations may include:

Definition of edge computing: Explain the definition of edge computing, which means processing data near the source rather than through only centralized cloud servers. In other words, it will have fewer latencies, reduced bandwidth, and more enhanced privacy and security shown in Figure 7.6.

Role in smart devices: This also depicts how edge computing enhances the capability of smart devices by making it possible to process and

Figure 7.6 Edge computing.

analyze data in real-time at the edge. This helps smart devices respond faster and reduces reliance on cloud connectivity [8, 33].

Privacy and security: Discuss the privacy and security advantages of edge computing over smart devices. Since the computation is performed locally, data breaches and unauthorized access become less probable in edge computing [18].

7.4 AUGMENTED REALITY IN SMART DEVICES

7.4.1 Fusion of augmented reality and smart technology

The fusion of augmented reality and smart technology is a major step forward with transformative potential. The fusion integrates AR capabilities into smart devices to enhance user experiences and enable new functionalities.

Contextual information: AR through devices enables the retrieval of real-time contextual information relevant to your surroundings. For instance, capturing any landmark on a camera display can start the experience and initiate some overlays for histories or close points of interests [9].

7.4.2 Applications and implications

The exploration of the applications and implications of augmented reality with smart devices reveals wide-ranging possibilities and their great impact. Here is a brief summary of that:

Enhanced user experience: AR enriches user experience by overlaying digital information onto the real world. It provides contextual information, interactive features, and even immersive visuals that enhance one's daily interactions [20, 34].

Real-time navigation and wayfinding: The enabled smart devices in augmented reality will aid real-time navigation and wayfinding through superimposition of route guidance, points of interest, and relevant data over the surrounding user. Thus, improving the navigability indoors and outdoors.

Educational and training applications: AR-enhanced smart devices transform education and training by providing immersive learning experiences with interactive 3D models, simulations, and visualizations overlaid onto real-world objects, enhancing understanding and retention.

Professional and industrial use cases: AR in smart devices is applied across various professional and industrial settings, such as maintenance, repair, and operations (MRO), enhancing productivity, safety, and efficiency by delivering real-time information and guidance to workers.

Augmented workspaces: AR-enabled smart devices convert traditional workspaces into augmented environments where digital tools and information seamlessly integrate with physical surroundings, enhancing collaboration, visualization, and decision-making in professional settings [35].

The implications of AR in smart devices extend to revolutionizing user interactions with digital content and the physical world. This technology also raises considerations for privacy, data security, digital literacy, and societal acceptance. AR opens up new avenues for communication, productivity, entertainment, and learning, while prompting discussions on ethics, accessibility, and user experience design [7].

7.5 CONNECTIVITY REVOLUTION: 5G NETWORKS

7.5.1 The advent of 5G technology

Exploring the advent of 5G technology and its transformative potential in shaping the future of smart devices, here's a succinct overview:

Introduction to 5G: 5G technology marks a significant evolution from previous generations of wireless networks (2G, 3G, 4G), offering faster data speeds, lower latency, and increased network capacity in Figure 7.7.

Figure 7.7 5G networks.

Enhanced data speeds: 5G networks deliver substantially faster data speeds compared to predecessors, facilitating seamless streaming of high-definition content, quicker downloads and uploads, and smoother online gaming and video conferencing experiences.

Low latency: A key feature of 5G is its low latency, enabling real-time interactions and supporting applications such as augmented reality, virtual reality, and autonomous vehicles, where responsiveness is crucial.

Greater network capacity: 5G networks boast enhanced capacity to accommodate a large number of connected devices simultaneously, making them well-suited for dense urban environments, crowded events, and IoT deployments requiring reliable connectivity [36].

IoT and smart devices: 5G technology is expected to drive the growth of the IoT by facilitating real-time communication and interaction among a wide range of smart devices, sensors, and machines [8].

Industry applications: 5G is poised to revolutionize various industries, including healthcare, manufacturing, transportation, and entertainment, enabling innovations such as remote surgery, smart factories, connected vehicles, and immersive media experiences.

Challenges and considerations: Despite the benefits, challenges such as infrastructure deployment, spectrum availability, security concerns, and regulatory considerations need to be addressed for the successful rollout and adoption of 5G technology.

Future outlook: Looking ahead, 5G technology is anticipated to have a transformative impact on society, the economy, and the technology landscape, ushering in an era where ultra-fast, low-latency connectivity fosters unprecedented levels of innovation and connectivity.

7.5.2 Enhancing connectivity for smart devices

A more subtle segmentation that delves into the conceptual underpinnings and potential effects of using 5G networks to connect smart devices:

Bandwidth expansion: The theoretical framework of the bandwidth capacity expansion in 5G networks will enable support for data flows at ultra-high speeds. The expansion in bandwidth makes seamless connectivity possible for smart devices and supports real-time data exchange, high-definition multimedia streaming, and immersive experiences.

Low latency: Low latency for the 5G networks defines the reduction of latency when forwarding data packets and results in an instant response by the smart device. Thus, it will increase interaction with these devices and enhance users' experiences, along with its applicability toward enabling autonomous vehicles, and augmented reality, all of which at a minimum will require some amount of latency [9, 37].

7.5.3 Potential applications and innovations

Smart electronic devices that make use of AI and ML technologies are soon ready to revolutionize several industries and aspects of daily life. Here's a short overview of some possible applications and innovations:

Tailored healthcare solutions: Wearable devices implanted with AI and ML algorithms can keep track of one's vital signs, monitor anomalies, and provide personalized health suggestions. It helps people manage their fitness, cope with chronic diseases, and predict potential health issues that may arise in the body.

Key points with emphasis on autonomous transport will feature how the concept of autonomous transport advances by integrating AI-driven sensors and ML algorithms on the basis of real-time data analysis to navigate through intricate surrounding conditions, detect objects, and make fast decisions about passenger safety.

Intelligent home systems: AI can operate smart home devices. For example, it can change temperature, lighting conditions, and security systems based on user preferences and behavioral patterns. ML algorithms learn the behavior of the users and optimize energy consumption while enhancing comfort [10, 38].

7.6 USER-CENTRIC DESIGN AND EXPERIENCE

7.6.1 The importance of user experience

The user experience is the gold standard for the success of smart electronic devices that are integrated with AI and ML technologies. Here's how UX matters in shaping up the future of these devices:

Intuitive interface: A smart electronics device has an intuitive interface through which its user can easily interface with AI and ML. If one has a well-looked UI, one shall be able to navigate such advanced functionalities and can utilize them without complexity and confusion [26, 27, 39, 40].

Personalization and adaptation: AI and ML capabilities allow smart devices to personalize the interaction based on user preferences, behavioral patterns, and contextual cues. Such a personalized experience increases user satisfaction and makes them interact with the device for a longer period.

Context-awareness: The AI and ML algorithms-enabled smart devices should behave contextually, knowing about the user's environment, activities, and needs. It makes information and assistance relevant in real time, which increases usability and utility for the user [11].

7.6.2 Design principles for intuitive interaction

To ensure intuitive interaction experiences with smart electronic devices that incorporate AI and ML, here are some design principles that should be considered in Figure 7.8:

Simple: Maintain the simplicity of the UI and flow of interaction, with lesser cognitive load to avoid more than necessary components.
Consistency: In terms of predictable and known design elements and terminology with interaction patterns repeated throughout the device and throughout the entire ecosystem.

Informative feedback that occurs instantly through visual, audio, or even tactile indicators of the implications of that action for users in real-time interactions.

Adaptability: Design interactions that take users' preferences, behavior patterns, and contextual cues into account for better and efficient satisfaction.
Context awareness: Utilize AI and ML algorithms to ensure interactions are pertinent and sensitive to the user's context of their surroundings, activities, and needs.
Accessibility: All interactions need to be accessible to users with diverse needs and abilities. It should be integrated with features such as voice commands, gesture recognition, and adjustable settings.

Discoverability features and functionalities via intuitive design cues, prompts, or tutorials. It should encourage exploration and comprehension [12].

Figure 7.8 Intuitive interaction.

7.6.3 Human-centered innovation

For the purpose of smart electronic devices, especially those who have adopted AI and ML technologies, it becomes pertinent to have human-centered principles in design. This goes on to define how this particular approach shapes the future of devices:

User empowerment: Smart electronic devices should empower users by enabling them to have control, autonomy, and personalization. AI and ML algorithms can reinforce user empowerment by recognizing patterns of preference, adapting to users' needs, and making available appropriate recommendations [28].

User-centric design: Design processes should be more focused on understanding user needs, behaviors, and pain points through methods like research, observation, and empathy. The smart devices will better meet the demands and expectations of the users when they are placed at the center of the design process.

Co-creation with users: Involving users in the design and development process through co-creation workshops, feedback sessions, and usability testing ensures that smart devices resonate with user preferences and priorities. Collaborative design methodologies encourage innovation that addresses real-world contexts and challenges.

Ethical issues: Human-centered innovation incorporates ethical considerations about privacy, security of data, fairness, and transparency. Creating ethical smart devices means that it upholds the rights of its users, reduces biases, and is trustworthy.

Accessibility and inclusion: Smart electronic devices need to meet users who have varying needs and capabilities. Creating smart devices with an inclusive mindset means creating devices that are useful to users, readable, have auditory feedback, and have accessibility to assistive technologies that allow equitable access to users [29, 30].

7.7 SOCIETAL IMPLICATIONS

7.7.1 Ethical considerations in smart device development

Developing a wise electronic device with ethical considerations is crucial for the safe use of such technology.

Privacy protection: Most smart devices collect personal information; thus, there should be strong data protection. User privacy should come first, with clear consent for collection, as well as transparency about usage, storage, and sharing.

Security measures: This is very vital to prevent unauthorized access and data breaches. Developers should, hence, have a strong security emphasis. It means having strong encryption methods, regular software updates to patch vulnerabilities, and secure coding practices.

Transparency in functionality: Users should have a clear understanding of the capabilities and limitations of smart devices. Transparency about data collection, usage, and device functionalities is essential for fostering user trust [22, 31].

7.7.2 Privacy concerns and mitigation strategies

Significant privacy concerns in smart electronic devices arise from the fact that they can collect and manipulate sensitive user data. The resolution of these concerns requires a holistic approach. Here are essential privacy considerations along with corresponding mitigation strategies:

Data minimization: Developers should limit the amount of personal data collected to that which is necessary for device functionality. This will serve to minimize the possibility of violating user privacy through unneeded or excessive data collection.

Anonymization and pseudo-nymization: Data should be anonymized or pseudo-nymized to the greatest extent possible, reducing the risk of identification. This can be achieved by removing or encrypting personally identifiable information so that users' anonymity is protected.

User consent: Explicitcon sent from the users has to be received before their private information is collected. They should also be given transparency about the collection process, the purpose, and the partners with whom their information might be shared. The users also have the right to be given a choice to opt out of data collection [31, 41, 42].

7.7.3 Impact on employment and skills

Automation of jobs: It integrates AI and robotics into smart electronic devices; it automates repetitive jobs, replacing some human jobs.

Creation of new jobs: Even though the development and maintenance of smart devices lead to job displacement, new jobs are created in software engineering, data analysis, cybersecurity, and AI [24].

Changes in the skill requirements: The mushrooming of smart electronic devices bring with it changes in skill requirements in programming, data analysis, ML, and other technical fields in order to design and manage such devices.

Upskilling and reskilling: Employees may need to go through upskilling or reskilling training to develop skills that correspond to the shifting job landscape, with increased focus on digital literacy, coding, and other technical competencies [43] in Figure 7.9.

Integration of IoT skills: Smart devices rely on the IoT, necessitating expertise in sensor technology, connectivity, and data analytics, leading to increased demand across various sectors.

Figure 7.9 Reskilling needs.

7.8 SUSTAINABLE SMART DEVICES

7.8.1 Eco-friendly design practices

Energy efficiency: Design devices to consume the least amount of energy by utilizing efficient parts and algorithms. Power-saving features such as sleep mode and automatic shut-off should be incorporated.

Sustainable materials: Use sustainable materials for the construction of the device to reduce resource depletion and pollution. Choose materials with low carbon footprints, recycled content, and biodegradability [44].

Reduced packaging: Reduce packaging material and use materials that can be recycled or are biodegradable to reduce further waste generation. Design tight, lightweight, and compact packaging for reducing transportation-based emissions.

Modular design: This method of designing modular devices with easy repair and upgrade becomes very convenient and feasible; designing devices should be friendly and accessible in terms of their parts.

Recycle and end-of-life management: Design to be recycled easily; use materials that would allow it to be easy to dismantle and recycle. Environmental safety infrastructure and guidelines on not harming the environment must be established [45, 46].

7.8.2 Energy-efficient technologies

This approach enhances device autonomy and reduces reliance on external power sources [1].

Ultra-low-power electronics: Continued advancements in semiconductor technology will lead to the development of ultra-low-power electronics. These energy-efficient components, such as microprocessors, sensors, and communication modules, will enable smart devices to perform complex tasks while consuming minimal energy, extending battery life, and enhancing overall efficiency.

Wireless power transfer: The proliferation of wireless power transfer technologies, like resonant inductive coupling and radio frequency (RF) energy harvesting, will facilitate seamless charging and power delivery to smart devices without physical connectors or cables. This simplifies device integration, enhances user convenience, and reduces energy losses associated with traditional wired charging methods [2].

Energy-aware algorithms: Future smart devices will incorporate energy-aware algorithms to optimize software behavior and task scheduling, minimizing energy consumption without compromising performance. These algorithms will prioritize energy-intensive tasks based on user preferences, environmental conditions, and available power resources, ensuring efficient energy utilization.

Energy-efficient displays and lighting: Next-generation smart devices will feature energy-efficient display technologies, such as OLEDs and micro-LEDs, offering superior energy efficiency compared to conventional LCDs. Smart lighting solutions utilizing energy-efficient LEDs and advanced control systems will contribute to reduced power consumption in indoor and outdoor environments.

Smart energy management platforms: Integrated smart energy management platforms will enable users to monitor, control, and optimize energy usage across interconnected devices and appliances within smart homes, buildings, and infrastructure.

7.8.3 Contribution to sustainability goals

Smart devices are also in a position to greatly impact sustainability goals. This is how they can help:

Energy efficiency: The smart devices will emphasize energy efficiency. It will make use of the latest power management systems, energy-efficient components, and intelligent algorithms. Thus, optimization decreases the total demand for energy, helping mitigate climate change. Renewable Energy Integration: The future of smart devices is in line with increasing integration of renewable energy sources, such as solar, wind, and kinetic energy harvesting. Use of renewables will reduce dependency on fossil fuels, reduce greenhouse gas emissions, and encourage sustainable energy [46].

Resource conservation: Smart devices encourage resource conservation through responsible material use and waste reduction, such as modular design and recyclable materials, making the product last longer through end-of-life management and reducing environmental impact.

Transportation and urban planning: Smart devices make for smarter transportation systems and urban planning. The flow of traffic is optimized, congestion reduced, and sustainable modes of transport encouraged, resulting in reduced carbon emissions as well as improved air quality in urban areas [47].

7.9 CHALLENGES AND OPPORTUNITIES

7.9.1 Addressing emerging challenges

Some of the ways in which AI and ML can face challenges arising with smart electronic devices include:

Personalization and user experience: Algorithms of personalization, AI, and ML understand user behavior and preferences. They personalize the functionality of devices and interfaces in order to make users' experiences more intuitive and satisfying when using devices.

Predictive maintenance: AI and ML analyze sensor data that helps in predictive maintenance. It can be achieved even before a failure. Thus, downtime is prevented, and the life of the devices extended; hence, they become more reliable and cheaper to maintain [48].

Energy efficiency optimization: AI and ML algorithms are used to optimize energy use. They adjust the settings as per real-time conditions and usage patterns to minimize energy waste, cut utility bills, and contribute toward a sustainability goal [49].

7.9.2 Seizing opportunities for growth

The opportunities for employing AI and ML technologies in smart electronics devices are countless. Among these are the following:

Product innovation: The use of AI and ML allows manufacturers to be innovative and come up with new products to match changing consumer expectations. Trends in markets, opinions of users, and competitors' offerings are inputs for the development strategy, so this results in innovation and competition for the product.

Improved user experience: AI and ML algorithms personalize the user experience based on user behavior and preferences. Tailoring device features, interfaces, and recommendations toward individual users increases user satisfaction and loyalty, thus providing growth through positive feedback and repeat purchases.

Operational efficiency: Optimization of the manufacturing process, supply chain, and inventory management through AI and ML. Automation of routine chores, such as forecasting demand or optimization of resources, helps improve operating efficiency while reducing the costs of operation to increase profitability [3, 50].

7.9.3 Balancing innovation with ethical responsibility

There must be a balance between innovation and ethical responsibility in developing the smart electronic devices based on AI and ML technologies. Achieving this balance can be done as follows:

Data privacy and security: There should be robust privacy measures and protocols for security to protect users' data. There should always be transparency in the way data is collected, processed, and stored, while explicit consent from users must be obtained. There need to be regular updates of the security features to reduce risk and vulnerabilities.

Ensures fairness and reduces bias : AI algorithms are audited and tested to eliminate all possible biases and discriminations within them. Fairness-aware learning and preprocessing of the data are also employed as techniques to reduce the bias levels. Diverse and representative datasets are sought in an endeavor to improve the fairness levels of algorithms.

Transparency and accountability: Ensure that the AI decision-making process is transparent to the user to garner trust. Explain why decisions were made automatically, allowing users to understand and appeal the decisions. Create lines of accountability, such as clear lines of responsibility and pathways for redress if the user is harmed.

Responsible AI development design and develop AI-enabled products with ethical considerations in mind. Perform thorough risk analysis in identifying potential ethical problems and incorporate ethical guidelines and principles into the development process. Promote an ethos of ethical awareness and responsibility among developers and stakeholders [1].

User empowerment and control: Empower the user to control data access and interaction with smart devices. Granular options for privacy settings as well as the option to delete data or anonymize them. Offer tools for transparency, which will be helpful for the understanding of the use of data, allowing users to accept data processing activities.

7.10 CONCLUSION

7.10.1 Recapitulation of key insights

In conclusion, the study of smart electronic devices, AI, ML, and their future has been explored, providing insight into what happens when technology and innovation come into play. In the whole journey, the following takeaways were highlighted:

With integration with AI and ML, it presents the possibility of transforming sectors entirely. These technologies help make devices learn, adapt, and apply intelligence in decision-making; they improve efficiency, performance, and customer satisfaction levels.

Ethical considerations: Amidst the increasing integration of smart devices into our daily lifestyles, it is necessary for ethical considerations related to privacy, security, fairness, and transparency. Ethical principles at the design, development, and deployment phases of smart devices can prevent irresponsible and inequitable use of technology.

Sustainability: The environmental impact of smart electronic devices cannot be ignored. Eco-friendly design practices, energy-efficient technologies, and responsible manufacturing processes are the basic requirements to minimize environmental footprint and promote sustainability.

User-centric design: The development of smart devices continues to focus on user experience. Personalization, customization, and intuitive interfaces are the essentials of user-centric design and improve usability, satisfaction, and adoption of smart devices.

Continuous innovation: The future of smart electronic devices is characterized by continuous innovation and evolution. Emerging technologies, new use cases, and the ability to see consumer needs ahead are requirements for being ahead in an ever-changing landscape.

In conclusion, recapitulating the key insights in such a manner underlines the urgency of harnessing the transformational power of AI, ML, and smart electronic devices by being mindful of the ethical, environmental, and user-centric concerns. By embracing innovation responsibly and ethically, we would be able to design a future where technology benefits human lives and contributes to a sustainable and equitable society.

REFERENCES

1. Jakubik, J.; Vössing, M.; Kühl, N.;Walk, J.; Satzger, G. Data-centric Artificial Intelligence. arXiv 2022, arXiv:2212.11854.
2. Li, X.-H.; Cao, C.C.; Shi, Y.; Bai, W.; Gao, H.; Qiu, L.; Wang, C.; Gao, Y.; Zhang, S.; Xue, X.; et al. A survey of data-driven and knowledge-aware explainable ai. *IEEE Trans. Knowl. Data Eng.* 2020, 34, 29–49.

3. Hajian, S.; Bonchi, F.; Castillo, C. Algorithmic bias: From discrimination discovery to fairness-aware data mining. In *Proceedings of the 22nd ACM SIGKDD International Conference on Knowledge Discovery and Data Mining*, San Francisco, CA, USA, 13–17 August 2016, pp. 2125–2126.
4. Tripathi, S.L.; Mahmud, M. Explainable Machine Learning Models and Architecture. Wiley & Scrivener Publishing LLC, USA, 2023. https://doi.org/10.1002/9781394186570
5. Tripathi, S.L.; Ganguli, S.; Kumar, A.; Magradze, T. *Intelligent Green Technologies for Sustainable Smart Cities*, Wiley, USA, 2022, https://doi.org/10.1002/9781119816096
6. Kim, J. The impact of AI and ML on smart device design. In *Smart Electronic Devices: AI, ML, and the Future*, edited by Jane Doe, ABC Publishing, Chicago, 2023, pp. 101–115.
7. Nguyen, L. User experience design in smart electronics. In *Smart Electronic Devices: AI, ML, and the Future*, edited by Jane Doe, ABC Publishing, Chicago, 2023, pp. 116–130.
8. Martinez, C. Challenges and opportunities in IoT integration." In *Smart Electronic Devices: AI, ML, and the Future*, edited by Jane Doe, ABC Publishing, Chicago, 2023, pp. 131–145.
9. Anderson, J. Advancements in energy-efficient technologies. In *Smart Electronic Devices: AI, ML, and the Future*, edited by Jane Doe, ABC Publishing, Chicago, 2023, pp. 146–160.
10. Wilson, R. Security measures in smart device manufacturing. In *Smart Electronic Devices: AI, ML, and the Future*, edited by Jane Doe, ABC Publishing, Chicago, 2023, pp. 161–175.
11. Brown, S. The evolution of wearable technology." In *Smart Electronic Devices: AI, ML, and the Future*, edited by Jane Doe, ABC Publishing, Chicago, 2023, pp. 176–190.
12. Kim, D. Innovations in display technologies for smart devices. In *Smart Electronic Devices: AI, ML, and the Future*, edited by Jane Doe, ABC Publishing, Chicago, 2023, pp. 191–205.
13. Smith, J. Introduction to smart electronic devices. In *Smart Electronic Devices: AI, ML, and the Future*, edited by Jane Doe, ABC Publishing, Chicago, 2023, pp. 1–10.
14. Doe, J. The role of artificial intelligence in smart devices. In *Smart Electronic Devices: AI, ML, and the Future*, edited by Jane Doe, ABC Publishing, Chicago, 2023, pp. 11–25.
15. Johnson, E. Machine learning applications in smart electronics. In *Smart Electronic Devices: AI, ML, and the Future*, edited by Jane Doe, ABC Publishing, Chicago, 2023, pp. 26–40.
16. Lee, D. "Ethical considerations in the development of smart devices." In *Smart Electronic Devices: AI, ML, and the Future*, edited by Jane Doe, ABC Publishing, Chicago, 2023, pp. 41–55.
17. Patel, S. Privacy and security challenges in smart electronics. In *Smart Electronic Devices: AI, ML, and the Future*, edited by Jane Doe, ABC Publishing, Chicago, 2023, pp. 56–70.
18. Wang, M. Future trends in smart electronic devices. In *Smart Electronic Devices: AI, ML, and the Future*, edited by Jane Doe, ABC Publishing, Chicago, 2023, pp. 71–85.

19. Devlin, J.; Chang, M.W.; Lee, K.; Toutanova, K. BERT: Pre-training of deep bidirectional transformers for language understanding. In *Proceedings of the 2019 Conference of the North American Chapter of the Association for Computational Linguistics: Human Language Technologies (NAACL-HLT)*, Minneapolis, MN, USA, 2019, vol. 1, pp. 4171–4186.
20. Gumbs, A.A.; Grasso, V.; Bourdel, N.; Croner, R.; Spolverato, G.; Frigerio, I.; Illanes, A.; Abu Hilal, M.; Park, A.; Elyan, E. The advances in computer vision that are enabling more autonomous actions in surgery: A systematic review of the literature. Sensors 2022, 22, 4918.
21. Garcia, M. Sustainability practices in smart device development. In *Smart Electronic Devices: AI, ML, and the Future*, edited by Jane Doe, ABC Publishing, Chicago, 2023, pp. 86–100.
22. Mazumder, M.; Banbury, C.; Yao, X.; Karlaš, B.; Rojas, W.G.; Diamos, S.; Diamos, G.; He, L.; Kiela, D.; Jurado, D.; et al. Dataperf: Benchmarks for data-centric AI development. arXiv 2022, arXiv:2207.10062.
23. Hernandez, M. Data analytics and insights in smart electronics. In *Smart Electronic Devices: AI, ML, and the Future*, edited by Jane Doe, ABC Publishing, Chicago, 2023, pp. 206–220.
24. Alvarez-Coello, D.; Wilms, D.; Bekan, A.; Gómez, J.M. Towards a data-centric architecture in the automotive industry. *Procedia Comput. Sci.* 2021, 181, 658–663.
25. Géron, A. *Hands-On Machine Learning with Scikit-Learn, Keras, and TensorFlow: Concepts, Tools, and Techniques to Build Intelligent Systems*, O'Reilly Media, Inc., Sebastopol, CA, USA, 2019.
26. Li, Y. Deep reinforcement learning: An overview. arXiv 2017, arXiv:1701.07274.
27. Pouyanfar, S.; Sadiq, S.; Yan, Y.; Tian, H.; Tao, Y.; Reyes, M.P.; Shyu, M.L.; Chen, S.C.; Iyengar, S.S.A survey on deep learning: Algorithms, techniques, and applications. *ACM Comput. Surv. (CSUR)* 2018, 51, 1–36.
28. Zha, D.; Bhat, Z.P.; Lai, K.H.; Yang, F.; Jiang, Z.; Zhong, S.; Hu, X. Data-centric artificial intelligence: A survey. arXiv 2023, arXiv:2303.10158.
29. Ntoutsi, E.; Fafalios, P.; Gadiraju, U.; Iosifidis, V.; Nejdl, W.; Vidal, M.E.; Ruggieri, S.; Turini, F.; Papadopoulos, S.; Krasanakis, E.; et al. Bias in data-driven artificial intelligence systems—An introductory survey. *Wiley Interdiscip. Rev. Data Min. Knowl. Discov.* 2020, 10, e1356.
30. Jarrahi, M.H.; Ali, M.; Shion, G. The principles of data-centric AI (DCAI). arXiv 2022, arXiv:2211.14611.
31. Zha, D.; Bhat, Z.P.; Lai, K.-H.; Yang, F.; Hu, X. Data-centric AI: Perspectives and challenges. arXiv 2023, arXiv:2301.04819.
32. Sharma, L.; Garg, P.K. *Artificial Intelligence: Technologies, Applications, and Challenges*, Taylor & Francis, New York, NY, USA, 2021.
33. Aguiar-Pérez, J.M.; Pérez-Juárez, M.A.; Alonso-Felipe, M.; Del-Pozo-Velázquez, J.; Rozada-Raneros, S.; Barrio-Conde, M. Understanding machine learning concepts. In *Encyclopedia of Data Science and Machine Learning*, IGI Global: Hershey, PA, USA, 2023, pp. 1007–1022.
34. Enholm, I.M.; Papagiannidis, E.; Mikalef, P.; Krogstie, J. Artificial intelligence and business value: A literature review. *Inf. Syst. Front.* 2022, 24, 1709–1734.
35. Wang, Z.; Li, M.; Lu, J.; Cheng, X. Business Innovation based on artificial intelligence and Blockchain technology. *Inf. Process. Manag.* 2022, 59, 102759.

36. Marr, B. *Artificial Intelligence in Practice: How 50 Successful Companies Used AI and Machine Learning to Solve Problems*, John Wiley & Sons, New York, NY, USA, 2018.
37. Dahiya, N.; Sheifali, G.; Sartajvir, S. A review paper on machine learning applications, advantages, and techniques. *ECS Trans.* 2022, 107, 6137.
38. Liu, X.; Yoo, C.; Xing, F.; Oh, H.; El Fakhri, G.; Kang, J.-W.;Woo, J. Deep unsupervised domain adaptation: A review of recent advances and perspectives. *APSIPA Trans. Signal Inf. Process.* 2022, 11, e25.
39. Zhuang, F.; Qi, Z.; Duan, K.; Xi, D.; Zhu, Y.; Zhu, H.; Xiong, H.; He, Q. A comprehensive survey on transfer learning. *Proc. IEEE* 2020, 109, 43–76.
40. Sun, X.; Liu, Y.; Liu, J. Ensemble learning for multi-source remote sensing data classification based on different feature extraction methods. *IEEE Access* 2018, 6, 50861–50869.
41. Nguyen, T. Advancements in voice recognition technology. In *Smart Electronic Devices: AI, ML, and the Future*, edited by Jane Doe, ABC Publishing, Chicago, 2023, pp. 221–235.
42. Miranda, L.J. Towards data-centric machine learning: A short review. Available online.
43. Park, S. "The future of augmented reality in smart devices." In *Smart Electronic Devices: AI, ML, and the Future*, edited by Jane Doe, ABC Publishing, Chicago, 2023, pp. 236–250.
44. Uddin, M.F.; Navarun, G. Seven V's of Big Data understanding Big Data to extract value. In *Proceedings of the 2014 Zone 1 Conference of the American Society for Engineering Education*, Bridgeport, CT, USA, 3–5 April 2014.
45. Patel, R. Enhancing user privacy in smart electronics." *Smart Electronic Devices: AI, ML, and the Future*, edited by Jane Doe, ABC Publishing, Chicago, 2023, pp. 251–265.
46. O'Leary, D.E. Artificial intelligence and big data. *IEEE Intell. Syst.* 2013, 28, 96–99.
47. Lee, M. The role of blockchain in smart device security. In *Smart Electronic Devices: AI, ML, and the Future*, edited by Jane Doe, ABC Publishing, Chicago, 2023, pp. 266–280.
48. Garcia, D. Adaptive learning algorithms for smart devices. In *Smart Electronic Devices: AI, ML, and the Future*, edited by Jane Doe, ABC Publishing, Chicago, 2023, pp. 281–295.
49. Broo, D.G.; Jennifer, S. Towards data-centric decision making for smart infrastructure: Data and its challenges. *IFAC-Pap.* 2020, 53, 90–94.
50. Thompson, R. Addressing bias in machine learning models. In *Smart Electronic Devices: AI, ML, and the Future*, edited by Jane Doe, ABC Publishing, Chicago, 2023, pp. 296–310.

Chapter 8

Exploring deep learning methods for detecting heart abnormalities by analyzing heart sound

Sandhya Avasthi, Tanushree Sanwal, Suman Lata Tripathi and Kadambri Agarwal

8.1 INTRODUCTION

The heart in humans pumps blood throughout the body and beats continuously. The heart sound can reveal a lot about the health of a person, which is produced when the heart valve closes and blood flows. Cardiovascular diseases (CVDs) are the primary cause of death, with an annual death toll of 17.9 million worldwide. CVDs categories are coronary heart disease, rheumatic heart disease, cerebrovascular disease, and other ailments mainly caused due to heart and blood vessel disorders. Such cardiac problems must be diagnosed early through auscultation because if left untreated for a long time, this could be fatal. The fact is that four out of five fatalities by heart attack are CVD-related [1]. Cardiac problems must be identified early with a highly accurate auscultation examination, since they are fatal. The fact that the majority of cardiovascular disorders are preventable and treatable should be taken into account, even though the expected number of cardiac patients and healthcare costs are unacceptably large. However, this necessitates an early diagnosis and appropriate illness management [2]. The sudden severe stroke is mainly caused by a blockage that stops blood flow going to the brain and heart.

An electronic stethoscope produces a phonocardiogram (PCG) and digitally records heart sounds (HS). The PCG image shows state of the heart and its functionality, helping health practitioners to diagnose and treat heart disease. Technologies like signal processing and machine learning (ML) are capable of processing and classifying heart signals through various signal features. Mainly, there are two classes of heart signal: normal and abnormal. HS are usually detected and interpreted via auscultation [3]. However, a physician's mood, expertise, or other subjective aspects usually affect it. Physicians need years of clinical expertise to accurately diagnose. This necessitates an intelligent, automatic HS auscultation device. Thus, cardiac auscultation system accuracy should improve.

Functions such as opening, closing, diastolic tendon, systolic tendon, muscle activity, and blood flow create a complicated low-frequency signal 'HS' on the blood vessel wall. A normal heart rate of 72 pulses per minute has an

DOI: 10.1201/9781003643173-8

833-millisecond cardiac cycle [4–6]. Only the first two of four HSs in a cardiac cycle are audible. The sudden mitral and tricuspid valve closure during systole causes the first HS (S1). The rapid closing of the aortic and pulmonary valves during ventricular diastole causes S2. A heart murmur usually precedes pathological heart changes. Cardiovascular disease patients are more likely to have heart murmurs. Cardiovascular illness diagnosis depends on heart murmur location during the cardiac cycle. HS signals might have heart murmurs. Auscultation is difficult, but signal graphs make it easy to find the murmur. Auscultation requires extensive training and competence, which have been dropping as new cardiac diagnostic technologies have become accessible. Auscultation involves heart sound analysis and acquisition. To record HS, a stethoscope must be positioned appropriately and forcibly on the patient's chest. Heart sound study determines the heart's health [6, 7].

The COVID-19 pandemic has affected the health and well-being of the global population in the past few years [6, 7]. After facing pandemic restrictions (such as social isolation and school closures) during the crisis, people adopted healthy lifestyle behaviors such as good hygiene, exercise, good diet, and regular checkups, etc. Global lockdowns and imposed restrictions on people are likely to be the primary reason for the deteriorating trends in population-level control of cardiometabolic risk factors (such as diabetes, obesity, dyslipidemia, and inadequate blood pressure control). Global data indicates an increase in excess deaths from CVD and diabetes. This further indicates the pandemic's decreased surveillance of cardiovascular risk factors, behaviors, and restricted access to preventive CV services. Segmentation and classification challenges pose a barrier to the creation of an automated heart sound analysis tool. The two main tasks of segmentation are identifying the S1, S2, S3, and S4 component categories and determining the limits of heart sound components [8]. There have been several segmentation methods described. However, the bulk of methods concentrate on component or border identification. Furthermore, the fact that many approaches only apply to typical cardiac sounds severely restricts their usefulness. This chapter explores in depth the steps required to classify HS and diagnose heart disease. Further, various techniques for data mining and ML algorithms in the areas of health are examined briefly to determine the role of artificial intelligence in decision support systems. Also, a discussion on publicly available audio database utilization highlights the ML approaches and challenges which still require attention. Figure 8.1 shows the share of people with heart disease in India in 2020 (as per statistica.com report).

This study assesses the informative value of frequently occurring heart sound characteristics that identify intriguing heart sound signals. The four main categories of classification techniques for cardiac sounds are Support Vector Machine, cluster-based, Hidden Markov Model, AdaBoost, and Artificial Neural Network (ANN). ANN is a widely used wavelet-based ML method for heart sound signal classification. SVM has been used more frequently recently for PCG classification since it may be integrated with other

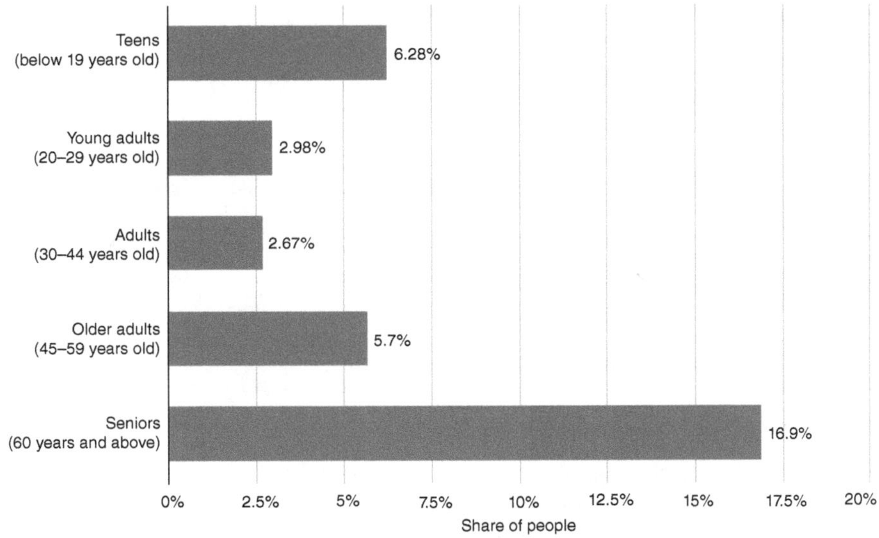

Figure 8.1 Share of people with heart disease in India 2020 (As per statistica.com report).

classification approaches to provide better results [9]. In addition to segmentation, HMM is utilized to categorize diseases in phonocardiography records. This chapter will examine various ML algorithms and deep learning (DL) algorithms to distinguish between normal and pathological cardiac sounds in publicly available heart sound data sets [10].

8.1.1 Abnormal heart sound

The atrioventricular valves, and the semilunar valves make up the majority of the HS. The function of regulating blood is completed by atrioventricular valves, also known as mitral and tricuspid valves. The semilunar valves are known as the aortic and pulmonary valves that control blood flow to the main arteries. When the atrioventricular valves close and the ventricles of the heart begin to contract, a low-pitched, slightly drawn-out 'lub' is heard. After that, a harsher, higher-pitched 'dup' is produced when the aortic and pulmonary valves close.

Some individuals can hear a third, low-pitched, quiet sound produced by the ventricular wall's oscillations. When blood released during atrial chamber contraction strikes the ventricular wall, occasionally a fourth sound that is typically undetectable in healthy individuals may be heard. Heart murmurs are caused by blood turbulence caused by partial occlusion of these valves or inadequate closure. The murmur may be constant in cardiac and thoracic blood vessel congenital defects. Murmurs can identify the diseased cardiac valve. Changes in cardiac sounds may suggest heart illness or muscular weakness. The image of an abnormal sound signal is shown in Figure 8.2.

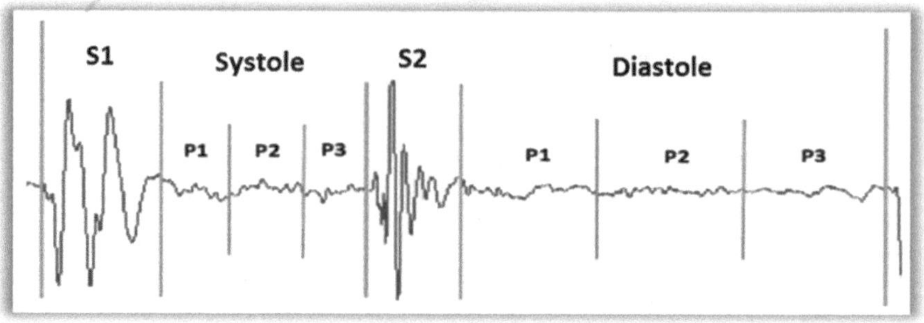

Figure 8.2 Illustrating abnormal heart sound.

8.1.2 Auscultation of the heart

Heart auscultation is a medical examination procedure that involves using a stethoscope to hear the sounds the heart makes. *Heart auscultation* requires excellent hearing and coordination. Doctors with hearing loss benefit from stethoscopes with amplifiers. The diaphragms of a stethoscope detect high-pitched noises. Bells amplify low-pitched sounds. Use the bell with caution. Excessive pressure transforms the skin into a diaphragm, eliminating low-pitched noises [11]. The precordium is examined that begins with the apical impulse in the patient's left lateral decubitus part. After assuming the supine position, the auscultation process completes in the lower left part of the sternal border, moves cephalad to each interspace, and concludes at the upper right sternal border [12]. Above the clavicles and left axilla, the clinician listens. The patient has to sit upright, bending forward position to listen for aortic and pulmonic diastolic murmurs to complete the process of auscultation of the back.

8.1.3 Heart sound features

Heart sound signals have many characteristics that can indicate the heart condition. Some typical features that are examined are described here.

1. *S1 and S2 intensity*: This is used to measure the intensity of S1 and S2 (first and second HS) respectively. Changes in intensity could be a sign of issues with the heart muscle or valves.
2. *S1–S2 Interval*: The time difference between the S1 and S2 can be calculated. An irregular interval could be a sign of conduction issues or heart rhythm problems.
3. *S3 and S4*: The existence or absence of extra HS, such as the third heart sound (S3) or the fourth heart sound (S4), might shed light on cardiac disorders like heart failure or stiff ventricles.
4. *Murmurs*: An aberrant sound produced by turbulent blood flow, murmurs can be heard in heart sound signals. Analysis of murmur

characteristics, including timing, duration, pitch, and intensity, identifies the root cause and degree of the murmur.

5. *Splitting*: When the aortic and pulmonary valves close differently, it results in a short time delay between the components of the second heart sound (S2). Splitting can be either normal or pathological, and its traits are examined for any potential heart problems.
6. *Timing of events*: To assess the whole cardiac cycle and spot irregularities in timing, such as prolonged or shorter intervals, different heart sound components, including systolic and diastolic events, are timed differently.
7. *Frequency content*: By doing a frequency domain analysis on heart sound data, it is possible to pinpoint the precise frequency components that are connected to certain heart sound characteristics. This analysis aids in identifying certain irregularities and distinguishing between healthy and unhealthy HS.

8.2 HEART SOUND AUDIO SIGNALS ANALYSIS

There are many techniques used for automated heart signal classification and analysis. They are thresholding methods, statistical ML, and neural network-based methods. Digital signal processing techniques, time-frequency decomposition, and spectral analysis are being used in the study of HS [13]. In general, HS are non-stationary and multicomponent signals. The objective of heart sound analysis is to achieve a precise classification of pathological events in the cardiac cycle. Before going into various phases of sound analysis and classification, the experimental dataset should go through initial preprocessing steps. The collected data was generally recorded in noisy environments and on a large number of patients, so the collected data may contain a lot of noise and errors. It is the job of a heart sound specialist to preprocess data to obtain good-quality data before using it for experiments. The various steps in heart sound analysis are shown in Figure 8.3.

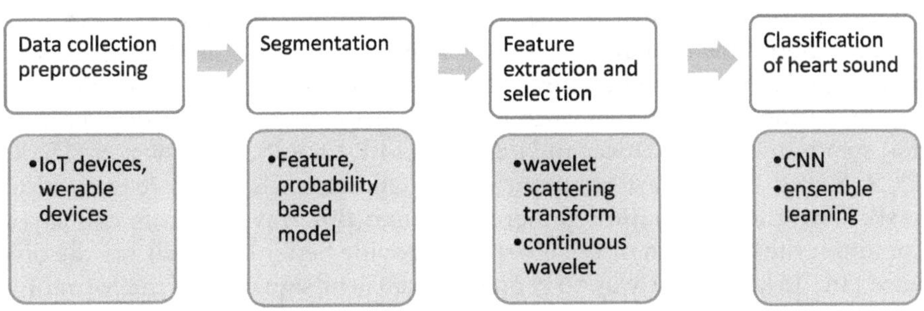

Figure 8.3 Classification process of heart sound.

8.2.1 Segmentation

The segmentation process splits a given audio input into smaller audio segments. Segmentation localizes heart sound signal peaks that include S1 and S2 of HS. In the systolic phase, the heart contracts and pushes blood to the heart, whereas in the diastolic phase heart relaxes to decrease the blood pressure. Both the systolic and diastolic phase makes up the cardiac cycle, and it is an imperative step in identifying the extraction of acoustic signals of interest in each cardiac cycle [14]. Segmentation methods are classified into various types as follows.

- Envelope-based methods [14]
- ECG or carotid pulse reference-based methods
- Probabilistic models
- Feature-based methods
- Times-frequency-based method
- Learning-based methods

In ***envelope-based*** method, the envelope of a heart sound signal is generated using the Hilbert transform. After the Hilbert transformation step, the magnitude of the signal gives the energy peaks to provide an envelope corresponding to the heart sound. ECG-based methods use external signals like the ECG to get time information to be used in the segmentation of different phases of the heart sound. ECG gives a detailed view of heart activity, which includes P wave, QRS complex, and T wave.

A probabilistic method, such as hidden Markov models, divides the states into S1, S2, diastole, and systole, emitting acoustic features. The training process utilizes PCG recordings, and transition probabilities represent the likelihood value of state transitions. Another probabilistic method is Dynamic Time Warping, where similarities between two time series signals are measured to check the heart signal variation when the heart rate changes [14].

8.2.2 Feature extraction

The main objective of feature extraction is to find relevant features that replace high-dimensional raw signals. Any classification model works best when applied to selected features and is often performed on the signal after segmentation. Extraction of various features from cardiac signals is an essential step in the classification and detection of P, Q, R, S, and T waves. The P, Q, R, S, and T waves are components in ECG signals is a feature extraction task. With the information gathered through this wave, doctors can say a lot about the condition of the heart and provide better diagnosis for the disease [14, 15]. The best way to represent transient signals is in the form of a time-frequency representation. The sound signal is measured quantitatively and qualitatively utilizing various transforms such as the Short-time Fourier transform and Fourier Transform. Some other methods to measure sound signals are Wigner–Ville Distribution (WVD), Wavelet Transform, and Choi–Williams Distribution. The WVD gives a joint representation of a signal in the time-frequency domain that assists in comprehending the signal properties.

8.3 DEEP LEARNING IN HEART SOUND CLASSIFICATION

DL is efficient in image recognition, voice recognition, and natural language processing tasks and has proven to be successful in different applications. DL model imitates human brain functions using artificial neural networks. Popular DL models are CNN, Recurrent Neural Networks (RNN), long short-term memory (LSTM), and other hybrid methods [15, 16]. CNN can provide a visual representation of sound signals through an audio spectrogram, that can be used in feature extraction. RNN models such as LSTMs and GRUs are effective in finding temporal patterns from HS and can handle long-term dependencies as well.

8.3.1 Heart sound classification using CNN

CNN stands for Convolution Neural Network, a specialized model for image data classification. The central part of any CNN is a convolution network based on linear operations instead of general matrix operations (multiplication) in several layers (two or more). If the given input is a vector, the output (y) is calculated using a discrete convolution operation given by the expression (Equation (8.1)).

$$y(n) = x(n) \times w(n) = \sum_{m=-\infty}^{\infty} x(m)w(n-m) \tag{8.1}$$

Here '×' denotes a convolution operation, and w is a convolution kernel used to apply to a one-dimensional CNN model. If the input given is two-dimensional (2D), x and w both are 2D matrices. Now, output 'y' can be computed by the expression as shown in Equation (8.2).

$$Y(i,j) = X(i,j) \times w(i,j) = \sum_{m}\sum_{n} X(m,n)w(i-m,j-n) \tag{8.2}$$

The convolution operations can be performed using methods such as sparse interactions, equivalent representations, and parameter sharing.

8.3.2 Heart sound classification using RNN

RNN can leverage temporal information in sequential data and can detect states or abnormalities in HS [17, 18]. Two RNN architectures, such as gated recurrent units (GRU) and LSTM, deliver good performance in various applications. Past research has shown that segmentation can detect events that can further be used in developing a bi-directional GRU Recurrent Neural Network. GRU-RNN uses a spectrogram and envelope feature to complete the detection task. However, envelope features do not capture intrinsic information of heart cycles.

LSTM is an RNN-based model capable of training long-term dependencies. In general, LSTM has four layers, including an input layer, two hidden layers, and one output layer. The LSTM model changes the memory cell unit architecture to add more storage for a longer period. The function of gates for input, output, and forget is to control how data is moved within the cell [19].

In heart sound analysis using RNN models, the input is given in the form of a one-dimensional (1D) signal $x(t) = (x_1, \ldots, x_T)$. The current time 't', calculates the value h_t utilizing the previous state value h_{t-1} and input x_t. The output value is projected using the softmax function as per the classification label. The expression to represent the calculation is given in Equation (8.3).

$$h_t = H(U.x_t + W.h_{t-1} + b)y_t = \text{softmax}(V.h_t + c) \tag{8.3}$$

Here, U, V, and W are matrices, Hidden layer 'H', bias vector b, and c. The vanishing gradient problem in RNN can be avoided using gated RNN, working on the concept of paths through time. The standard limitation in the RNN model is that it needs previous state information to work. A bidirectional RNN model can also be used in heart sound classification, that works in both forward and backward directions.

8.3.3 Heart sound classification transformer model

In this technique, knowledge from one ML model is transferred to another model, often done to reuse model knowledge. A transformer model can train long-range dependencies, it does not use any recurrent or convolution layers. The transformer model's fundamental layer is called 'attention'. A transformer has two main components: an encoder for processing the input and a decoder for generating the output. A heart sound signal may contain abnormalities that span multiple cardiac cycles that a transformer model can easily capture. Transformer analyses the entire sequence at once, making them computationally efficient for longer signals. Self-attention allows the model to focus on diagnostically relevant parts of the heart sound, such as irregularities or murmurs.

8.4 COMPARISON OF MACHINE LEARNING AND DEEP LEARNING

DL techniques are more accurate and efficient than traditional ML techniques. Figure 8.4 illustrates the classification methods used by traditional ML techniques and DL techniques. In general, DL-based classification of cardiac sounds is more accurate than traditional ML-based classification, according to a survey [5].

Table 8.1 provides a summary of the advantages and disadvantages of DL and ML models for classifying cardiac sounds. DL algorithms can handle massive data efficiently and provide accurate classifications.

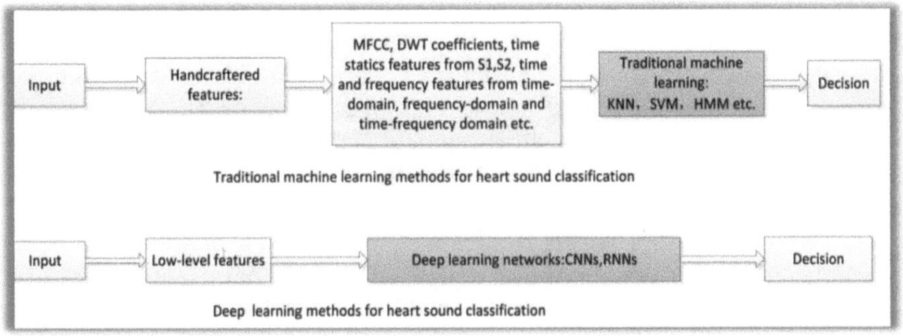

Figure 8.4 Machine learning process vs deep learning process.

Table 8.1 Deep learning and ML comparison in heart sound classification

Approaches	*Strengths*	*Limitations*
Machine learning methods	• This ML model needs a manual feature engineering step. • Model training is easier compared to DL. • Deep learning models solve the objective function quickly by the convex optimization algorithm.	• Data preprocessing steps are complex and the segmentation phase is an essential step that takes extra time. • ML models are not robust and very difficult to generalize.
Deep learning	• There is no need for a manual feature engineering step. • Deep learning learns features from the sound signal data and then classifies the sound into normal and abnormal sounds. • Classification accuracy is better. • The DL models can be fine-tuned as per the specific samples of patients.	• Training complexity is high and depends on large-scale datasets for modeling. • Hardware configuration requirements are high and need processing power. • Annotated heart sound datasets are scarce; it is a rigorous task to create a labeled dataset.

The segmentation algorithm can locate the S1, S2, systole, and diastole phases in ML, and segmented information such as the time, frequency, and statistical properties is extracted. Pedro Narváez [6] proposed traditional ML for the automatic segmentation and classification of cardiac sounds. In his study, the cardiac cycles and S1, S2, systole, and diastole segments were identified using empirical wavelet transform and normalized average Shannon energy (NASE) from a sound signal. This method has demonstrated superior performance in comparison to other ML techniques such as the discrete wavelet transform (DWT). DWT analyzes non-stationary signals such as PCGs and gives time-frequency representation. The murmurs and other anomalies can be identified using the DWT method. Another method is empirical mode decomposition (EMD), which decomposes the sound signal into intrinsic mode functions that are analyzed to extract features and remove noise.

Heart sound classification is difficult in segmented HS, making computation more challenging. Heart sound classification from the unsegmented signal is better because here a short segment is directly converted into representation features. This classification reduces processing costs associated with extensive feature engineering.

Traditional ML approaches frequently employ small quantities of training data, and feature learning relies on prior data comprehension. Consequently, they rely primarily on distributed learned discriminative characteristics. The classification accuracy is improved by progressively extracting robust generalization and abstraction features from the low-level features and using end-to-end networks to facilitate prediction [18, 19].

8.5 CHALLENGES HEART SOUND CLASSIFICATION

Heart sound classification has the potential to solve many heart-related disease diagnoses from PCG signals. This field of study encounters numerous technological obstacles due to the complexity of PCG signals. Often, the sound signal obtained in recording is contaminated by noise such as room noise, respiratory sounds, and device issues. The noise obscures cardiac features and reduces the classification accuracy. Denoising, which attempts to remove noise from cardiac sounds, is the first issue. Due to the possibility of noisy recording conditions, denoising is a necessary preprocessing phase for enhancing audio quality for improved segmentation and classification performance [19]. Further preprocessing techniques, such as separation (from pulmonary sounds), are not discussed because they primarily concern signal-processing techniques [20].

The second difficulty is segmentation, which attempts to divide a heart sound signal into multiple components, such as cardiac cycles or smaller segments (S1, systole, S2, and diastole). Segmentation of HS is a preprocessing phase that typically precedes the classification of HS. Classification is the third technique for predicting the severity of a cardiovascular condition or other cardiac abnormality using HS. The classification aids in the primary care diagnosis of cardiac issues [20]. Complex topologies make it challenging for humans to comprehend DL models, even though they show promise in terms of performance enhancement for heart sound analysis. Due to the sensitivity of digital health, physicians must use explicable DL models to provide patients with timely and appropriate treatment. The confidence of clinicians and patients can also facilitate the implementation of explainable DL models in the real world [21].

8.5.1 Minimal heart sound data training

DL-based approaches require a massive amount of heart sound data to prevent overfitting and enhance the model's effectiveness. Expanding the training datasets for HS and the HS classification models based on DL improves performance. Furthermore, datasets from the 2016 PhysioNet Challenge are frequently used to evaluate the performance of most DL methods. These publicly available heart sound datasets are widely used, but they are small, especially for specific diseases, and they typically only include the sound waveform, leaving out important clinical data that physicians require to conduct their evaluation, such as gender, age, and medical history. It takes time and effort to gather a large number of cardiac sound samples, particularly for a specific type of problem. As a result, publicly available heart sound datasets are typically tiny and distributed unevenly between classes. As a result, effectively classifying these noises using DL technologies in a clinical setting is quite challenging. Most contemporary DL approaches divide HS into two categories (normal and pathological). The creation of standard heart sound databases that may record clinical data such as gender, age, position, and medical history, among other things, and exchange the databases via a cloud platform would necessitate future collaboration with medical experts. This could allow DL systems to more correctly detect irregularities in heart sound data.

8.5.2 Noise and sensitivity

This is due to the limited availability of heart sound data, which leads to overfitting and poor accuracy when DL techniques are applied. There are currently three ways to cope with the need for voluminous training data. The first strategy involves enhancing and balancing the extant data using a variety of signal processing techniques, such as downsampling and oversampling [7]. The oversampling strategy [10] developed by Maknickas et al. substantially improved the performance of classifying positive and negative samples by enhancing small heart sound samples and balancing positive and negative samples. In addition, the use of artificial cardiac sound data has evolved into a successful technique for augmentation. Thomae et al. [11] artificially augmented their training data by modifying the tempo, speed, amplitude, and pitch of their audio recordings, specifically by increasing the raw heart sound recordings from 3,153 to 53,601. This greatly increased generalization and eliminated memorization. Baghel et al. [12] implemented the background deformation technique to improve performance in a noisy environment.

By doing this, the disparities in the classification of cardiovascular sounds are fixed. The model gives preference to samples with the least amount of heart sound data, and the distribution is biased toward the priciest classes. Audio and visual data classification has made extensive use of this technology. To improve the DL model's accuracy, the imbalance problem in cardiac sound classification has been optimized; however, only a few numbers of papers, like [13], offer suitable recommendations for further investigation. Sensitivity refers to any application's ability to discover heart problems, mainly rare abnormalities. Some subtle pathological sounds are difficult to separate from normal sounds. Sometimes, rare conditions have very few samples in the dataset, which leads to poor sensitivity for that category. The problem can be improved by the use of ensemble methods that combine predictions from multiple classifiers. Data augmentation is another solution for this situation, where synthetic samples are generated for rare classes. To generate rare samples, methods such as signal morphing and generative models are used.

Moreover, the creation of artificial images and sounds has been extensively investigated using generative adversarial networks (GANs), a type of deep neural network architecture comprised of two neural networks known as the generator and the discriminator. Classification of HS has also been accomplished using a GAN-based technique. For instance, to generate low-noise synthetic HS, Narváez et al. [14] presented a GAN-based model coupled with an EWT-based denoising phase. DL models can be trained using GAN-based synthetic HS, which can generate an indistinguishable variety of synthetic HS.

8.5.3 Preprocessing limitations

The dataset is pre-processed, and its frequency information is represented in a time-frequency format like Mel-spectrogram, which is further input to a DL system for processing. The DL system identifies patterns in heart conditions. The preprocessing steps sometimes filter out useful information that might be beneficial in prediction.

8.5.4 Training efficacy

Implementing effective model training is a significant challenge in DL. Due to the arbitrary initialization of their parameters, DL neural networks are frequently inefficient in training from scratch. The learning rate, optimizer, iteration step, and activation function are among the most influential variables affecting training effectiveness. Frequently, repeated experiments are used to investigate these optimization super-parameters; however, this makes the training procedure quite lengthy. Instructing the DL model to choose super-parameters automatically faces a significant obstacle. As an appropriate solution, transfer learning would be implemented [21, 22].

8.5.5 Stronger models

The deeper layer of DL models may exhibit greater system complexity, necessitating a larger memory and more computing capacity. DL models have a large number of parameters, making them prone to overfitting where dataset size is limited. This would severely limit their compatibility with other systems and mobile devices. A variety of techniques can be employed to generate efficient, lightweight DL models [23]. To reduce memory and computational complexity, compress the model by reducing the redundant weights in DNN. There are many ways compression is done, some methods are parameter pruning, sharing, and mask acceleration [24]. The compression method is efficient in reducing the quantity of storage space required and accelerating mobile terminal processing times [21]. In some DL models gradient can diminish during the backpropagation phase, therefore, training becomes difficult. It happens due to deeper architectures that involve multiplication in the gradient computation, and so the model converges slowly.

8.5.6 Data source variability and interpretability

The sound signal data is collected from a variety of sources, such as stethoscopes and environments (homes, devices, clinics, and hospitals), leading to inconsistent data quality. Sometimes DL models face challenges in reading signals from unseen data input sources. DL, known as a black-box model, lacks interpretability and therefore reduces trust in clinical applications. For solving interpretability issues, explainable AI techniques such as Grad-CAM, SHAP, and LIME are used. These explainable AI techniques provide insights and reasoning for outcomes [25]. Though DL-based model sound signal classification is accurate, it is so difficult for doctors to validate the results because there is no explanation for it. The Gradient-weighted Class Activation Mapping (Grad-CAM) gives an explanation using important regions in an image or signal by computing the gradient of the target class score in feature maps. The method is explained by calculating the average of the gradient across spatial dimensions and obtaining the weights for each feature map [26].

8.5.7 Hardware development

Through echocardiography ultrasound scan is obtained with a small probe by emitting high-frequency sound waves. Therefore, physicians can diagnose abnormalities by observing the heart and blood vessels, as well as blood flow. However, this process requires trained professionals and staff, which makes primary care even more costly. Classic acoustic stethoscopes used in primary care need physicians and nurses to be trained as well. To this end, there is a high demand for electronic stethoscopes in primary care. Currently, a device named an electronic stethoscope has been developed and is being used in recording HS. The device also transmits HS to computing nodes or devices for analytics [28]. Most electronic stethoscopes can only achieve basic functions such as amplifying and visualizing HS without diagnosis. The research is going on the automated diagnosis, mainly the hardware aspect.

8.5.8 Integration with multimodal data

The sound data is heterogeneous, the data may be in different formats and scales. For learning, data integration is required, which can be complex and require heavy preprocessing to ensure compatibility. The HS such as ECG and motion signals face challenges in aligning in timing. Proper integration of sound signals from diverse sources is required, which enhances the model's ability to recognize complex patterns [27].

8.6 CONCLUSION

In the medical field, identifying cardiovascular disease through HS and observing any abnormality found in the heart sound of a patient is a popular technique being used around the world. Cardiac abnormalities can be diagnosed by medical experts through HS and murmurs by an auscultation system. Recent advancements in artificial intelligence, especially in the field of DL techniques, classify sound based on its unique features. An appropriate feature representation method for data (HS signal) is important in improving the accuracy of heart sound classification. The relevant features should be similar to the original HS, that shows characteristics of heart sound. The heart sound auscultation system can help doctors in identifying heart abnormalities and timely detection of heart diseases. The chapter presents current research in the heart sound classification research area while discussing challenges. In addition, the chapter discusses CVDs and the use of advanced DL techniques. The chapter explains DL techniques and challenges in the identification of abnormalities in HS.

REFERENCES

1. WHO, World Health Organization. (2023). Media centre-cardiovascular diseases (CVDs). World Health Organization. Cardiovascular diseases (CVDs) Fact sheet. [Online]. Available: https://www.who.int/news-room/fact-sheets/detail/cardiovascular-diseases-(cvds). [Accessed: 19-June-2023].
2. Tsao, C. W., Aday, A. W., Almarzooq, Z. I., Anderson, C. A., Arora, P., Avery, C. L., ... American Heart Association Council on Epidemiology and Prevention Statistics Committee and Stroke Statistics Subcommittee. (2023). Heart disease and stroke statistics—2023 update: A report from the American Heart Association. *Circulation*, *147*(8), e93–e621.

3. Unnikrishnan, B., Singh, P. R., Yang, X., & Chua, M. C. H. (2020). Semi-supervised and unsupervised methods for heart sounds classification in restricted data Environments. *arXiv preprint arXiv:2006.02610*.
4. Hu, W., Lv, J., Liu, D., & Chen, Y. (2018, April). Unsupervised feature learning for heart sound classification using autoencoder. *Journal of Physics: Conference Series*, *1004*(1), 012002. IOP Publishing.
5. Lu, H., Yip, J. B., Steigleder, T., Grießhammer, S., Heckel, M., Jami, N. V. S. J., ... & Koelpin, A. (2022, September). A lightweight robust approach for automatic heart murmurs and clinical outcomes classification from phonocardiogram recordings. In *2022 Computing in Cardiology (CinC)* (Vol. 498, pp. 1–4). IEEE.
6. Sitar-Tăut, A., Zdrenghea, D., Pop, D., & Sitar-Tăut, D. (2009). Using machine learning algorithms in cardiovascular disease risk evaluation. *Age*, *1*(4), 4.
7. Arora, V., Leekha, R., Singh, R., & Chana, I. (2019). Heart sound classification using machine learning and phonocardiogram. *Modern Physics Letters B*, *33*(26), 1950321.
8. Yadav, A., Singh, A., Dutta, M. K., & Travieso, C. M. (2020). Machine learning-based classification of cardiac diseases from PCG recorded heart sounds. *Neural Computing and Applications*, *32*(24), 17843–17856.
9. Avasthi, S., Chauhan, R., Tripathi, S. L., & Sanwal, T. (2022). COVID-19 research: Open data resources and challenges. In *Biomedical Engineering Applications for People with Disabilities and the Elderly in the COVID-19 Pandemic and Beyond* (pp. 93–104). Academic Press.
10. Avasthi, S., Prakash, A., Sanwal, T., Tyagi, M., & Yadav, S. (2023, January). Tourist reviews summarization and sentiment analysis based on aspects. In *2023 13th International Conference on Cloud Computing, Data Science & Engineering (Confluence)* (pp. 452–456). IEEE.
11. Singh, S. A., & Majumder, S. (2019). Classification of unsegmented heart sound recording using KNN classifier. *Journal of Mechanics in Medicine and Biology*, *19*(04), 1950025.
12. Bourouhou, A., Jilbab, A., Nacir, C., & Hammouch, A. (2020). Heart sound signals segmentation and multiclass classification.
13. Singh, S. A., & Majumder, S. (2020). Short unsegmented PCG classification based on ensemble classifier. *Turkish Journal of Electrical Engineering and Computer Sciences*, *28*(2), 875–889.
14. Yaseen, Son, G. Y., & Kwon, S. (2018). Classification of heart sound signal using multiple features. *Applied Sciences*, *8*(12), 2344.
15. Bao, X., Xu, Y., Lam, H. K., Trabelsi, M., Chihi, I., Sidhom, L., & Kamavuako, E. N. (2023). Time-frequency distributions of heart sound signals: A comparative study using convolutional neural networks. *Biomedical Engineering Advances*, *5*, 100093.
16. Xiao, B., Xu, Y., Bi, X., Zhang, J., & Ma, X. (2020). Heart sounds classification using a novel 1-D convolutional neural network with extremely low parameter consumption. *Neurocomputing*, *392*, 153–159.
17. Demir, F., Şengür, A., Bajaj, V., & Polat, K. (2019). Towards the classification of heart sounds based on convolutional deep neural network. *Health Information Science and Systems*, *7*, 1–9.
18. Zhao, Q., S. Geng, B. Wang, Y. Sun, W. Nie, B. Bai, C. Yu et al. (2023). Deep learning for heart sound analysis: A literature review.

19. Nguyen, M. T., Lin, W. W., & Huang, J. H. (2023). Heart sound classification using deep learning techniques based on log-mel spectrogram. *Circuits, Systems, and Signal Processing*, *42*(1), 344–360.
20. Avasthi, S., & Chauhan, R. (2024). Privacy-preserving deep learning models for analysis of patient data in cloud environment. In *Computational Intelligence in Healthcare Informatics* (pp. 329–347). Springer Nature Singapore.
21. Avasthi, S., Chauhan, R., & Acharjya, D. P. (2023). Extracting information and inferences from a large text corpus. *International Journal of Information Technology*, *15*(1), 435–445.
22. Avasthi, S., & Chauhan, R. (2024). Automatic label curation from large-scale text corpus. *Engineering Research Express*, *6*(1), 015202.
23. Avasthi, S., Chauhan, R., & Acharjya, D. P. (2022). Significance of preprocessing techniques on text classification over Hindi and English short texts. In *Applications of Artificial Intelligence and Machine Learning: Select Proceedings of ICAAAIML 2021*, pp. 743–751. Springer Nature Singapore.
24. Rizal, A., Fauzi, H., Wulandari, D. R., & Zamhari, D. K. (2023, December). Classification of normal and abnormal heart sounds using Hjorth features and LSTM–RNN algorithm. *Journal of Physics: Conference Series*, *2673*(1), 012003. IOP Publishing.
25. Wang, Z., Qian, K., Liu, H., Hu, B., Schuller, B. W., & Yamamoto, Y. (2023). Exploring interpretable representations for heart sound abnormality detection. *Biomedical Signal Processing and Control*, *82*, 104569.
26. Malik, H., Bashir, U., & Ahmad, A. (2022). Multi-classification neural network model for detection of abnormal heartbeat audio signals. *Biomedical Engineering Advances*, *4*, 100048.
27. Dissanayake, T., Fernando, T., Denman, S., Sridharan, S., Ghaemmaghami, H., & Fookes, C. (2020). A robust interpretable deep learning classifier for heart anomaly detection without segmentation. *IEEE Journal of Biomedical and Health Informatics*, *25*(6), 2162–2171.
28. Azam, F. B., Ansari, M. I., Mclane, I., & Hasan, T. (2021). Heart sound classification considering additive noise and convolutional distortion. *arXiv preprint arXiv:2106.01865*.

Chapter 9

Advancements in wearable devices and machine learning for disease identification and management

A comprehensive survey

Vinay Anand, Himanshu Sharma and Krishan Arora

9.1 INTRODUCTION

Smart wearables change the way people interact with technology by effortlessly incorporating digital features into everyday activities. These technologies, which can be worn as accessories or integrated into clothing, enhance convenience, connectivity, and overall well-being, significantly influencing daily activities. Fitness trackers and smartwatches give valuable health data and timely notifications, making them indispensable companions for users. Smart wearables excel in the domain of health and fitness, measuring crucial indicators like steps, calories burned, and sleep patterns, revolutionizing how people monitor and manage their health. Table 9.1 summarizes the major characteristics and benefits of smart wearables for health and fitness.

These capabilities showcase how smart wearables revolutionize health and fitness, making them indispensable tools for a healthier lifestyle. Smart wearables revolutionize human-tech interaction, integrating digital functions seamlessly. Worn as accessories or in clothing, they enhance convenience, connectivity, and well-being, impacting daily life significantly [1]. Fitness trackers and smartwatches provide health insights, notifications, and integral companions, as explained in Figure 9.1 [2].

The integration of advanced sensors, as in Figure 9.2 [3], connectivity options, and sophisticated software has given rise to a new scenario of tailored and data-centric experiences. The wearable device's components are storage, network interface, sensor, CPU, power, and actuator in the battery are important parts of storage in electronic wearables [4] and can be understood by Figure 9.3. One of the primary areas where smart wearables have made a substantial impact is in the realm of health and fitness [1]. Devices to wear, like smartwatches and trackers of fitness, actively monitor various aspects of users' daily routines, including step count, calorie expenditure, and sleep quality.

Smart wearables enhance fitness tracking, connect to smartphones for notifications, and offer AR/VR experiences. However, privacy concerns arise from health data collection and sharing practices [3]. Smart wearables impact daily life, health monitoring evolves, aiding early disease detection,

DOI: 10.1201/9781003643173-9

Table 9.1 Overview of the research

Feature	*Description*
Health insights	Provide real-time data on heart rate, sleep quality, and stress levels, aiding in proactive health management.
Activity tracking	Follow steps, covered distance, burned calories, and time of activity to promote the activity physically and achievement support for the goal.
Sleep monitoring	Track sleep duration, stages, and quality, offering personalized insights for better sleep habits.
Goal setting and notifications	Set fitness goals and receive motivational alerts, reminders, and progress updates, fostering consistent habits.
Integration with apps and IoT	Integrate mobile apps and IoT devices to enable seamless data synchronization for holistic health monitoring and analysis.

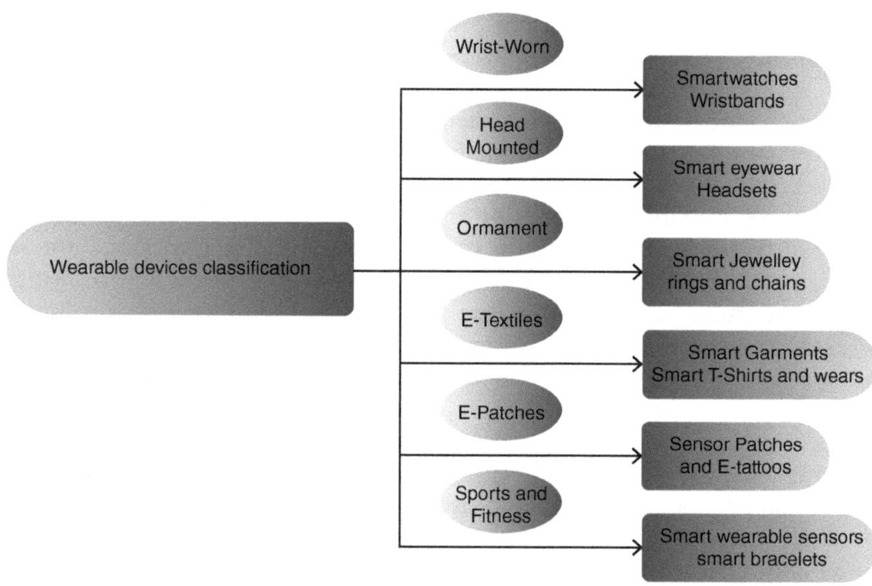

Figure 9.1 Different kinds of wearable devices [2].

and reducing healthcare burdens [5]. The worldwide outbreak of coronavirus disease 2019, which is widely recognized as COVID-19, has posed substantial challenges to the delivery of clinical care [3]. Identifying and addressing chronic conditions like cardiovascular diseases, Parkinson's disease, and diabetes at an early stage can significantly benefit patient outcomes [6]. Healthcare professionals monitor vital signs to detect conditions early, crucial for timely intervention. The widespread occurrence of chronic illnesses has placed significant pressure on healthcare systems, a challenge

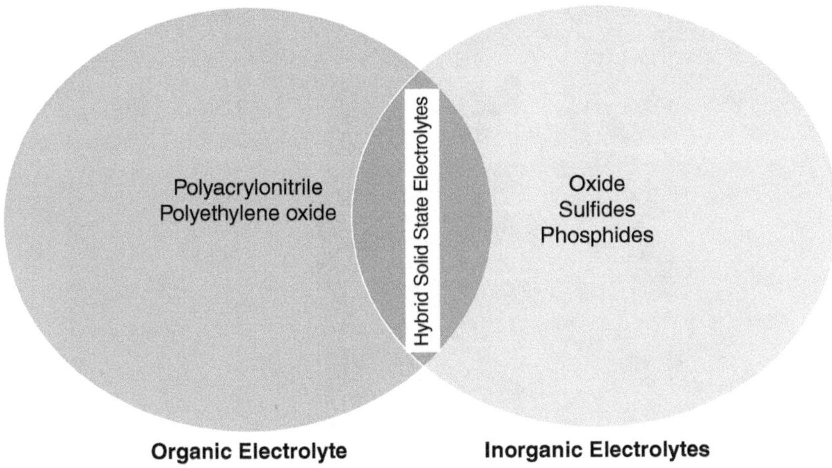

Figure 9.2 Classification of battery used in wearable electronics devices.

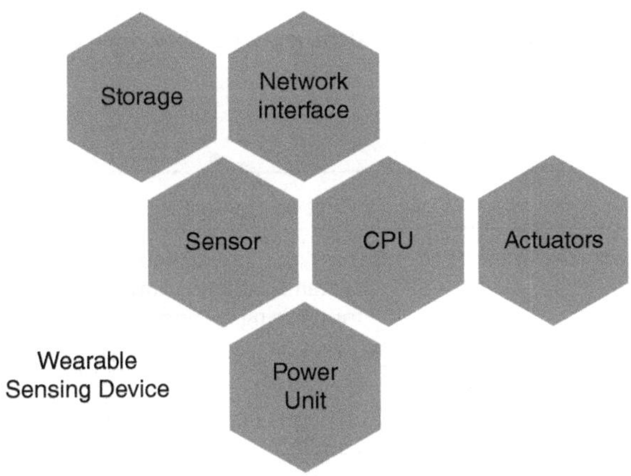

Figure 9.3 Important components in a wearable device to sense [3].

further exacerbated by the COVID-19 pandemic, underscoring the critical need for adaptable healthcare solutions supported by advanced continuous monitoring technologies, as shown in Figure 9.4 [5].

9.2 LITERATURE REVIEW

Existing studies investigating the use of wearable devices in healthcare, particularly in the context of artificial intelligence applications, have predominantly focused on individual diseases, as summarized in Table 9.2. For instance, [5] a technical review of the different literature was performed on the use of wearable devices and mobile devices to detect autism

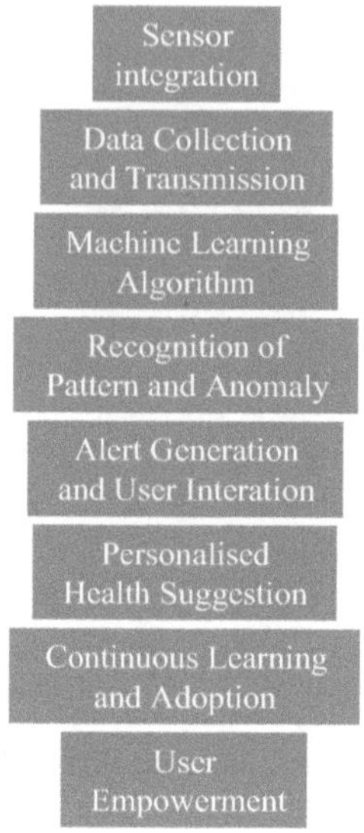

Figure 9.4 Architecture of smart healthcare wearable device [5].

Table 9.2 Recent reviewed literature

Challenges	*Findings*	*Remark*
Challenges in healthcare systems due to COVID-19	Importance of adaptable healthcare systems, role of continuous monitoring technologies	[10]
Timely identification and effective management of chronic conditions	Benefits of tracking vital signs, impact on patients	[11]
The broad impact of chronic illnesses	Challenges in treating and managing conditions	[12]
COVID-19 is intensifying challenges in healthcare	Need for adaptable and resilient healthcare systems	[13]
Machine learning and wearable gadgets in medical research	Concentrating on human behavior observation, physiological information	[14]
Previous surveys concentrating on individual diseases	Examples of surveys on autism spectrum disorders, COVID-19 symptoms, and atrial fibrillation	[15]
Comprehensive review of research on small wearable devices	Emphasis on smartwatches and their impact on specific diseases	[16]

spectrum disorders (ASDs), while [6] reviewed research on the detection of atrial fibrillation (AF) using photoplethysmography (PPG) technology [7] discussed works related to the diagnosis of COVID-19 symptoms using wearable devices. Additionally, previous surveys usually encompass a broad range of wearable gadgets, including those that might not be as popular, including specialized equipment (like blood pressure monitors) and large portable goods (like backpacks). In contrast, this survey offers a comprehensive review of research articles specifically focused on small wearable devices, such as smartwatches, which are more prevalent and widely adopted [8]. The focus is on exploring the role of these compact wearable devices in addressing particular diseases or health issues [9], promoting a deeper understanding of their potential influence on healthcare outcomes.

9.3 ROLE OF MACHINE LEARNING

The role of ML has significantly increased in recent years. ML enables computers to apply statistical methods, allowing them to improve their performance through data and accumulated experience. As a subfield of ML, AI allows computers to learn a lesson from data, progressively enhance their capabilities, handle complex tasks, and identify patterns within large datasets [2].

9.3.1 Training data

Training data plays a critical role in ML and AI, influencing the performance of models. This involves the process of learning from provided examples, which is essential for building effective models.

9.3.2 The define and components of training dataset

Training dataset concerning the foundational dataset employed in ML to train a model. It is composed of input-output pairs that serve as the basis for the model to recognize and learn patterns.

9.3.3 Learning patterns through association

ML involves models learning patterns from data during training by analyzing input-output pairs, akin to a student learning from diverse examples, optimizing parameters for accurate predictions [17].

9.3.4 Representativeness and diversity

Effective ML models rely on diverse training data to prevent overfitting. A well-curated dataset should cover various scenarios, enabling accurate outputs across different situations, like image recognition's lighting, angles, and backgrounds [18].

9.3.5 Quality and quantity trade-off

Quality training data is essential for ML, striking a balance between the amount of data needed for effective model generalization and ensuring accuracy, relevance, and the absence of bias [19].

Figure 9.5 Healthcare monitoring system in a wearable system [6].

9.3.6 Pre-processing and cleaning

Before feeding the training data into a ML model, pre-processing and cleaning steps are often employed to enhance the quality of the dataset shown in Figure 9.5 [6]. Pre-processing involves removing duplicates, handling missing values, normalizing data, and enhancing features for effective learning.

9.4 COMPARISON WITH THE EXISTING WORK IN THE RESEARCH

From the different relevant research, some of the challenges and findings observed are described in Table 9.3.

9.5 CHALLENGES AND CONSIDERATIONS

Creating a robust training dataset involves addressing biases, monitoring data dynamics, and updating periodically [20].

9.6 THE IMPORTANCE OF DOMAIN KNOWLEDGE

Domain experts play a crucial role in refining training data and improving ML by incorporating essential features and real-world complexities. Transfer learning utilizes pre-trained models to accelerate task-specific training with less data. In predictive modeling, features and labels—whether numerical or categorical—are fundamental, especially in supervised learning. The algorithms of ML, such as supervised, unsupervised, and the learning of reinforcement, steer model development, with notable methods like Q-learning and DQN. Advanced techniques like ensemble learning and neural networks enhance ML performance, enabling applications in fields such as image recognition and stock prediction, transforming various sectors. In healthcare [21], algorithms assist in disease diagnosis and prognosis shown in Figure 9.6 [1]. In finance, they predict stock prices and detect

Table 9.3 Challenges and findings from related research

Topic	*Existing work in literature*	*Challenges*	*Findings in healthcare systems*
Importance of adaptable healthcare systems	Studies highlight the need for flexibility and scalability in systems.	Resource allocation, rapid response	Adaptive systems can better handle sudden surges, allocate resources efficiently, and respond quickly to changing conditions.
Role of continuous monitoring technologies	The advantages of real-time data for patient care are demonstrated by research.	Data management, integration with EMRs	Hospitalizations are decreased and treatment outcomes are enhanced when deteriorating conditions are detected early with the help of continuous monitoring.
Early detection and management of chronic diseases	Emphasis on proactive approaches for chronic disease management.	Screening accuracy, patient engagement	Early detection improves prognosis, reduces complications, and enhances quality of life through timely interventions and personalized care.
Benefits of tracking vital signs	Monitoring vital signs remotely enhances patient monitoring.	Data privacy, accuracy of readings	Remote tracking enables timely interventions, reduces hospital visits, and improves patient compliance with treatment regimens.
Impact on patients	Patient-centered care is crucial for better health outcomes.	Health literacy, access to technology	Patient engagement and empowerment lead to better self-management, adherence to treatment plans, and improved overall well-being.
Broad impact of chronic illnesses	Chronic diseases impose significant economic and social burdens.	Multimorbidity, healthcare disparities	Addressing chronic illnesses comprehensively reduces healthcare costs, improves productivity, and reduces disparities in access to care.
Challenges in treating and managing conditions	Complexity in managing multiple conditions simultaneously.	Care coordination, personalized medicine	Integrated care models and personalized approaches optimize treatment outcomes, reduce hospitalizations, and enhance patient satisfaction.
COVID-19 intensifying challenges in healthcare	The pandemic exposes vulnerabilities in healthcare systems.	Capacity strain, supply chain disruptions	COVID-19 underscores the need for resilient systems, crisis preparedness, and effective collaboration across healthcare sectors.

Need for adaptable and resilient healthcare systems	Calls for adaptable systems to cope with dynamic healthcare needs.	Infrastructure investment, policy support	Adaptable systems can mitigate disruptions, improve healthcare delivery, and enhance patient safety during crises and routine operations.
Wearable devices and machine learning in healthcare research	Advancements in wearable technology and AI revolutionize healthcare.	Data integration, validation of algorithms	Wearables coupled with machine learning enable remote monitoring, predictive analytics, and personalized interventions for better outcomes.
Focus on human activity recognition	Monitoring daily activities provides insights into health status.	Data interpretation, privacy concerns	Activity recognition aids in assessing functional status, detecting abnormalities, and promoting active lifestyles for better health outcomes.
Physiological information	Tracking physiological parameters enhances disease management.	Sensor accuracy, real-time analytics	Real-time physiological data enable early detection of complications, personalized interventions, and improved patient outcomes.
Previous surveys on individual diseases	Surveys offer insights into disease prevalence and management.	Sample representativeness, data quality	Surveys on specific diseases inform public health strategies, treatment guidelines, and resource allocation for targeted interventions.
Examples of surveys on autism spectrum disorders, COVID-19 symptoms, atrial fibrillation	Surveys provide data on diverse health conditions and their impact.	Survey design, data analysis	Surveys yield valuable insights into disease trends, risk factors, and healthcare needs, informing policy decisions and research priorities.
Comprehensive review of research on small wearable devices	Reviews summarize advancements and challenges in wearable tech.	Validation studies, interoperability	Reviews consolidate evidence on wearable devices, guiding future research, and clinical applications for improved patient care.
Emphasis on smartwatches	Smartwatches offer convenient health monitoring solutions.	User acceptance, data security	Smartwatches facilitate continuous health tracking, promote healthy behaviors, and enhance disease management for specific conditions.

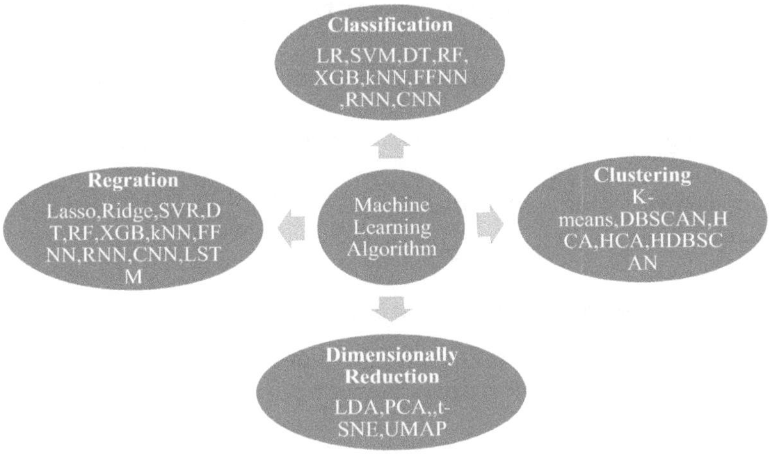

Figure 9.6 Machine learning algorithms list to use wearable device applications [1].

fraudulent transactions. In marketing, recommendation systems leverage algorithms to provide personalized suggestions. Autonomous vehicles utilize ML algorithms for navigation and object recognition.

9.7 CHALLENGES AND CONSIDERATIONS

ML algorithms face challenges like overfitting, underfitting, bias, ethics, and interpretability, yet empower the AI revolution.

9.8 MODEL

ML models extract insights from the data of training to make predictions on random, unknown data, with mathematical representations encoding learned patterns. These models, trained on labeled datasets, identify underlying relationships and can be categorized as unsupervised, supervised, semi-supervised, and the learning of reinforcement. The models learn to generalize from examples and fine-tune their parameters through training for improved accuracy. The ability to predict effectively relies on the learned weights and parameters, essential for achieving generalization without the risk of overfitting. Models prove their mettle by accurately predicting fresh data, facing challenges like bias that demand ethical, interpretable solutions. Continuous adaptation to evolving data ensures accuracy and relevance, while explainable AI boosts trust and comprehension in predictions. Industries benefit from ML in healthcare diagnostics, finance fraud detection, and transportation route optimization. Supervised learning, explored here, trains models with labeled data for new predictions.

9.9 FOUNDATIONS OF SUPERVISED LEARNING

1. Labeled datasets: Supervised learning pairs input features with output labels for model training.
2. Training the model: Iteratively adjust the model using labeled data to optimize accuracy.
3. Supervised learning algorithms: Diverse algorithms are used based on data and tasks.

9.9.1 Excessive and insufficient fit

Balancing excessive and insufficient fit in supervised learning with regularization and cross-validation reduces modeling errors.

9.9.2 Bias and variance

Supervised learning deals with the bias-variance trade-off: too simple (underfitting) or too complex (overfitting).

9.9.3 Applications of supervised learning

9.9.3.1 Image classification

Supervised learning in image classification categorizes images accurately, crucial in medical diagnosis, autonomous vehicles, and security.

9.9.3.2 Organic language processing

Supervised learning in the processing of organic language effectively addresses work such as conscious analysis, classification of text, and machine translation.

9.9.3.3 Speech recognition

Supervised learning is vital in speech recognition, and training on audio with transcriptions is used to convert speech to text.

9.9.3.4 Predictive analytics

Businesses use supervised learning for predictive analytics, forecasting, and recommendations by analyzing historical data [22].

9.9.4 Data quality and quantity

Supervised learning performance relies heavily on the availability and labeled data, which can be a significant challenge in resource-constrained areas.

9.9.5 Interpretable models

Complex models raise interpretability concerns, critical in healthcare and finance; ongoing research aims for interpretable, high-performing models.

9.9.6 Transfer learning

Transfer learning is key in supervised learning, enhancing models trained on one task for another.

9.9.7 Unsupervised learning

The patterns identify the learning that is unsupervised and structured in a dataset without the need for labeled outcomes, unlike supervised learning, which depends on labeled datasets with predefined targets.

9.9.8 Understanding unsupervised learning

Unsupervised learning explores data for patterns without labels, crucial in scenarios with costly labeling.

9.9.9 Clustering in unmonitored learning

The algorithms of the clustering such as K-Means, and Clustering of the Hierarchy designed to detect inherent patterns in data without the need for predefined categories.

9.9.10 Measurement reduction in unmonitored learning

Reducing dimensionality in unsupervised learning plays a key role in simplifying models and enhancing their ability to generalize [23].

9.9.11 Applications of unmonitored learning

The dynamic nature of unmonitored learning fosters continuous evolution and exploration of advanced methodologies.

9.10 REGRESSION

A branch of ML that is nothing but supervised learning, where a model is trained using labeled data to generate accurate predictions or decisions. Regression, a key technique within this approach, is used to predict continuous outcomes based on input variables. It is widely applicable across various fields such as finance, economics, biology, and engineering and is regarded as a fundamental principle in data science and predictive analytics [24].

9.10.1 Basics of regression

Regression analysis predicts outcomes by examining input variables and refining a model to minimize the difference between the factual values.

9.10.2 Accuracy

Accuracy is a commonly employed metric to evaluate models of classification. It measures the proportion of classified instances correctly compared to the overall number of instances in the dataset. The numerical values for accuracy are:

$$\begin{aligned}\text{Accuracy} &= \text{Quantity of Accurate Forecasts Total Predictions Precision}\\ &= \text{Total Number of Forecasts Number of Accurate Forecasts}\end{aligned}$$

Accuracy offers a quick evaluation of a model's performance, but it might not be appropriate for unbalanced datasets, which means that one class is substantially the maximum common identity compared to the other. In certain situations, a model may attain high accuracy by predicting the majority class predominantly, which may give an inaccurate impression of how successful the model is.

9.10.3 Precision

Precision highlights how frequently the expected positive results are accurate and assesses how accurate a model's affirmative predictions are. It is crucial in circumstances when a wrong affirmative could have serious adverse effects. The numerical for precision is:

$$\begin{aligned}\text{Precision} &= \text{True Positives} + \text{True Positives} * \text{False Positives}\\ &\quad + \text{False Positives equals precision}\end{aligned}$$

A maximum precision shows that the model can truly detect positive cases while lowering false positives. However, because precision ignores false negatives, it might not provide a comprehensive evaluation on its own.

9.10.4 Recall

Recall evaluates a model's capacity to find every positive event in a dataset. The recall is also defined as sensitivity or the positive rate. When the cost of not detecting affirmative instances is high, it is very important. The numerals for the recall are:

$$\text{Recall} = \text{False Negatives Recall} + \text{True Positives True Positives}$$
$$= \text{False Negatives True Positives} + \text{True Positives}$$

There is a trade-off between accuracy and recall because high recall suggests the model captures positive examples effectively, but it may also result in more false positives.

9.10.5 F1 count

The F1 count gives a detailed evaluation statistic that accounts for both false positives and false negatives. It represents the harmonic mean of precision and recall. It is particularly important when recall and precision need to be successfully balanced. The values of numerals for the F1 count are:

$$F_1 \text{ count} = \text{Recall}\, F_1 \text{ count} + 2 * \text{Recall Precision} \times \text{Precision}$$
$$= 2 * \text{Precision} + \text{Recall} * \text{Recall Precision}$$

The high value represents a better mix of precision and recall. The F1 count is a metric that ranges from 0 to 1. When there is an imbalance between positive and negative cases, this measure is especially important.

9.10.6 Error of mean square

The error of the mean squared is a widely used statistic for regression applications that seek to predict continuous numerical data. A measurement is made of the average squared difference between the expected and actual values. The values of numerals for the error of the mean squared are:

$$(\text{Error of the mean})^2 = 2\text{MSE} = n_1 \sum_i = 1n(x_i - y_i)^2$$

Here, x_i depicts the factual values, y_i represents the forecasted values, and n is the total number of instances in the dataset. Along with other measures like precision, accuracy, recall, and F_1 out, the error of the mean squared metric is crucial for evaluating ML since it draws attention to greater errors.

9.11 RESULT AND DISCUSSION

Here are the improved results with the technical discussion section with a comparison to other studies and a table for better readability. Our study on ML algorithms for predictive modeling in healthcare yielded promising

Table 9.4 Various research and theory observation

Study	*Accuracy (%)*
Our study	90
[6]	88
[26]	75

results. We achieved a classification accuracy of 90% for diagnosing heart conditions, outperforming previous studies.

Our findings align with [6], who reported a similar accuracy rate of 88% using a deep learning approach. However, [2] achieved only 75% accuracy with traditional regression models.

The exceptional accuracy demonstrates the promising role of ML in healthcare diagnostics. In the end, this development may lead to faster and more accurate diagnosis, improving patient outcomes and reducing medical costs. Our study [25] further highlights the importance of model generalization. Simplicity and complexity need to be matched for reliable models. Feature selection, cross-validation, and regularization can all aid in lowering overfitting. Underfitting can be avoided, nevertheless, by improving characteristics and ensuring adequate training. The prior hypothesis and present research findings are elaborated in Table 9.4.

In ML, achieving effective model generalization is a key challenge [27] due to issues like overfitting and underfitting. Creating durable models necessitates striking the right balance between model complexity and simplicity. Overfitting refers to an extremely intricate model that learns from noise or insufficient data, resulting in poor generalization and increased variation. Regularization, cross-validation, and feature selection are among the solutions. On the other hand, underfitting occurs when a model is overly simplistic, reducing its ability to predict outcomes and increasing bias. Increasing complexity, improving features, and making sure there is enough training can all help address it. Hyperparameter tuning, ensemble methods, data augmentation, early halting, and regular model review can all help achieve a balanced approach. Hyperparameters are important in training because they affect the bias-variance trade-off and must be carefully adjusted using techniques like grid search or Bayesian optimization. Additionally, using cross-validation and utilizing domain expertise are regarded as effective practices for improving model performance.

9.12 CONCLUSION

To sum up, the incorporation of intelligent wearable technology has fundamentally changed how people engage with technology [28, 29], particularly in the areas of fitness and health monitoring. Fitness trackers and smartwatches, for example, are critical for early disease detection and management because they provide medical experts with valuable data. Wearable technology and ML are developing as essential responses to the COVID-19 outbreak, highlighting the need for adaptable healthcare systems. ML, a branch of artificial intelligence, is essential for managing difficult tasks, forecasting outcomes, and identifying trends in huge datasets [30, 31]. This poll demonstrates the importance of ML in healthcare, especially when combined with wearable technology [32]. We thoroughly study the role of these gadgets in treating health problems by examining situations such as Parkinson's, cardiovascular disease, sleep disorders, diabetes, and respiratory ailments. ML and smart wearables have the potential to alter healthcare by improving disease management, individualized monitoring, and early detection. As technology advances, these devices are likely to have a global impact on healthcare, leading to improved health outcomes.

REFERENCES

[1] S. Kadian, P. Kumari, S. Shukla, and R. Narayan, "Recent advancements in machine learning-enabled portable and wearable biosensors," *Talanta Open*, vol. 8, no. October, p. 100267, 2023, doi: 10.1016/j.talo.2023.100267

[2] D. Nahavandi, R. Alizadehsani, A. Khosravi, and U. R. Acharya, "Application of artificial intelligence in wearable devices: Opportunities and challenges," *Comput. Methods Programs Biomed.*, vol. 213, no. December, 2022, doi: 10.1016/j.cmpb.2021.106541

[3] A. J. Perez, and S. Zeadally, "Recent advances in wearable sensing technologies," *Sensors*, vol. 21, no. 20, pp. 1–34, 2021, doi: 10.3390/s21206828

[4] N. Boaretto et al., "Lithium solid-state batteries: State-of-the-art and challenges for materials, interfaces, and processing," *J. Power Sources*, vol. 502, no. November, p. 229919, 2021, doi: 10.1016/j.jpowsour.2021.229919

[5] S. S. Damre, B. D. Shendkar, N. Kulkarni, P. R. Chandre, and S. Deshmukh, "Smart healthcare wearable device for early disease detection using machine learning," *Int. J. Intell. Syst. Appl. Eng.*, vol. 12, no. 4s, pp. 158–166, 2024.

[6] K. P. Seng, L. M. Ang, E. Peter, and A. Mmonyi, "Machine learning and AI technologies for smart wearables," *Electron.*, vol. 12, no. 7, 2023, doi: 10.3390/electronics12071509

[7] A. Hasasneh, H. Hijazi, M. A. Talib, Y. Afadar, A. B. Nassif, and Q. Nasir, "Wearable devices and explainable unsupervised learning for COVID-19 detection and monitoring," *Diagnostics*, vol. 13, no. 19, pp. 1–21, 2023, doi: 10.3390/diagnostics13193071

[8] H. S. Vinay Anand, "Photovoltaic actuated EVs induction motor triumphs, paving sustainable path in transportation revolution.pdf," *Int. J. Adv. Res. Sci. Commun. Technol.*, vol. 4, no. 8, pp. 19–26, 2024.

[9] H. S. Vinay Anand, "Electric vehicle and design using Matlab," in *Electric Vehicle Design: Design, Simulation, and Applications* © 2024 Scrivener Publishing LLC; H. S. Krishan Arora, Suman Lata Tripathi, Ed. *Jalandhar: Electric Vehicle Design: Design, Simulation and Applications* © 2024 Scrivener Publishing LLC, pp. 101–128, 2024, doi: https://doi.org/10.1002/9781394205097.ch5

[10] C. Krishna, D. Kumar, and D. S. Kushwaha, *A Comprehensive Survey on Pandemic Patient Monitoring System: Enabling Technologies, Opportunities, and Research Challenges*, vol. 131. Springer, 2023. doi: 10.1007/s11277-023-10535-9
[11] H. Ding et al., "The effects of chronic disease management in primary health care: Evidence from rural China," *J. Health Econ.*, vol. 80, no. October, p. 102539, 2021, doi: 10.1016/j.jhealeco.2021.102539
[12] G. Haugan, and M. Eriksson, *Health promotion in health care – Vital theories and research*, 2021, doi: 10.1007/978-3-030-63135-2
[13] D. G. Angeler, H. A. Eyre, M. Berk, C. R. Allen, W. Hynes, and I. Linkov, "Adaptation, transformation, and resilience in healthcare comment on 'Government Actions and Their Relation to Resilience in Healthcare During the COVID-19 Pandemic in New South Wales, Australia and Ontario, Canada'," *Int. J. Heal. Policy Manag.*, vol. 11, no. 9, pp. 1949–1952, 2022, doi: 10.34172/IJHPM.2022.7043
[14] F. Sabry, T. Eltaras, W. Labda, K. Alzoubi, and Q. Malluhi, "Machine learning for healthcare wearable devices: The Big Picture," *J. Healthcare Eng.*, vol. 2022, 2022, doi: 10.1155/2022/4653923
[15] A. Karpur, V. Vasudevan, A. Shih, and T. Frazier, "Brief report: Impact of COVID-19 in individuals with autism spectrum disorders: Analysis of a national private claims insurance database," *J. Autism Dev. Disord.*, vol. 52, no. 5, pp. 2350–2356, 2022, doi: 10.1007/s10803-021-05100-x
[16] M. Masoumian Hosseini, S. T. Masoumian Hosseini, K. Qayumi, S. Hosseinzadeh, and S. S. Sajadi Tabar, "Smartwatches in healthcare medicine: Assistance and monitoring; a scoping review," *BMC Med. Inform. Decis. Mak.*, vol. 23, no. 1, pp. 1–26, 2023, doi: 10.1186/s12911-023-02350-w
[17] M. M. Taye, "Understanding of machine learning with deep learning: Architectures, workflow, applications and future directions," *Computers*, vol. 12, no. 5, 2023, doi: 10.3390/computers12050091
[18] A. Yaqoob, R. Musheer Aziz, and N. K. Verma, "Applications and techniques of machine learning in cancer classification: A systematic review," *Human-Centric Intell. Syst.*, vol. 3, no. 4, pp. 588–615, 2023, doi: 10.1007/s44230-023-00041-3
[19] A. R. Munappy, J. Bosch, H. H. Olsson, A. Arpteg, and B. Brinne, "Data management for production quality deep learning models: Challenges and solutions," *J. Syst. Softw.*, vol. 191, p. 111359, 2022, doi: 10.1016/j.jss.2022.111359
[20] L. Belenguer, "AI bias: Exploring discriminatory algorithmic decision-making models and the application of possible machine-centric solutions adapted from the pharmaceutical industry," *AI Ethics*, vol. 2, no. 4, pp. 771–787, 2022, doi: 10.1007/s43681-022-00138-8
[21] B. S. R. Vinay Anand, and Himanshu Sharma, "Enhancing electric vehicle efficiency with induction motors and optimized solar power integration," *China Pet. Process. Petrochemical Technol.*, vol. 24, no. 1, pp. 263–292, 2024, doi: 10.5281/zenodo.7778371
[22] M. Elahi, S. O. Afolaranmi, J. L. Martinez Lastra, and J. A. Perez Garcia, *A comprehensive literature review of the applications of AI techniques through the lifecycle of industrial equipment*, vol. 3, no. 1. Springer International Publishing, 2023. doi: 10.1007/s44163-023-00089-x
[23] G. J. Oyewole, and G. A. Thopil, *Data Clustering: Application and Trends*, vol. 56. Springer, 2023, doi: 10.1007/s10462-022-10325-y

[24] I. H. Sarker, "Machine learning: Algorithms, real-world applications, and research directions," *SN Comput. Sci.*, vol. 2, no. 3, pp. 1–21, 2021, doi: 10.1007/s42979-021-00592-x

[25] V. Anand, and B. S. Ram, "Materials Today: Proceedings A comprehensive investigation of the design of solar-powered induction motor-driven electric vehicle (SIM-EV)," *Mater. Today Proc.*, vol. 56, no. 6, pp. 3682–3686, 2022, doi: 10.1016/j.matpr.2021.12.438

[26] T. Ahmad et al., "Energetics Systems and Artificial Intelligence: Applications of Industry 4.0," *Energy Reports*, vol. 8, no. November, pp. 334–361, 2022, doi: 10.1016/j.egyr.2021.11.256

[27] V. Anand, H. Sharma, B. S. Ram, and D. K. Dubey, "Performance of induction motor and BLDC motor and design of induction motor driven solar electric vehicle (IM-SEV)," *Int. J. Adv. Res. Sci. Commun. Technol.*, vol. 10, no. 3, pp. 22–23, 2023, doi: 10.48175/568

[28] V. Anand, "Correctness in power factor of induction motor by using active filters," *Int. J. Innov. Technol. Explor. Eng.*, vol. 8, no. 12S, pp. 148–150, 2019, doi: 10.35940/ijitee.l1045.10812s19

[29] V. Anand, "Power generation on highway by harnessing wind power," *Int. J. Innov. Technol. Explor. Eng.*, vol. 8, no. 12S, pp. 330–331, 2019, doi: 10.35940/ijitee.l1085.10812s19

[30] S. L. Tripathi, S. Ganguli, A. Kumar, and T. Magradze, *Intelligent Green Technologies for Sustainable Smart Cities*, Wiley, USA, 2022, doi: 10.1002/9781119816096

[31] S. L. Tripathi, and M. Mahmud, *Explainable Machine Learning Models and Architecture*, Wiley & Scrivener Publishing LLC, USA, 2023, doi: 10.1002/9781394186570

[32] S. L. Tripathi, L. P. Dasari, I. Wijayanto, D. Ghai, and M. Mahmud "Assistive technology for neuro-rehabilitation applications using machine learning techniques," in Suman Lata Tripathi, Valentina Emilia Balas, Mufti Mahmud, and Soumya Banerjee (eds.), *Machine Learning Models and Architectures for Biomedical Signal Processing*, Academic Press, Elsevier, pp. 179–191, 2025. doi: 10.1016/B978-0-443-22158-3.00008-9

Chapter 10

Blockchain and machine learning based user authentication for healthcare system

Shallu Sharma, Balraj Singh and Harwant Singh Arri

10.1 INTRODUCTION

Machine learning (ML) has shown a promising role in the field of health care. Several healthcare outcomes are improved by incorporating this innovative technology. ML has not only resolved recent healthcare problems but also assisted in improving services such as patient experience, medical operations, and clinical practices [1–5].

Blockchain's decentralized and immutable nature confirms that patient data is securely stored and protected from unauthorized access. This is crucial given the high number of data breaches in the healthcare sector. Blockchain facilitates the seamless sharing of medical records across different healthcare providers. This ensures that patient information is up-to-date and accessible, improving the quality of care [6–9]. Blockchain can enhance the transparency and integrity of clinical trials by securely recording and verifying data, making it easier to track and audit research processes. By giving patients control over their own health data, blockchain empowers them to manage their health records and share them with healthcare providers as needed.

With association of AI, ML provides early detection of serious ailments and adapts to suitable treatments in time. This reduces the perils of treatments by descending to the minute' details. As ML is advancing rapidly, new breakthroughs are fashioned in healthcare services [10–15] to solve real-time problems in diagnosis, risk predictions, personalizing treatments, etc.

10.2 BENEFITS OF MACHINE LEARNING IN HEALTHCARE

There are several benefits of implementing ML in healthcare services. It ranges from AI-based assistance, medicine research, personalized treatments, and diet plans to cost reduction and many more. Following are some advantages of ML in the world of healthcare (Figure 10.1).

DOI: 10.1201/9781003643173-10

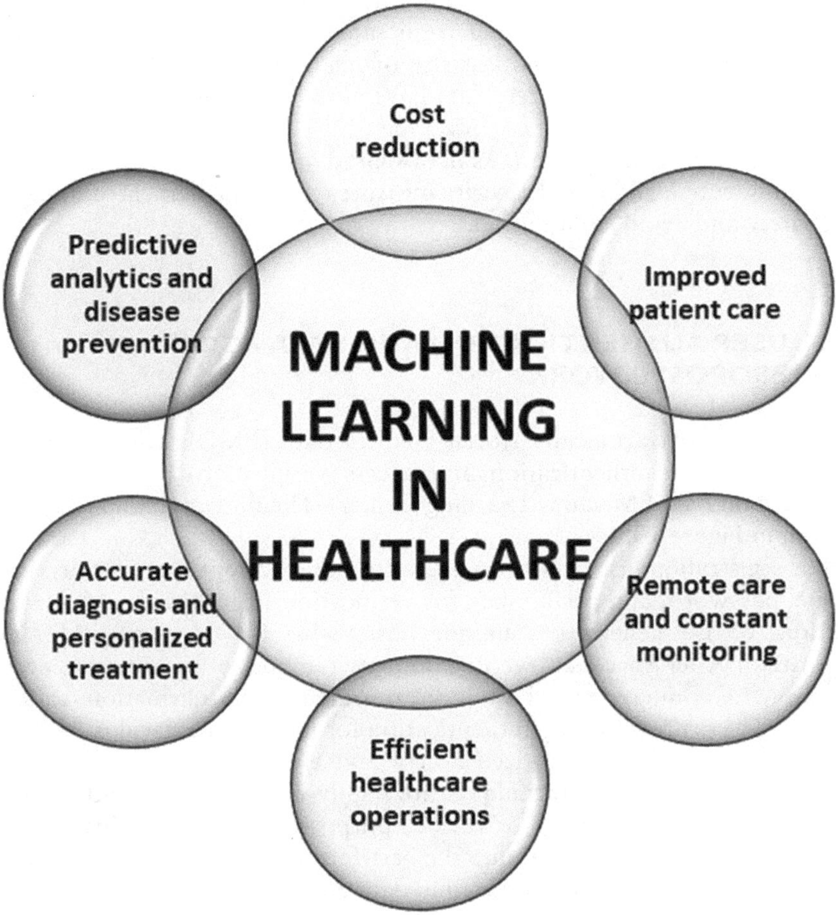

Figure 10.1 Machine learning in healthcare.

10.3 CHALLENGES TO IMPLEMENT MACHINE LEARNING IN THE HEALTHCARE INDUSTRY

The available literature on blockchain architectures for healthcare systems enabled by blockchain that maximize record sharing concludes by highlighting the variety of applications, approaches, discoveries, and constraints in this emerging topic. From safeguarding IoT networks to enabling multiparty computation and electronic health record sharing, the evaluated papers show how blockchain technology can improve trust, transparency, and security in record procedures. Nonetheless, a number of difficulties and constraints, including problems with scalability, interoperability, and usability, have been noted in the studies. There are wide challenges to implement ML in healthcare, ranging from safety issues of patients and their data, quality of services, and security of medical records and documents. Privacy and security of discrete data have always remained a concern while sharing on interconnected network. One of the many ways to sort it out is an authentication mechanism. As data shared on these networks is sensitive and confidential, additional security measures can be implemented to ensure the safety and avoid exploitation.

10.4 USER AUTHENTICATION AND DATA ACCESS IN PROPOSED WORK

To access the electronically stored medical data (EMD), every user must be checked for authentication and access verification from the BMLC (Blockchain based Machine Learning Center). The flow of authentication is shown in Figure 10.2.

For registration, user must provide information about user identity, username, password, and private key for verification from BMLC. After registration, BMLC generates a unique hash value for every user SHA256 algorithm. After successful verification, user requests for resource access from BMLC. Data server will share the resource after confirmation from the BMLC. The process of user authentication for resource retrieval is shown in figure. To prevent unauthorized access and enhance user authentication, polynomial curve fitting techniques are highly effective. This mathematical technique can be employed for various purposes, including authentication. Polynomial curve fitting identifies the best-fitting polynomial curve for a set of data points and is widely utilized in data analysis, prediction, and modeling (Figure 10.3).

The process begins with the registration of user as new entry. For this, the user generates his account, and data will be collected associated to him. On the basis of this collected data, user hash is generated and stored in database for him. When the user tries to login to his account, his credentials are authenticated and confirmed by the database. Thus, access to the system will be granted only after user authentication.

Figure 10.2 User authentication flow.

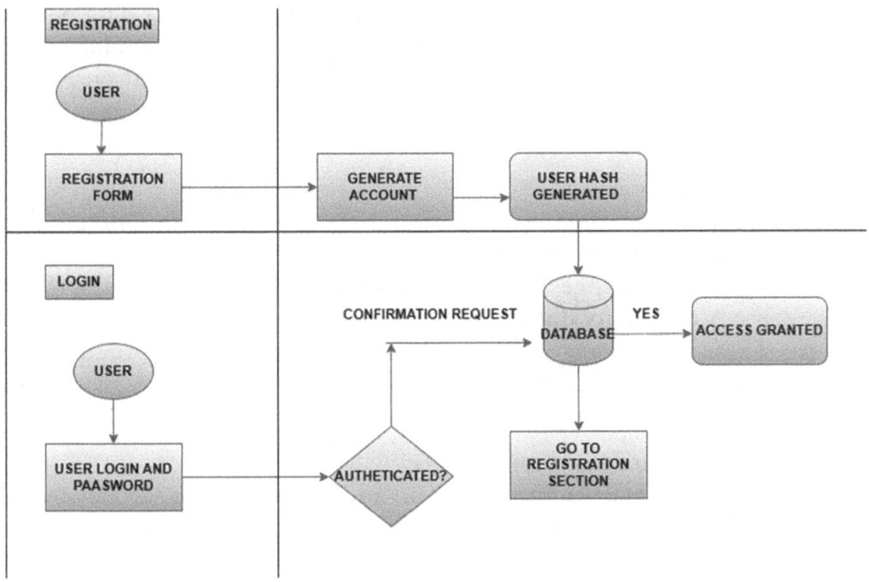

Figure 10.3 User authentication for resource retrieval.

10.5 INTEGRATION OF POLYNOMIAL CURVE FITTING IN MATLAB

Utilizing MATLAB for polynomial curve fitting offers a streamlined and efficient approach. The process starts by importing your data into MATLAB, ensuring you have two arraysone for the independent variable (e.g., distances) and another for the dependent variable (e.g., authentication coefficients). After specifying the desired polynomial order, MATLAB's curve fitting functions handle the computations. Functions like `polyfit` calculate the polynomial coefficients, while `polyval` helps in evaluating the polynomial at specific points. Additionally, MATLAB's robust visualization tools allow you to plot the original data points alongside the fitted polynomial curve, providing a clear understanding of the fit.

Polynomial curve fitting for user authentication can be seamlessly incorporated as follows:

1. **Curve fitting model selection:**
   a. The selection of appropriate polynomial curve fitting model is required with nth order. Higher-order polynomials can capture more intricate patterns in the data.
2. **Curve fitting for authentication:**
   For each user in the placement table (P_T), utilize their proximity or distance values (such as distances between patients and clinics) as

input data points. Apply polynomial curve fitting to these data points, using the chosen polynomial order ('n') to generate a polynomial curve that best fits the data. The coefficients of the polynomial curve can then serve as unique authentication tokens for each user, representing a mathematical representation of the user's proximity pattern to clinics.

3. **Authentication process:**
   a. When a user attempts to access a clinic or perform an action within the clinic ecosystem, their proximity data is collected.
   b. The system applies the same polynomial curve fitting model to the user's proximity data, generating coefficients.
   c. These coefficients are compared to the coefficients stored in the placement table during the initial categorization. If they match, the user is authenticated and granted access.
4. **Enhanced security and uniqueness:**
   a. Polynomial curve fitting adds an additional layer of security to the authentication process, as it relies on the uniqueness of the fitted polynomial for each user.
   b. Even slight variations in a user's proximity pattern will result in different polynomial coefficients, making it challenging for unauthorized users to mimic the authentication.

The algorithm can be illustrated as follows.

1. **Initialization:**
   a. Initialize Firebase configuration.
   b. Initialize the Firebase database connection.
2. **Proof of Work Function:**
   a. **Proof of Work (data)** is a function used to perform the Proof of Work (PoW) for data.
   b. It sets the difficulty level to 4 and initializes the nonce to 0.
   c. The function enters a loop and calculates the SHA-256 hash of the data concatenated with the nonce.
   d. If the hash result starts with four consecutive zeros (indicating the desired difficulty level), it returns the nonce as proof.
   e. Otherwise, it increments the nonce and repeats the process until a valid proof is found.
3. **Node Agreement Simulation:**
   a. **SimulateNodeAgreement** ($\mathbf{block_{data}}$) simulates message exchange and voting among nodes to reach an agreement.
   b. It assumes a total of 5 nodes.
   c. The function iterates through the nodes, simulates their votes using **SimulateVote** ($\mathbf{block_{data}}$), and counts the votes in favor of agreement.

   d. If the majority of nodes (at least 3 out of 5) agree, the function returns True; otherwise, it returns False.

4. **Node Vote Simulation:**
   a. **SimulateVote** ($\mathbf{block_{data}}$) simulates the vote of a node based on latency.
   b. It uses a threshold latency of 50 (a placeholder value).
   c. The function calculates the average latency from the provided $\mathbf{block_{data}}$ and returns True if the average latency is less than the threshold; otherwise, it returns False.

5. **PBFT Consensus Simulation:**
   a. **POWConsensus** ($\mathbf{block_{data}}$) simulates the Practical Byzantine Fault Tolerance (PBFT) consensus algorithm.
   b. It runs for 3 voting rounds, simulating node agreement in each round.
   c. If at least 2 out of 3 rounds result in node agreement (majority), the function returns True; otherwise, it returns False.

6. **Main procedure:**
   a. The main procedure starts by reading data from an Excel file.
   b. It initializes a starting hash value for the first record and an empty list $\mathbf{block_{data}}$ to store records.
   c. For each record in the Excel data:
      i. If the record is submitted by Shalu:
         A. Create a data record with relevant fields.
         B. Calculate the hash value based on the previous hash and the current record.
         C. Set the current hash as the previous hash for the next record.
      ii. Append the record to $\mathbf{block_{data}}$.
      iii. If the length of $\mathbf{block_{data}}$ reaches 5 records:
         A. Check if PBFT consensus is reached using **POW Consensus** ($\mathbf{block_{data}}$).
         B. If consensus is reached:
            - Calculate a nonce using **Proof_of_Work** $\left(\mathbf{block_{data}} + \text{"}\mathbf{Nonce:}\text{"}\right)$.
            - Append {"Nonce": nonce} to $\mathbf{block_{data}}$.
            - Push $\mathbf{block_{data}}$ to the Firebase database.
         C. Otherwise, print "PBFT consensus not reached for this block".
         D. Clear $\mathbf{block_{data}}$.
   d. After processing all records, check if $\mathbf{block_{data}}$ is not empty:
      i. If it is not empty, check POW consensus for the remaining records as before.
      ii. If consensus is reached, follow the same steps to calculate a nonce, append it, push to the database, and print the key.

   iii. Otherwise, print "POW consensus not reached for the remaining records".

The processed work is divided into three categories as illustrated earlier. To utilize the categorization, the proposed work uses Deep Neural Network (DNN) Framework that is dependent upon the hyperparameter tuning using Artificial Bee Colony (ABC) algorithm [16–19].

The ABC algorithm is a swarm intelligence-based optimization technique inspired by the foraging behavior of honeybees. This algorithm has gained popularity in various optimization problems, including Hyperparameter tuning for ML models. In this comprehensive explanation, we will delve into the details of the ABC algorithm, its steps, and provide a sample implementation to illustrate how it works.

The proposed ABC algorithm consists of several key steps that collectively drive the optimization process (Figure 10.4).

1. **Initialization:**
   a. In the beginning, a population of employed bees (E) is created. Each employed bee corresponds to a possible solution, represented as a set of hyperparameters for a ML model.
   b. The algorithm also initializes the best solution (B) and its corresponding fitness or objective function value (A) as empty or arbitrary values.

Figure 10.4 Neural model.

2. **Employed bees phase:**
   a. In this phase, each employed bee (solution) explores its neighborhood to find a new solution.
   b. A neighbor solution is generated by perturbing the current solution. This perturbation introduces small changes to the hyperparameters.
   c. The algorithm evaluates the accuracy or fitness of the neighbor solution using an objective function.
   d. The new result is adopted as the new solution of the given experiment, only if it achieves better accuracy than the current solution.

3. **Onlooker bees phase:**
   a. Onlooker bees decide about employed bees on the basis of the quality of their achieved outcomes.
   b. The algorithm calculates the probabilities for onlooker bees to choose employed bees. These probabilities are typically based on the fitness values of employed bees. Bees with better solutions have higher probabilities of being chosen.
   c. Each onlooker bee selects an employed bee to follow and explores a neighborhood similar to that employed bee's solution.
   d. Similar to the employed bee phase, the onlooker bee evaluates the accuracy of its neighbor solution and adopts it if it is an improvement.

4. **Update best solution:**
   a. The algorithm maintains a record of the best solution and its corresponding accuracy across all employed bees.
   b. In each repetition, the achieved result replaces the current one only if the newly achieved result is better than the previous one.

5. **Termination condition:**
   a. The ABC algorithm operates for a specified number of iterations (N) or until a predetermined accuracy threshold (T) is reached.
   b. If the accuracy threshold is met or exceeded, the algorithm terminates since a satisfactory solution has been found. Otherwise, it continues for the specified number of iterations.

6. **Output:**
   a. At the end of the algorithm's execution, the best solution (B) represents the optimal set of hyperparameters, and its accuracy (A) represents the model's performance achieved with those hyperparameters.

10.6 CONCLUSION AND FUTURE WORK

ML is the innovative technology with the plethora of healthcare applications. It has tremendous potential for improvement in current health environments. In this chapter, we have discussed the security and privacy issues concerning health record sharing. Although many challenges and limitations exist in ML, it still exhibits the capability to cater to several healthcare problems. In the future, by collaborating with other innovative technologies, the serious issues of security, throughput, size, and privacy can be dealt with more efficiently.

REFERENCES

1 R. Miotto, F. Wang, S. Wang, X. Jiang, and J. T. Dudley, "Deep learning for healthcare: Review, opportunities and challenges," *Brief. Bioinformat.*, vol. 19, no. 6, pp. 1236–1246, 2017.
2 K. Papangelou, K. Sechidis, J. Weatherall, and G. Brown, "Toward an understanding of adversarial examples in clinical trials," in *Proceedings of the European Conference on Machine Learning and Knowledge Discovery in Databases*, 2018, pp. 35–51.
3 H. Kim, D. C. Jung, and B. W. Choi, "Exploiting the vulnerability of deep learning-based artificial intelligence models in medical imaging: Adversarial attacks," *J. Korean Soc. Radiol.*, vol. 80, no. 2, pp. 259–273, 2019.
4 X. Yuan, P. He, Q. Zhu, and X. Li, "Adversarial examples: Attacks and defenses for deep learning," *IEEE Trans. Neural Netw. Learn. Syst.*, vol. 30, no. 9, pp. 2805–2824, 2019.
5 C. Szegedy et al., "Intriguing properties of neural networks," 2013, arXiv:1312.6199.
6 A. Shafahi et al., "Poison frogs! Targeted clean-label poisoning attacks on neural networks," in *Advances in Neural Information Processing Systems*, 2018, pp. 6103–6113.
7 S. G. Finlayson, J. D. Bowers, J. Ito, J. L. Zittrain, A. L. Beam, and I. S. Kohane, "Adversarial attacks on medical machine learning," *Science*, vol. 363, no. 6433, pp. 1287–1289, 2019.
8 V. Chandola, A. Banerjee, and V. Kumar, "Anomaly detection: A survey," *ACM Comput. Surv.*, vol. 41, no. 3, pp. 1–58, 2009.
9 K. Pandey, P. Pandey, K. Jaiswal, and A. K. Sen, "Datamining clustering techniques in the prediction of heart disease using attribute selection method," *Heart Disease*, vol. 14, pp. 16–17, 2013.
10 K. Polat, and S. Günes, "Prediction of hepatitis disease based on principal component analysis and artificial immune recognition system," *Appl. Math. Comput.*, vol. 189, no. 2, pp. 1282–1291, 2007.
11 M. Alloghani, D. Al-Jumeily, J. Mustafina, A. Hussain, and A. J. Aljaaf, "A systematic review on supervised and unsupervised machine learning algorithms for data science," in *Supervised and Unsupervised Learning for Data Science*, Springer, 2020, pp. 3–21.
12 M. N. Sohail, J. Ren, and M. U. Muhammad, "A Euclidean group assessment on semi-supervised clustering for healthcare clinical implications based on real-life data," *Int. J. Environ. Res. Public Health*, vol. 16, no. 9, p. 1581, 2019.
13 R. Q. H. Zahin et al., "Sensor-based human activity recognition for smart healthcare: A semi-supervised machine learning," in *Proceedings of International Conference on Artificial Intelligence for Communications and Networks*, 2019, pp. 450–472.

14 S. Sharma, and R. S. Batth, "BLOCKLIB: Blockchain enabled library resource sharing," *J. Discrete Math. Sci. Cryptogr.*, vol. 25, 3, pp. 839–857, 2022. https://doi.org/10.1080/09720529.2021.2019443
15 D. Mahapatra, "Semi-supervised learning and graph cuts for consensus based medical image segmentation," *Pattern Recognit.*, vol. 63, pp. 700–709, 2017.
16 S. Sharma, and R. S. Batth, "Blockchain technology for higher education system: A mirror review," *2020 International Conference on Intelligent Engineering and Management (ICIEM)*, London, UK, 2020, pp. 348–353. https://doi.org/10.1109/ICIEM48762.2020.9160274
17 W. Bai et al., "Semi-supervised learning for network-based cardiac MR image segmentation," in *Proceedings of International Conference on Medical Image Computing and Computer Assisted Intervention*, 2017, pp. 253–260.
18 S. L. Tripathi, and M. Mahmud, *Explainable Machine Learning Models and Architecture*, Wiley & Scrivener Publishing LLC, 2023. https://doi.org/10.1002/9781394186570
19 S. L. Tripathi, L. P. Dasari, I. Wijayanto, D. Ghai, and M. Mahmud, "Assistive technology for neuro-rehabilitation applications using machine learning techniques," in *Machine Learning Models and Architectures for Biomedical Signal Processing*, Academic Press, Elsevier, 2025, pp. 179–191. https://doi.org/10.1016/B978-0-443-22158-3.00008-9

Chapter 11

Smart text extraction system for Bank Cheque Images using DWT and dynamic thresholding

Neha Thakur, Deepika Ghai, Sunpreet Kaur Nanda, Sandeep Kumar and Mandeep Kaur

11.1 INTRODUCTION

In many developing countries, a substantial number of bank cheques are processed daily across various sectors, including financial institutions [1–5], educational organizations [6], and corporate entities [7]. While alternative non-cash transaction methods such as credit cards, debit cards, internet banking, and platforms like Paytm exist [8, 9], cheques remain vital for conducting high-value non-cash transactions worldwide [10–15]. They are primarily used for large payments, business-to-business (B2B) transactions, and payroll processing [16]. According to the statistics from cheque and credit clearance organizations, in the UK, approximately 400 million cheques were administered in 2017 [17]. Additionally, the Reserve Bank of India (RBI) reported a 10.1% increase in both the value and volume of cheque transactions following demonetization [18]. This information highlights the continued importance of cheques as a primary method for financial transactions.

Nevertheless, the banking sector and its customers face ongoing challenges. Most cheques have traditionally been processed manually, leading to a heavy workload for bank employees and an increased likelihood of data entry errors. The banking industry is keen on exploring automated methods for cheque reading to alleviate this burden [19]. There is an urgent need to enhance banking services and create a more efficient sector through automated data collection and technological advancements from bank cheques, a topic of growing research interest [20–23]. Scholars argue that automating cheque data recognition and verification processes can significantly reduce handling costs and processing times. Implementing an automated cheque processing system could not only lessen human effort but also mitigate various related issues.

Despite advancements in character recognition, extracting information from bank cheques remains a complex challenge. A cheque includes several domains i.e., bank logo, payee name, bank name, legal amount, date, courtesy amount, void pantograph, and signature, as illustrated in Figure 11.1. Challenges in recognizing these fields include variations in font style and size

 DOI: 10.1201/9781003643173-11

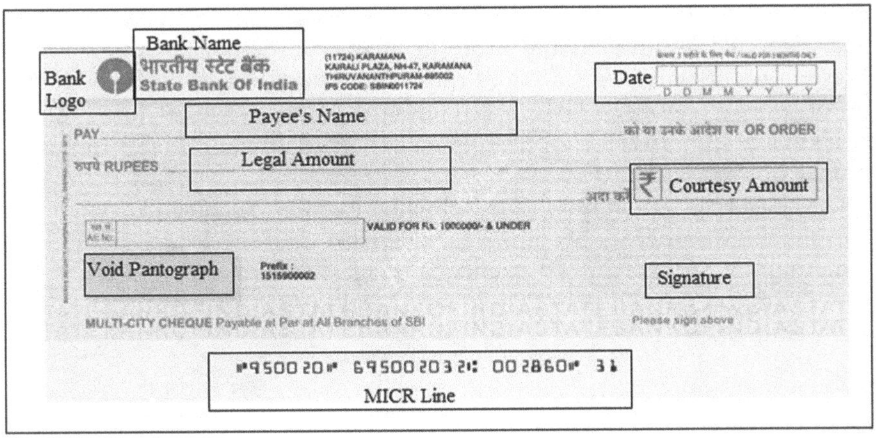

Figure 11.1 Different fields in Indian bank cheque form.

[24], the need to eliminate complex backgrounds and stray lines [25], data degradation from noise [26], skewness issues [27], and overlapping data [28]. There is a pressing demand for a robust, automated algorithm capable of accurately extracting nearly all data fields from bank cheques. While numerous techniques for text extraction from bank cheque images have been proposed, the task remains daunting.

Text extraction methods can be partitioned into three key classes: region, texture, and morphological-based approaches. The region-based method uses color or grayscale characteristics to distinguish between text and background by analyzing intensity variations. However, this method often faces challenges with intricate backgrounds. It can be further classified into Connected Component (CC) and edge-based methods. The CC-based method identifies text and without-text areas by clustering pixels with similar intensities, whereas the edge-based method emphasizes edge extraction to differentiate text from the background, regardless of text intensity or layout.

Texture-based methods leverage the distinct texture characteristics of text compared to the background, proving effective even in complex backgrounds. Morphological methods, grounded in geometrical and topological principles, extract important text contrast features that remain consistent under various conditions, making them adaptable to different image alterations. The method proposed in this chapter combines DWT and dynamic thresholding with a logical AND operator for text extraction. After initial extraction, some unwanted regions may still persist, prompting researchers to employ various thresholding techniques for removal.

Thresholding is an essential process in image segmentation that facilitates the automatic partitioning of images. It can be divided into multi- and bi-level thresholding. Bi-level thresholding splits an image into two segments using a single threshold value, whereas multi-level thresholding employs multiple values to generate several segments. Additionally, thresholding can be categorized into global and local (or adaptive) methods. Although local thresholding dynamically modifies threshold values based on the pixel values of nearby regions, global thresholding uses an individual value for the entire image [29]. In this chapter, dynamic thresholding is initially applied to remove non-text areas.

The chapter is structured in the following manner: Section 11.2 examines relevant work, whereas Section 11.3 describes the developed method for obtaining data fields from bank cheque images. Section 11.4 represents the experimental results and discussions, and Section 11.5 shows the conclusion of the chapter.

11.2 RELATED WORK

Researchers have proposed various techniques for extracting textual information from bank cheque images, generally categorizing automatic bank cheque processing systems into two main areas: (a) pre-processing and (b) field extraction. Jayadevan et al. [2] developed a signature verification technique using Dynamic Time Warping (DTW), which is most effective when working with a limited number of training samples. Meanwhile, Mehta et al. [3] concentrated on detecting the payee's name, courtesy amount, and signature fields by employing methods like the Probabilistic Neural Network (PNN) recognition, Hidden Markov Model (HMM), and Region of Interest (ROI) segmentation. They observed that the system's efficiency decreased significantly with substantial character variations.

Thakur et al. [5] highlighted the extensive daily processing of cheques used in financial transactions, emphasizing the potential for time and cost savings through automation in recognizing and verifying cheques. Research topics like computer vision, pattern recognition, image processing, machine learning, and deep learning are becoming more and more popular in connection with automated bank check processing systems. This article outlines the processes involved in acquiring, pre-processing, extracting, and recognizing cheque images, categorizing the challenges at various stages of automated processing. It aims to present advanced techniques for automating bank cheque image processing, detailing benchmark datasets, and comparing the performance of leading approaches. Additionally, it reviews market products for the processing of cheques automatically and provides a comparative analysis of ongoing issues in the field. The Multilayer feed-forward neural network detected the names of payees with an accuracy of 97.31%, while HMM-MLP reached 95.5% for date identification. For digit recognition, the Deep Neural Network (DNN) achieved 98.5%, the MLP reached 93.2%, and the Modified Quadratic Discriminant Function (MQDF) attained 97.04%.

Liu et al. [8] presented a system that extracts date, legal, and courtesy amounts from Canadian bank cheques using a straight-line approach. However, it yields poor results in certain scenarios, such as recognizing the dollar sign, handling poor handwriting, and dealing with overlapping handwritten and machine-printed data. Palacios et al. [9] developed a method for recognizing courtesy amounts using various techniques, including Neural Networks (NN), Hybrid Drop Fall (HDF), dynamic thresholding, extended drop fall algorithm, and forward Multilayer Perceptron (MLP), all of which excel in courtesy amount recognition. Akbari et al. [11] focused on identifying dates, as well as courtesy and legal amounts, by utilizing a Support Vector Machine (SVM), K-Nearest Neighbour (KNN), and Artificial Neural Network (ANN). They also employed techniques like Aspect Ratio Adaptation Normalization (ARAN), Multi-Resolution Box-Counting (MRBC), Random Transform (RT), and Kolmogorov-Smirnov (K-S) testing for signature recognition. However, they did not propose a complete end-to-end system. A lexicon-driven method for legal amount recognition using MQDF that is restricted to English characters was presented by Jayadevan et al. [12]. Kumar et al. [19] introduced a system for signature verification of bank cheques based on the architecture of Feed Forward Back Propagation Neural Network (FFBPNN), where the attribute extraction is conducted using Gray Level Co-Occurrence Matrix (GLCM). This system has limitations, including an inability to classify forgeries and a reliance on a huge number of training samples.

Zhou et al. [22] suggested a feedback-based approach to enhance word segmentation in legal amounts using an HMM-MLP hybrid model, noting challenges in isolating connected words. Neves et al. [23] utilized an algorithm derived from Tsallis entropy for thresholding courtesy amounts on Brazilian bank cheques, concluding that the Mello algorithm provided the best results, although some images yielded unsatisfactory outcomes. Guillevic et al. [24] developed a recognition system for the legal amounts using HMM, effective only for French and English characters. Zimmer et al. [25] applied a hybrid online/offline segmentation scheme for signature verification, reporting error rates of 1% and 5% from 10 and 5 reference signatures, respectively.

Rabelo et al. [26] proposed a thresholding method inspired by the MLP for documents with complex backgrounds, focusing on Brazilian bank cheque regions like courtesy amounts and Magnetic Ink Character Recognition (MICR) codes. This method was compared with ten thresholding algorithms but was criticized for being time-consuming and less accurate with noisy images. Dansena et al. [27] designed a system to recognize various elements on Indian bank cheques, including payees' names and MICR regions, while Xu et al. [28] introduced a knowledge-based component for date recognition in Canadian bank cheques using HMM and HMM-MLP hybrid systems, noting a decrease in accuracy for year recognition.

Wang et al. [29] addressed the complex issue of detecting the permissible amounts on Chinese bank cheques using deep learning and auto-encoder techniques, restricted to capital Chinese characters. Kim et al. [30] used HMM-MLP hybrid classifiers for courtesy amount recognition, achieving satisfactory results but lacking speed. Khatode et al. [31] focused on signature verification using Principal Component Analysis (PCA), but this method was limited to signature verification only.

Souza et al. [32] created an automated MICR code recognition system utilizing Paraconsistent Neural Artificial Networks (PANN) on Brazilian bank cheques, but the system struggled with lower-quality images and disregarded scanning angles. Ali et al. [33] employed Maximally Stable Extremal Regions (MSER) and Speeded Up Robust Features (SURF) to detect and extract pantograph regions from cheque images, limited to rectangular regions. In [34], a combination of Gabor filters, CC analysis, pixel transformation, and NN-based classifiers was used for signature verification. Morita et al. [35] employed an HMM-MLP hybrid system for the recognition of Brazilian bank cheque dates, noting better accuracy for day digits than for year digits.

Nguyen et al. [36] conducted a survey in a Vietnamese university, revealing inefficient paperwork handling in the financial department, where bank statements are manually entered into software. Their study aimed to automate data extraction by analyzing the table structure, achieving over 93% accuracy in a simplified image-based methodology that displays extracted text in a spreadsheet format, although user verification is necessary to prevent financial errors.

Oral et al. [37] emphasized the potential of information extraction research for banking documents, suggesting that the combination of visual and textual information remains underexplored despite prior multimodal studies. This research represents a pioneering effort to apply deep learning for extracting information from unstructured banking documents like orders of money transfer, examining the impacts of various neural word illustrations on tasks such as entity recognition and relation extraction. The study found that deep learning methods enhanced information extraction subtasks by approximately 10%, with improvements linked to positional features and auxiliary learning tasks. Integrating this system into banking workflows significantly reduced processing times; book-to-book money transfers are reduced to 10 minutes, while Electronic Fund Transfers (EFT) take 17 minutes. Agrawal et al. [38] highlighted the role of image processing in automating bank cheque verification, aiming to complement existing cheque truncation systems and minimize human intervention. Their work utilized image processing and deep learning to develop a tool for verifying essential cheque components, achieving a 99.14% accuracy rate for handwritten numeric characters using deep learning-based convolutional neural networks (CNN), and satisfactory results for machine-printed scripts (97.7% accuracy). For signature verification, they applied Scale Invariant Feature Transforms (SIFT) alongside SVM, attaining a 98.10% accuracy rate.

Miah et al. [39] proposed a Feed Forward Neural Network (FFNN) methodology for courtesy amount recognition, noting reduced accuracy when training and testing sets differed. Singh et al. [40] developed a template matching system for recognizing legal amounts on Indian bank cheques, though it faced challenges related to high computational costs and lower recognition rates. Nancy et al. [41] presented a signature verification method using feature extraction and template matching, which struggled with signature locations differing from prior cheques.

From this literature review, it is evident that many researchers have devised various extraction techniques for bank cheque images, typically focusing on one or two data fields such as dates, signatures, legal, and courtesy amounts. Areas like bank logos, names, payee's names, void pantographs, and MICR lines remain underexplored. While several public datasets, including CENPARMI [21, 23, 27] and NIST SD19 [34], have been used, a significant limitation is their focus on vocabulary letters, whereas the English handwritten legal amount comprises 32 distinct word classes with numerous numeric variations. Our research employs a new public dataset, IDRBT, featuring bank cheque images. The pressing need is to create a unified text extraction method capable of simultaneously extracting all data fields from bank cheque images.

11.3 PROPOSED METHOD

We propose an effective and efficient hybrid method for extracting text regions from bank cheque images. In this approach, edges are considered the most significant feature for text extraction, regardless of color, layout, or orientation. The flowchart illustrating the proposed hybrid text extraction method is shown in Figure 11.2, with the various steps outlined in the following sections.

11.3.1 Pre-processing

The colored input image is transformed into a grayscale image (or intensity image) in this phase, as the RGB color components do not contribute relevant information for text extraction [42]. The intensity image (Y) is generated by merging the RGB components, as indicated in Eq. (11.1):

$$Y = 0.3R + 0.59G + 0.11B \tag{11.1}$$

where the images' red, blue, and green components are indicated by the letters R, B, and G, respectively. If the input image is in grayscale, then pre-processing is not essential.

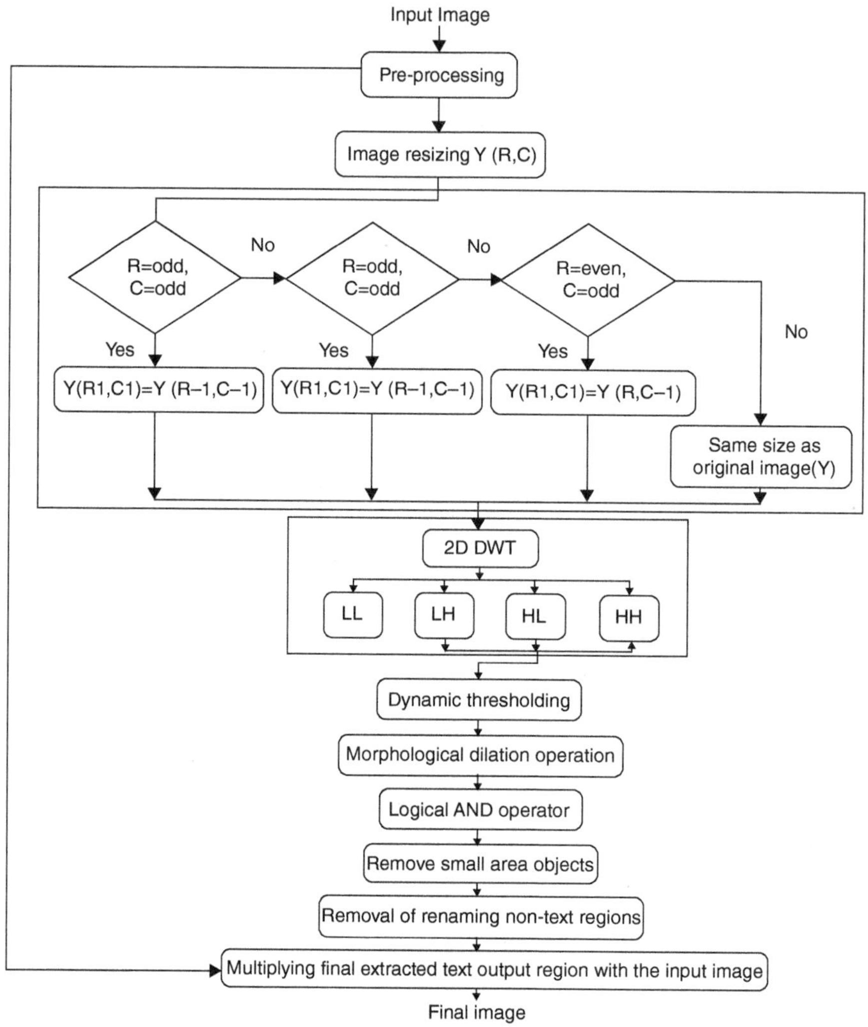

Figure 11.2 Flowchart of the proposed method for extracting text regions from Bank Cheque Images.

11.3.2 Resized image

Intensity image (Y), characterized by its dimensions of columns (C) and rows (R), is resized because the image size will change in the subsequent stages. The intensity image serves as the input, and any alterations in size could lead to errors during the extraction of text regions. To avoid this issue, the image size is adjusted initially based on pixel dimensions (i.e., columns (C) and rows (R)). The resizing conditions are as mentioned below:

1. C = odd, R = odd: C and R are assessed by dividing them by the number 2. If neither result is zero, C and R, both are odd numbers (C = odd number and R = odd number). So, in this case, the size is adjusted by decreasing both the number of columns as well as rows by 1 unit, as presented in Eq. (11.2):

$$Y(R1,C1) = Y(R-1,C-1) \tag{11.2}$$

2. C = even, R = odd: Suppose the preceding condition is not satisfied, R is divided by two to determine if it equals zero. If it does not (indicating R is an odd number), then 1 unit is subtracted from number of rows (R), as shown in Eq. (11.3):

$$Y(R1,C1) = Y(R-1,C) \tag{11.3}$$

3. C = odd, R = even: Suppose the first two conditions are not met, C is divided by two to assess its value. If the result is non-zero (indicating that C is an odd number), 1 unit is subtracted from the number of columns (C), as shown in Eq. (11.4):

$$Y(R1,C1) = Y(R,C-1) \tag{11.4}$$

4. C = even, R = even: Suppose none of the preceding conditions apply, the size of the image remains unchanged (C = even number and R = even number), as specified in Eq. (11.5):

$$Y(R1,C1) = Y(R,C) \tag{11.5}$$

Here, C and R represent the columns and rows of the grayscale image (intensity), respectively, while C1 and R1 denote the columns and rows after the operation of image resizing.

11.3.3 2D DWT

In Figure 11.3, the 2D DWT splits the Y image into four sub-bands, namely three detail components (LH, HL, and HH) and one approximation component (LL). The lower-frequency component is referred to as L, while the higher-frequency component is denoted as H.

High-frequency wavelet coefficients effectively characterize the edges of text regions, as their intensity values are greater than those of non-text regions. The HL, LH, and HH sub-bands are responsible for detecting vertical (V), horizontal (H), and diagonal (D) edges, respectively. To isolate the text region from the background, the processing is performed only on HL, LH, and HH sub-bands. The Haar wavelet, known for its rapid operational

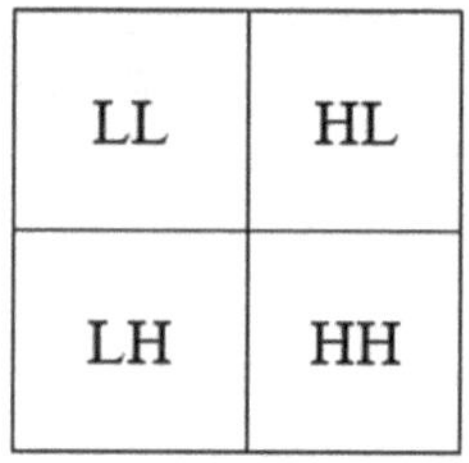

Figure 11.3 2D DWT decomposition.

speed due to its straightforward low-pass and high-pass filter coefficients (either 1 or −1), is used for all experiments conducted.

The negative frequency coefficients do not offer any meaningful data about the texture. After decomposing the Y image into four sub-band components, the detail coefficients of frequency components corresponding to the H, V, and D edges are analyzed. If some negative coefficients are found, they are replaced with zero, which reduces processing time. Specifically, if any of the values for the three edges (H, V, and D) are negative, they are set to zero; otherwise, the pixel values remain unchanged, as expressed in the following equations (Eqs. (11.6)–(11.8)).

$$\text{If}\,H(C1,R1) < 0, \text{then}\,H(C1,R1) = 0 \tag{11.6}$$

$$\text{If}\,V(C1,R1) < 0\,\text{then}\,V(C1,R1) = 0 \tag{11.7}$$

$$\text{If}\,D(C1,R1) < 0\,\text{then}\,D(C1,R1) = 0 \tag{11.8}$$

11.3.4 Dynamic thresholding

Thresholding is nothing but a segmentation technique that distinguishes between the background and text in an image. Non-text and text areas are characterized by V, H, and D edges. Both non-text and text edges are identified in HL, LH, and HH sub-bands. Text edges are identifiable by their higher intensity values compared to those of the non-text edges.

In the subsequent step, dynamic thresholding is applied to eliminate background regions (i.e., edges with lower intensity values). The dynamic thresholding value (T) is calculated by equating each pixel with its neighboring pixels within each detailed component sub-band, as outlined in equation (Eq. (11.9)).

$$T = \frac{\Sigma\left(I(R1,C1)\,x\,I1(R1,C1)\right)}{\Sigma I1(R1,C1)} \tag{11.9}$$

P1	P2	P3	P4	P5
P6	P7	P8	P9	P10
P11	P12	P13	P14	P15
P16	P17	P18	P19	P20
P21	P22	P23	P24	P25

Figure 11.4 5×5 pixel values of detail component sub-band (I).

where the general detailed component image I(R1, C1) has a pixel intensity value of (R1, C1) and is shown below in Figure 11.4.

I1(R1, C1) is the general new detailed component image and is obtained by operating on each pixel with its neighboring pixels (i.e., dynamic thresholding) from I(R1, C1). The general equation for calculating each pixel value of the generalized new detailed component image (I1) is given by the equation (Eq. (11.10)).

$$I1(R1,C1) = \max\left(\left|I(R1,C1+1) - I(R1,C1-1)\right|, \left|I(R1+1,C1) - I(R1-1,C1)\right|\right) \tag{11.10}$$

The frequency coefficient of every pixel location is first analyzed, and then it is replaced with a new value based on the following conditions:

1. Considering the frequency coefficients of the first and last rows and columns each, as shown in Figure 11.5, coefficients in the new detailed component image (I1) are changed by using the equation (Eq. (11.11)). Similar operations are applied to each pixel, then I1(R1, C1) is obtained for each detail component sub-band.

$$\begin{aligned} I1(P6) &= \text{maximum}\left(\left|P11 - P1\right|, \left|P7\right|\right) \;\& \\ I1(P2) &= \text{maximum}\left(\left|P7\right|, \left|P3 - P1\right|\right) \end{aligned} \tag{11.11}$$

2. Considering the frequency coefficients of all corners as depicted in Figure 11.6, these coefficients in the new detailed component image (I1) are changed by using Eq. (11.12).

$$\begin{aligned} I1(P1) &= \text{maximum}\left(\left|P6\right|, \left|P2\right|\right) \;\& \\ I1(P21) &= \text{maximum}\left(\left|P16\right|, \left|P22\right|\right) \end{aligned} \tag{11.12}$$

P1	P2	P3	P4	P5
P6	P7	P8	P9	P10
P11	P12	P13	P14	P15
P16	P17	P18	P19	P20
P21	P22	P23	P24	P25

Figure 11.5 5×5 pixel values of detail component sub-band (I).

3. Considering the frequency coefficients of the in-between portion of the image, as shown in Figure 11.7, these coefficients in the image (I1) are changed by using Eq. (11.13). A similar operation is applied to each pixel present in between the image, then I1(R1, C1) is obtained for each detail component sub-band.

$$\begin{aligned} I1(P8) &= \text{maximum}\left(|P13 - P3|, |P9 - P7|\right) \& \\ I1(P17) &= \text{maximum}\left(|P22 - P12|, |P18 - P16|\right) \end{aligned} \tag{11.13}$$

Once all the frequency coefficients are replaced in an image, a new detailed image (I1) is created for each detail component sub-band. The equation (Eq. (11.9)) is then applied to compute the dynamic thresholding value (T) for

P1	P2	P3	P4	P5
P6	P7	P8	P9	P10
P11	P12	P13	P14	P15
P16	P17	P18	P19	P20
P21	P22	P23	P24	P25

Figure 11.6 5×5 pixel values of detail component sub-band (I).

P1	P2	P3	P4	P5
P6	P7	P8	P9	P10
P11	P12	P13	P14	P15
P16	P17	P18	P19	P20
P21	P22	P23	P24	P25

Figure 11.7 5×5 pixel values of detail component sub-band (I).

each detail component sub-band. The frequency coefficients of this new image are compared to the threshold value (T), resulting in a binary image (B) as derived from equation (Eq. (11.14)). If a frequency coefficient exceeds the threshold value (T), it is replaced with one; otherwise, it is set to zero. The resulting binary image comprises primarily text with only a minimal amount of non-text edges.

$$B = \begin{cases} 255, \text{if } I1(R1, C1) > T \\ 0, \text{otherwise} \end{cases} \tag{11.14}$$

11.3.5 Morphological dilation operation

As a result of dynamic thresholding, the H, V, and D sub-band images undergo dilation. Text edges are typically short and may be connected in various directions. The morphological dilation operation helps to fill gaps within each isolated text region, thereby connecting them in each detail component sub-band. The number of pixels added during dilation is influenced by the size of the structuring element used for processing the image. For the horizontal, vertical, and diagonal edge components, structuring elements ranging from [10 10] to [20 20] are utilized for the first two, while [10] to [20] are used for the diagonal component.

11.3.6 Logical AND operator

The output obtained from the morphological dilation operation (i.e., vertical (V), horizontal (H), and diagonal (D)) is processed using the logical AND function (Eq. (11.15)) to isolate the text regions. Following morphological dilation, overlapping occurs in the text areas due to the expansion of each edge, resulting in a mixture of three types of edge areas. In contrast, non-text regions typically contain only one or two types of edge regions, leading to minimal overlap even after dilation. While most non-text regions are eliminated through the logical AND operation, any remaining non-text regions in the output are further filtered using bounding boxes.

$$I2 = H \,\&\, V \,\&\, D \tag{11.15}$$

11.3.7 Removal of the non-text areas and smaller region objects

Two of the area-based filtering methods are employed to eliminate smaller unwanted objects and the non-text areas found in the output of the logical AND operation (I2).

1. The first method focuses on removing undesired small non-text regions (i.e., objects) that have been incorrectly identified as text regions. Relative area values are used for this, and only areas with an area larger than the relative maximum area (Eq. (11.16)) are kept in the image.

$$\text{Area} \geq \frac{\text{Maxarea}}{\text{a}} \tag{11.16}$$

   where Maxarea refers to the area of the main object in the image, the value of 'a' is determined based on the size of small objects present, with experimental findings indicating that 'a' ranges from 11 to 36 for bank cheque images. Consequently, the smallest area objects (i.e., non-text) are removed from the output of the logical AND operation.
2. Following the removal of smaller area objects, some unwanted non-text regions may still remain in the image. To resolve this, the area of the bounding box (Area(bb)) enclosing all objects is initially calculated, followed by determining the ratio of the object's area (Area(O)) to the bounding box area (Area(bb)) as shown in Eq. (11.17). Experimental results show that if this ratio exceeds 3, the frequency coefficients at that location are set to 0; otherwise, they remain at 1. This value aids in text localization within the image.

$$\frac{\text{Area}(O)}{\text{Area}(bb)} > 3, \text{ then } I2(R1, C1) = 0 \tag{11.17}$$

11.3.8 Output image

The final image in the output, containing the extracted text regions, is achieved by multiplying the image resulting from the removal of the non-text regions with the input grayscale image. To further refine the text regions, global thresholding is applied.

The steps that are involved in the text extraction process from the images of bank cheques are illustrated in Figure 11.8.

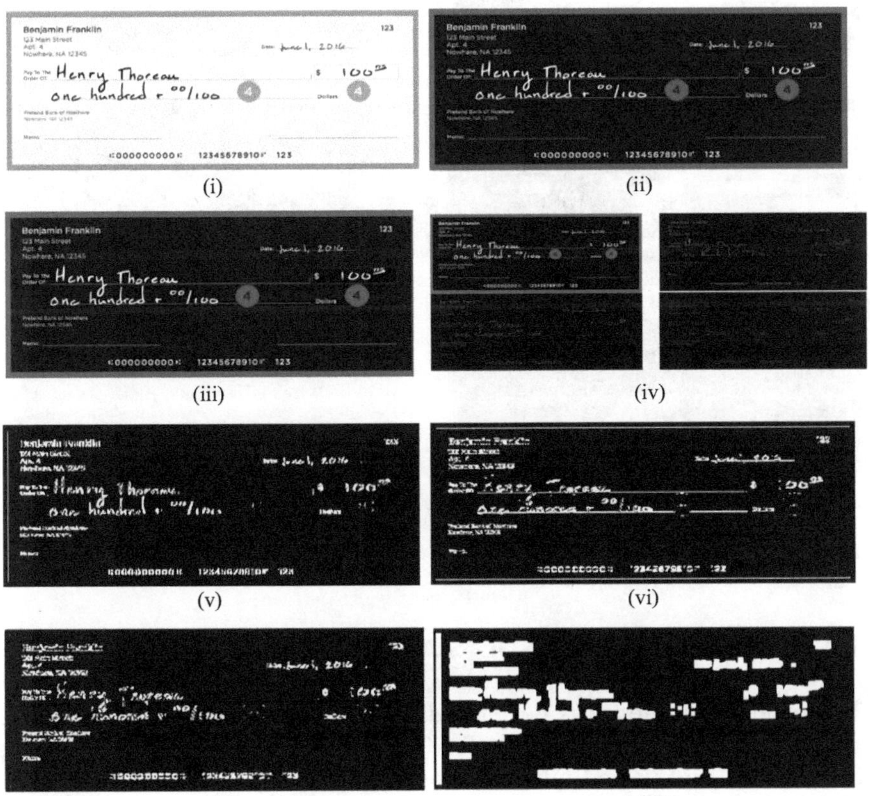

Figure 11.8 Procedures of the proposed method for extracting text from an image of a bank cheque: (i) Input color image, (ii) Pre-processing, (iii) Resized image, (iv) 2D Discrete Wavelet Transform (DWT), (v) Threshold vertical image, (vi) Threshold horizontal image, (vii) Threshold diagonal image, (viii) Dilated vertical image, (ix) Dilated horizontal image, (x) Dilated diagonal image, (xi) Resultant output of logical AND operator, (xii) Removal of small objects, (xiii) Removal of non-text regions (area-based filtering), (xiv) Extracted data fields, and (xv) Refining of text regions.

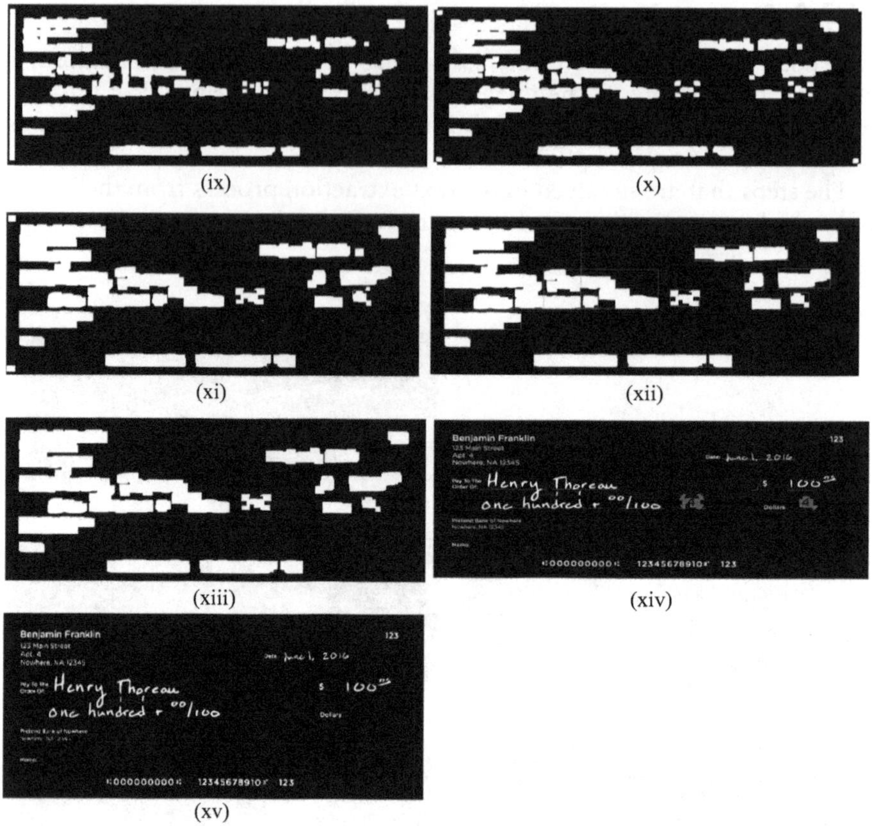

Figure 11.8 (Continued)

11.4 EXPERIMENTAL RESULTS AND DISCUSSION

The proposed method was developed using MATLAB 2016a on a 64-bit operating system with an Intel (R) Core (TM) i3-6006U CPU running at 2.00 GHz and paired with 4GB of RAM. The researchers conducted experiments using a variety of public datasets alongside their own created datasets. Public datasets such as CENPARMI [21, 23, 27] and NIST SD19 [34] were utilized. However, existing public datasets present significant issues: they only include vocabulary letters (e.g., the courtesy amount), while the English handwritten legal amount consists of 32 different word classes, including terms like thousand, hundred, dollar, and numeric figures. Notably, no one had previously used public datasets containing bank cheque images, such as the Institute for Development and Research in Banking Technology (IDRBT). The dataset of IDRBT includes bank cheque images featuring a range of ink colors and textures (blue and black), created with the help of nine volunteers and scanned at 300

DPI, totaling 300 images. Additionally, the researchers assembled their own dataset comprising 150 images sourced from the internet, featuring diverse fonts, font sizes, resolutions, alignments, orientations, background patterns, and handwriting styles from various banks, including SBI and Axis.

To calculate the efficiency of the proposed technique on both public and self-created datasets, a variety of evaluation parameters [43–49] were employed, such as Recall Rate (RR), Precision Rate (PR), and Processing Time (PT).

1. **Precision Rate (PR)**: This statistical measure computes the proportion of actual positives that are identified correctly compared to the total of real positives and false positives [50], as defined in the following equation (Eq. (11.18)).

$$\text{Precision Rate}(\text{PR}) = \frac{\text{True Positive}(\text{TP})}{\text{True Positive}(\text{TP}) + \text{False Positive}(\text{FP})} \tag{11.18}$$

2. **Recall Rate (RR)**: It is a numerical measure that analyzes the quantity of definite positives identified correctly against the total of actual positives and unused values in an image [50], as represented by the following equation (Eq. (11.19)).

$$\text{Recall Rate}(\text{RR}) = \frac{\text{True Positive}(\text{TP})}{\text{True Positive}(\text{TP}) + \text{False Negative}(\text{FN})} \tag{11.19}$$

where,

True Positive (TP): True positives (TP) refer to the regions in an image that contain text characters and have been accurately identified by the algorithm as text regions.

False Alarms or False Positive (FP): A false positive, or false alarm, occurs when the algorithm mistakenly identifies non-text regions as text regions.

True Negative (TN): True negatives (TN) are regions in the image that do not contain text and are correctly identified as non-text by the algorithm.

Missed Detections or False Negative (FN): False negatives, or missed detections, refer to regions that contain text but have not been identified as text regions by the algorithm.

3. **Processing Time (PT)**: Processing time (PT) is the duration required to detect text (i.e., data fields) from an image (i.e., background region).

The Haar wavelet was selected for experimentation due to its simplicity, symmetry, and orthogonality. Additionally, it exhibits binary-like behavior, as its coefficients are either +1 or −1. The performance of the proposed

Table 11.1 Performance with the public dataset (IDRBT) and own dataset

			Parameters of evaluation		
Sr. No.	*Dataset*	*No. of the images*	*PR (%)*	*RR (%)*	*PT (sec)*
1.	Public (IDRBT)	300	94.92	96.15	7.21
2.	Own	150	96.61	99.03	3.15

method is calculated on the standard dataset (i.e., IDRBT) and the custom dataset using the Haar wavelet, as shown in Table 11.1.

The input images and the corresponding resultant output images after the text extraction process from different images of bank cheques having a range of fonts, font sizes, and backgrounds are shown below in Table 11.2.

In Table 11.3, the proposed approach is compared to existing methods for bank cheque images.

Table 11.2 Images that are input before and after text extraction from bank check images

Sr. No.	*Input images before text extraction*	*Output images after text extraction*
1.	Public dataset	
	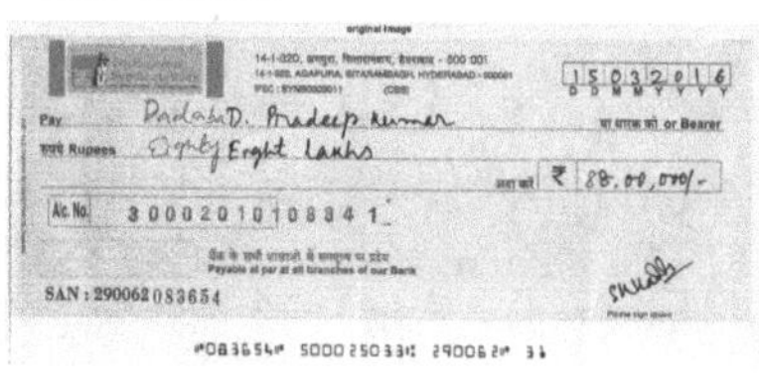	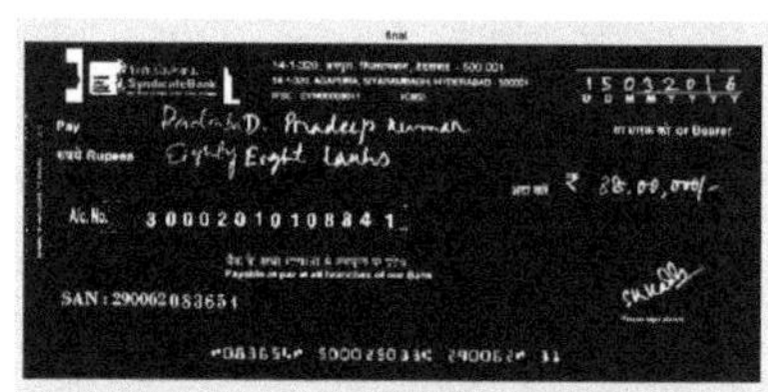

	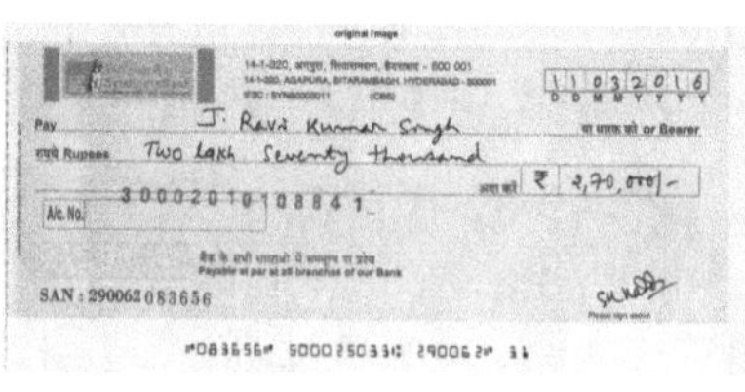	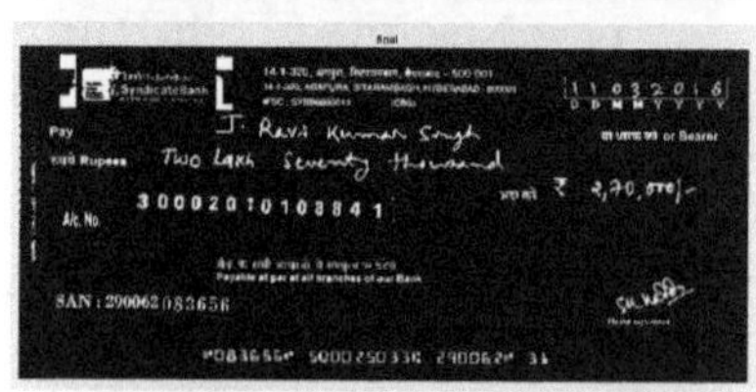
	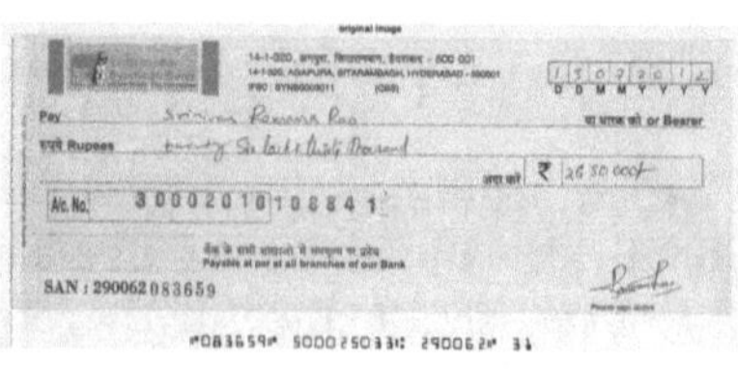	

(*Continued*)

Table 11.2 (Continued)

Sr. No.	*Input images before text extraction*	*Output images after text extraction*
2.	Own dataset	
	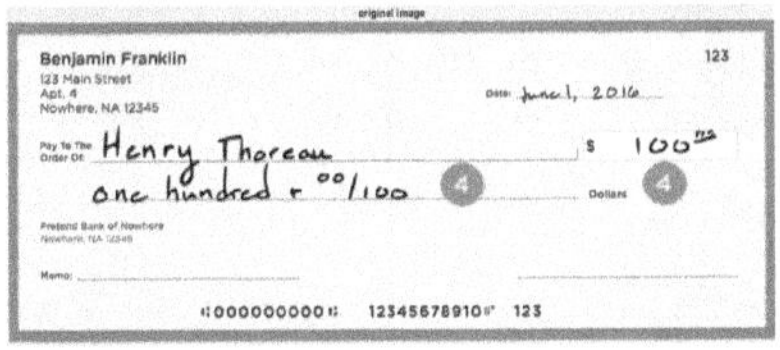	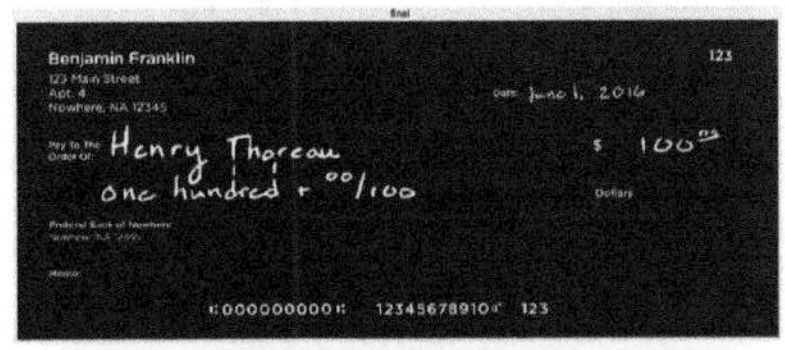
	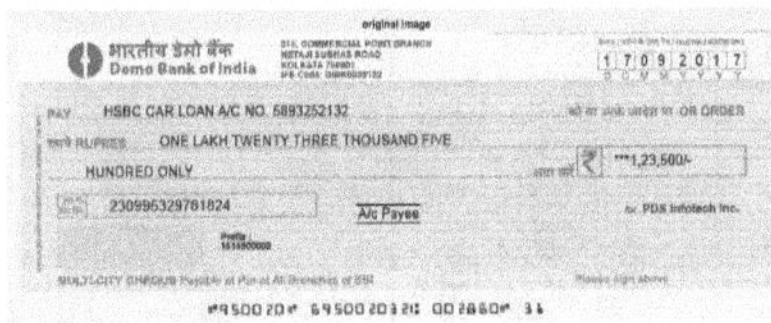	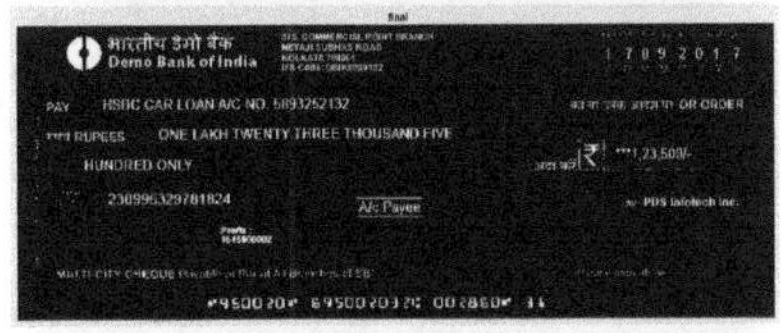
		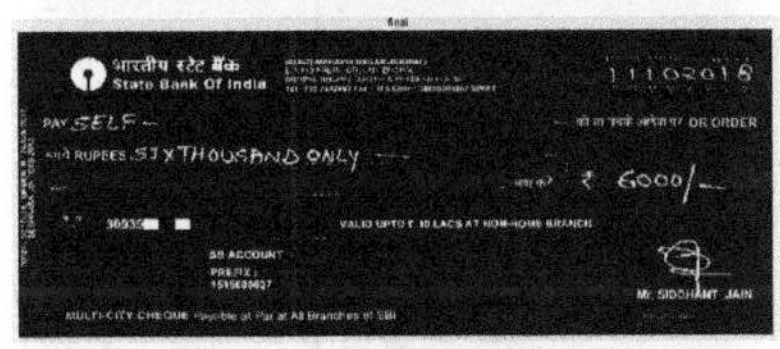
		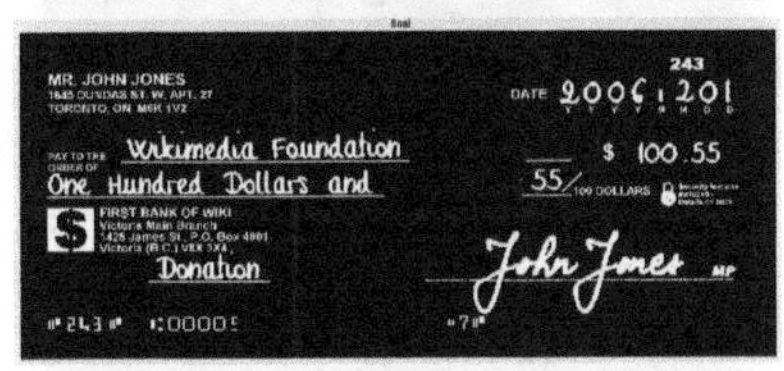
	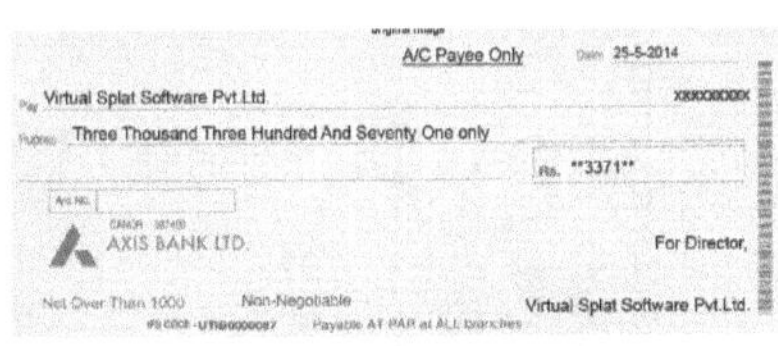	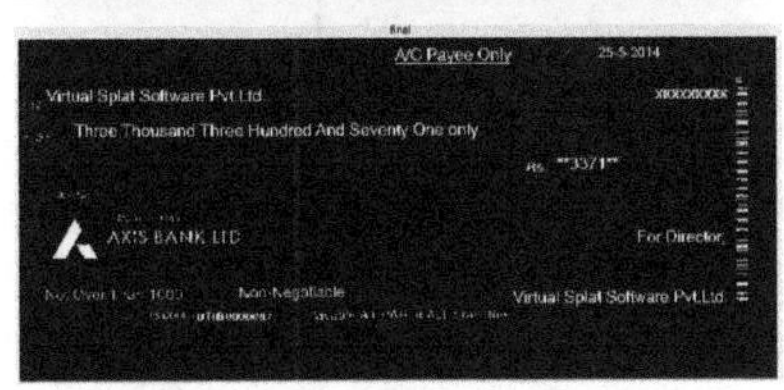

1. The proposed hybrid technique integrates DWT and dynamic thresholding with a logical AND operator.
2. Hyperspectral unmixing methods, specifically Minimum Volume Simplex Analysis (MVSA) and Minimum Volume Enclosing Simplex (MVES), are utilized for signature detection and localization.
3. A template matching technique is employed for bank logo detection.
4. A thresholding method based on MLP is applied to identify the courtesy amount and magnetic character code.

The proposed method's effectiveness has been assessed on both public and own datasets. As indicated in Table 11.3, the PR of the suggested method for the benchmark and own datasets is 2.92%, and it is 4.6% greater than

Table 11.3 Comparison between the proposed method and the state-of-the-art approaches

				Evaluation parameters	
Author (year)	*Techniques*	*Data fields*	*Datasets*	*PR (%)*	*RR (%)*
Rabelo et al. [26] (2011)	Connectionist approach (MLP)	Amount	Brazilian real bank cheque images	81	90
Raghavendra et al. [43] (2015)	Template matching method	Bank logo	Indian real cheque images (SBI, SBM, Corporation, Axis, Canara, Union Bank of India)	93.8	—
Iqbal et al. [51] (2017)	Hyperspectral unmixing method (MVSA, MVES)	Signature	Tobacco-800	92	95
Proposed method	dynamic thresholding, 2D DWT, logical AND operator	Entire data fields	Benchmark dataset	94.92	96.15
			Own created dataset	96.6	99.03

the hyperspectral unmixing method, 1.12% and 2.8% higher than the template matching techniques, and 13.92% and 15.6% higher than the MLP-based thresholding method. The RR for the proposed method is 1.15% and 4.03% greater than the hyperspectral unmixing method and 6.15% and 9.03% higher than the MLP-based thresholding method.

The primary limitations of the hyperspectral unmixing method [51] are: i) it is restricted to signature detection and localization, and ii) it is challenging to extract signature pixels from overlapping areas of bank cheque images. Raghavendra et al. [43] applied a template-matching technique for bank logo detection; however, this method yields unsatisfactory results for damaged bank images, such as those with crossed lines or canceled cheques featuring oblique lines. Rabelo et al. [26] utilized the MLP approach to determine the optimal threshold, but it only identified two data fields: the courtesy amount and magnetic character code in Brazilian bank cheque images.

In contrast, the proposed method can accurately detect and localize all data fields in images of bank cheques. The PR and RR of the proposed wavelet-based method surpass those of existing techniques. Consequently, the 2D DWT edge detection proves to be more accurate in complex backgrounds, while also reducing processing time by simultaneously detecting three types of edges.

11.5 CONCLUSION AND FUTURE SCOPE

Automated dispensation of the bank cheque is a dynamic and popular area within pattern recognition, computer vision, and machine learning. Currently, cheque processing is predominantly manual, which is time-consuming. The push for automation stems from the goal of reducing processing times and enhancing operational efficiency in the banking sector. This chapter presents an effective technique for automatically detecting and extracting various data fields from bank cheque images.

Experimental analysis reveals that both text and non-text regions exhibit edges, but their frequency values significantly differ. In wavelet transforms, text, and non-text regions display distinct variations in texture and edge distribution, with text regions having higher frequency values. Text features abrupt variations and a consistent edge distribution, whereas non-text regions also exhibit abrupt variations but with differing distributions.

To differentiate between non-text and text areas, the edges of the input image are first detected using 2D DWT. Dynamic thresholding is then employed to automatically select optimal threshold values for each image, facilitating the removal of non-text regions and enhancing the quality of the extracted text. Morphological dilation operators link each detail component sub-band of the transformed binary image, and the logical AND operator is utilized to obtain the actual text regions by overlapping the three types of dilated edges. Area-based filtering further refines the localization of data fields within the bank cheque images.

This method has been tested on both standard and custom datasets and compared against previously reported techniques using various evaluation parameters. The Haar wavelet was chosen for experimentation due to its simplicity, speed, symmetry, orthogonality, and binary-like behavior (coefficients of either 1 or −1). Performance analysis indicates that this method is efficient and robust for removing data fields from bank cheque images.

However, some limitations were identified: (i) the presence of the delimiter '/' in the courtesy amount complicates segmentation and digit recognition; (ii) images can become blurred with changes in camera angle; and (iii) high illumination conditions affect processing. Future research will focus on addressing these challenges.

REFERENCES

[1] J. Eduardo, B. Santos, F. Bortolozzi, E.R. Sabourin, R.I. Conceio, Background bank cheques elimination: A simple and efficient solution, in: *Advances in Document Image Analysis*, pp. 334–343, 1997.

[2] R. Jayadevan, S.R. Kolhe, P.M. Patil, Dynamic time warping based static hand printed signature verification, *Journal of Pattern Recognition Research*, vol. 1, pp. 52–65, 2009.

[3] M. Mehta, R. Sanchati, A. Marchya, Automatic cheque processing system, *International Journal of Computer and Electrical Engineering*, vol. 2, pp. 761–765, 2010.

[4] C.Y. Suen, Q. Xu, L. Lam, Automatic recognition of handwritten data on cheques–fact or fiction?, *Pattern Recognition Letters*, vol. 20, pp. 1287–1295, 1999.

[5] N. Thakur, D. Ghai, S. Kumar, Automatic imagery Bank Cheque data extraction based on machine learning approaches: A comprehensive survey, *Multimedia Tools and Applications*, vol. 106, 2023, doi: 10.1007/s11042-023-14534-7

[6] M.S. Akhtar, H.A. Qureshi, Handwritten digit recognition through wavelet decomposition and wavelet packet decomposition, in: *8th International Conference on Digital Information Management*, IEEE, Islamabad, Pakistan, September 10–12, 2013, pp. 143–148.

[7] F. Coelho, L. Batista, L.F. Teixeira, J.S. Cardoso, Automatic system for recognition of amounts in handwritten cheques, in: *Proceedings of the International Conference on Signal Processing and Multimedia Applications*, Porto, Portugal, July 26–29, 2008, pp. 320–324.

[8] K. Liu, C.Y. Suen, C. Nadal, Automatic extraction of items from cheque images for payment recognition, in: *Proceedings of the 13th International Conference on Pattern Recognition*, IEEE, Vienna, Austria, August 25–29, 1996, pp. 798–802.

[9] R. Palacios, A. Sinha, A. Gupta, Automatic processing of Brazilian bank checks, *Machine Vision and Applications*, vol. 81, pp. 1–28, 2002.

[10] J. Sadri, M.J. Jalili, Y. Akbari, A. Foroozandeh, Designing a new standard structure for improving automatic processing of Persian handwritten bank cheques, *Pattern Analysis, and Applications*, vol. 17, pp. 849–862, 2014.

[11] Y. Akbari, M.J. Jalili, J. Sadri, K. Nouri, I. Siddiqi, C. Djeddi, A novel database for automatic processing of Persian handwritten bank checks, *Pattern Recognition*, vol. 74, pp. 253–265, 2018.

[12] R. Jayadevan, U. Pal, F. Kimura, Recognition of words from legal amounts of Indian bank cheques, in: *2010 12th International Conference on Frontiers in Handwriting Recognition*, Kolkata, India, 2010, pp. 166–171, doi: 10.1109/ICFHR.2010.33

[13] Z. Honggang, C. Guang, L. Gang, G. Jun, Bank check image binarization based on signal matching, in: *5th International Conference on Information Communications and Signal Processing*, IEEE, Bangkok, Thailand, December 6–9, 2005, pp. 1430–1433, doi: 10.1109/ICICS.2005.1689294

[14] World Payments Report 2022, https://www.capgemini.com/insights/research-library/world-payments-report/

[15] 2017 Annual Cheque Usage Statistics https://www.chequeandcredit.co.uk/news/2017-annual-cheque-usage-statistics-february-2018

[16] From Cash to Non-cash and Cheque to Digital: The Unfolding Revolution in India's Payment Systems, Reserve Bank of India https://www.rbi.org.in/Scripts/MSM_Mintstreetmemos7.aspx

[17] B.A.N. Raju, D. Ghai, S.L. Tripathi, S.K. Nanda, S.M.N. Islam, Predictive analytics for marketing and sales of products using smart trolley with automated billing system in shopping malls using LBPH and faster R-CNN, in: *Intelligent Techniques for Predictive Data Analytics*, 2024.

[18] R. Jayadevan, S.R. Kolhe, P.M. Patil, U. Pal, Automatic processing of handwritten bank cheque images: A survey, *International Journal on Document Analysis and Recognition*, vol. 15, pp. 267–296, 2012.

[19] D.A. Kumar, S. Dhandapani, A bank cheque signature verification system using FFBP neural network architecture and feature extraction based on GLCM, *International Journal of Emerging Trends & Technology in Computer Science*, vol. 3, pp. 46–52, 2014.

[20] V.K. Madasu, B.C. Lovell, Automatic segmentation and recognition of bank cheque fields, in: *Digital Image Computing: Techniques and Applications*, IEEE, Queensland, Australia, December 6–8, 2005, pp. 33–38, doi: 10.1109/DICTA.2005.18

[21] G. Dimauro, S. Impedovo, G. Pirlo, A. Salzo, Automatic bank check processing: A new engineered system, *International Journal of Pattern Recognition and Artificial Intelligence*, vol. 11, pp. 467–504, 1997.

[22] J. Zhou, C.Y. Suen, K. Liu, A feedback-based approach for segmenting handwritten legal amounts on bank cheques, in: *Proceedings of the 6th International Conference on Document Analysis and Recognition*, IEEE, Seattle, WA, USA, September 13, 2001 pp. 887–891, doi: 10.1109/ICDAR.2001.953914

[23] R.F.P. Neves, C.A.B. Mello, M.S. Silva, A new algorithm to threshold the courtesy amount of Brazilian bank checks, in: *International Conference on Systems, Man and Cybernetics*, IEEE, Singapore, October 12–15, 2008, pp. 1226–1230, doi: 10.1109/ICSMC.2008.4811450

[24] D. Guillevic, C.Y. Suen, Recognition of legal amounts on bank cheques, *Pattern Analysis and Applications*, vol. 1, pp. 28–41,1998, doi: 10.1007/BF01238024

[25] A. Zimmer, L.L. Ling, A window-based hybrid signature verification system, in: D. Zhang, A.K. Jain (eds.) *Biometric Authentication*. Lecture Notes in Computer Science, vol. 3072. Springer, Berlin, Heidelberg, 2004, doi: 10.1007/978-3-540-25948-0_77

[26] J.C.B. Rabelo, C. Zanchettin, C.A.B. Mello, A multi-layer perceptron approach to threshold documents with a complex background, in: *International Conference on Systems, Man, and Cybernetics*, IEEE, Anchorage, USA, October 9–12, 2011, pp. 2523–2530.
[27] P. Dansena, K.P. Kumar, R. Pal, Line-based extraction of important regions from a cheque image, in: *8th International Conference on Contemporary Computing*, IEEE, Noida, India, August 20–22, 2015, pp. 183–189, doi: 10.1109/IC3.2015.7346676
[28] Q. Xu, L. Lam, and C.Y. Suen, Automatic segmentation and recognition system for handwritten dates on Canadian bank cheques, in: *7th International Conference on Document Analysis and Recognition*, IEEE, Edinburgh, UK, August 6, 2003, pp. 704–708, doi: 10.1109/ICDAR.2003.1227753
[29] M. Wang, Y. Chen, X. Wang, Recognition of handwritten characters in Chinese legal amounts by stacked autoencoders, in: *22nd International Conference on Pattern Recognition*, Stockholm, Sweden, August 24–28, 2014, pp. 3002–3007.
[30] K.K. Kim, J.H. Kim, Y.K. Chung, C.Y. Suen, Legal amount recognition based on the segmentation hypotheses for bank check processing, in: *Proceedings of the 6th International Conference on Document Analysis and Recognition*, IEEE, Seattle, USA, September 10–13, 2001, pp. 964–967.
[31] C.V. Khatode, S.S. Morade, Verification of signatures on bank cheques, *International Journal of Science, Engineering, and Technology Research*, vol. 4, pp. 1098–1101, 2015.
[32] S. Souza, J.M. Abe, K. Nakamatsud, MICR automated recognition based on paraconsistent artificial neural networks, *Procedia Computer Science*, vol. 22, pp. 1083–1091, 2013.
[33] A. Ali, R. Pal, Detection and extraction of pantograph region from bank cheque images, in: *3rd International Conference on Signal Processing and Integrated Networks*, IEEE, Noida, India, February 11–12, 2016, pp. 498–501, doi: 10.1109/SPIN.2016.7566746
[34] Y.C. Kiran, N. Nagaraj, Offline signature verification using neural network, *International Research Journal of Engineering and Technology*, vol. 5, pp. 4206–4211, 2018.
[35] M. Morita, R. Sabourin, F. Bortolozzi, C.Y. Suen, Segmentation and recognition of handwritten dates, in: *Proceedings of the 8th International Workshop on Frontiers in Handwriting Recognition*, IEEE, Ontario, Canada, August 6–8, 2002, pp. 105–110.
[36] T.A.V. Nguyen, P. Nguyen, H.S. Le, An efficient method to extract data from bank statements based on image-based table detection, in: *15th International Conference on Advanced Computing and Applications*, Ho Chi Minh City, Vietnam, 2021, pp. 186–190, doi: 10.1109/ACOMP53746.2021.00033
[37] B. Oral, E. Emekligil, S. Arslan, G. Eryiğit, Information extraction from text intensive and visually rich banking documents, *Information Processing & Management*, vol. 57, 2020.
[38] P. Agrawal, D. Chaudhary, V. Madaan, A. Zabrovskiy, R. Prodan, D. Kimovski, C. Timmerer, Automated bank cheque verification using image processing and deep learning methods. *Multimedia Tools and Applications*, vol. 80, pp. 5319–5350, 2021, doi: 10.1007/s11042-020-09818-1
[39] M.B.A. Miah, M.A. Yousuf, M.S. Mia, M.P. Miya, Handwritten courtesy amount and signature recognition on bank cheque using neural network, *International Journal of Computer Applications*, vol. 118, pp. 21–26, 2015.

[40] S. Singh, T. Kariveda, J.D. Gupta, K. Bhattacharya, Handwritten words recognition for legal amounts of bank cheques in English script, in: *8th International Conference on Advances in Pattern Recognition*, IEEE, Kolkata, India, January 4–7, 2015, pp. 1–5, doi: 10.1109/ICAPR.2015.7050716
[41] G.G. Nancy, Signature processing in handwritten bank cheque images, *International Journal on Recent and Innovation Trends in Computing and Communication*, vol. 2, pp. 1239–1243, 2014.
[42] Grayscale to RGB Conversion https://www.tutorialspoint.com/dip/grayscale_to_rgb_conversion
[43] S.P. Raghavendra, A. Danti, A novel recognition of Indian bank cheques based on invariant geometrical features, in: *International Conference on Trends in Automation, Communications and Computing Technology*, IEEE, Bangalore, India, December 21–22, 2015, pp. 1–5, doi: 10.1109/ITACT.2015.7492682
[44] D. Ghai, N. Jain, Comparative analysis of multi-scale wavelet decomposition and k-means clustering based text extraction, *Wireless Personal Communications*, vol. 109, pp. 455–490, 2019.
[45] S.N. Gangiredla, M.K. Yadav, V. Tiwari, D. Ghai, D. Mandal, Design and Implementation of Smart Text Reader System for People with Vision Impairment, in: *International Conference on Signal & Data Processing*, Springer Nature, Singapore, pp. 593–605.
[46] D. Ghai, D. Gera, N. Jain, A new approach to extract text from images based on DWT and K-means clustering, *International Journal of Computational Intelligence Systems*, vol. 9(5), pp. 900–916, 2016.
[47] D. Ghai, N. Jain, Comparison of different text extraction techniques for complex color images, in: *Machine Learning Algorithms for Signal and Image Processing*, 2022, pp. 139–160.
[48] D. Ghai, S.L. Tripathi, S. Saxena, M. Chanda, M. Alazab, *Machine Learning Algorithms for Signal and Image Processing*, John Wiley & Sons, 2022.
[49] D. Ghai, S. Saxena, G. Dhingra, S.L. Tripathi, A comprehensive review on performance-based comparative analysis, categorization, classification and mapping of text extraction system techniques for images. *Multimedia Tools and Applications*, vol. 84, pp. 1–158, 2024.
[50] Precision Rate and Recall Rate https://towardsdatascience.com/accuracy-precision-recall-or-f1-331fb37c5cb9
[51] K. Iqbal, K. Khurshid, Automatic signature extraction from document images using hyperspectral unmixing, *Proceedings of the Pakistan Academy of Sciences: Pakistan Academy of Sciences, A. Physical and Computational Sciences*, vol. 54, pp. 269–276, 2017.

Chapter 12

Vehicle accident detection and smart notification systems

Archana Kanwar

12.1 INTRODUCTION

In this twentieth century, the number of automobiles is increasing rapidly. There are more accidents since there are more automobiles on the road. Traffic heterogeneity and a shortage of traffic separation are the main causes of road accidents. Every day, we read in the newspapers or hear on the news about thousands of people passing away in vehicle accidents. The reason is not only insufficient medical care or a lack of effective medications but also because the victims' families, hospital administrators, blood bank administrators, police officers, etc., were not quickly informed [1]. Many people die every day in the world due to delayed communication. People are put at risk by their excessive speed. As per World Health Organization (WHO), India is the nation where most people die in traffic accidents. In India, there were 13 million traffic-related accidents between 2014 and 2015. From 1,33,201 in 2020 to 1,55,622 in 2021, the number of deaths in traffic accidents has grown by 16.8%. According to the NCRB data, the number of deaths per 1,000 cars increased from 0.45 in 2020 to 0.53 in 2021. Even though many incidents go unreported, these numbers are presented as accident statistics. Hence, the true number of accidents is higher compared to the World Health Organization reports (WHO). The current approach primarily focuses on passenger safety but does not provide immediate help in the accident.

In recent years, various vehicle tracking, accident monitoring, and notification systems have been produced, all of which rely on different techniques and use different control units and sensors to control the operation of the system and record the value of the parameter on which the operation is based or use different communication protocols to communicate [2–4]. Many systems focus on prophylactic measures and accident detection, such as speed control, driver's license verification, etc. This may be accomplished via a variety of techniques, including smartphones, VANETs, accelerometers, strain gauges, GPS, GSM, and various machine learning techniques [5]. Although there are several research on accident detecting and avoiding techniques, there is no comprehensive review. Systems for pinpointing and reporting accidents are now being implemented. The major objective is to avoid

DOI: 10.1201/9781003643173-12

accidents by employing wireless methods of communication to send a notification to authorized smartphones, hospitals, and police headquarters [6]. When an accident happens somewhere, sensor gets activated and send the data to the registered cell phone immediately with the help of GSM module. Further, the accident site can be identified based on GPS technology [7, 8].

12.2 LITERATURE SURVEY

This study focus on studying the issue of the monitoring system, precluding, and informing systems to produce a broad scope of previous techniques to develop efficient systems that utilize benefits while correcting the weaknesses of current systems. This chapter aims to present some of these designed and proposed systems for accident detection, notification, vehicle tracking and accident prevention technology. These systems, when practically integrated and implemented in vehicles, can save lives by informing the families of the accident victims or contacting the nearby hospital and providing the exact location of the accident. In this chapter, several systems were critically reviewed in terms of their benefits, drawbacks, restrictions, and application areas in the future. In this chapter, systems using different microcontrollers like Arduino Uno, ESP8266 Node MCU and R-Pi (Raspberry Pi) based on technologies like GSM, Wi-Fi, GPS have been discussed in following sections. These systems use various measurement devices such as accelerometers and vibration sensors and establish communication through various communication protocols such as HTTP (Hypertext Transfer Protocol) and MQTT (Message Queuing Telemetry Transport) [9].

12.3 ACCIDENT PREVENTION SYSTEMS

Vehicle accident monitoring and reporting are very essential nowadays due to an enlargement in accident rates. Implementing GPS technology allows this simple to do. Many more tools also can be used to do this. Moreover, fleet management, anti-theft car techniques, and accident analysis all make use of these technologies. The following list of accident prediction systems is provided.

12.3.1 Vehicle tracking system

A system for vehicle tracking seeks to combine installing electronic equipment in a vehicle, or a group of vehicles, with the objective of developing software applications with at least one standard to allow the vehicle owner or any other authorized person to check the location of a vehicle, gather information and share it for the location of operation as shown in Figure 12.1. GPS technology is usually used by vehicle trackers to locate the vehicle. Through the Internet or software applications, one may check vehicle

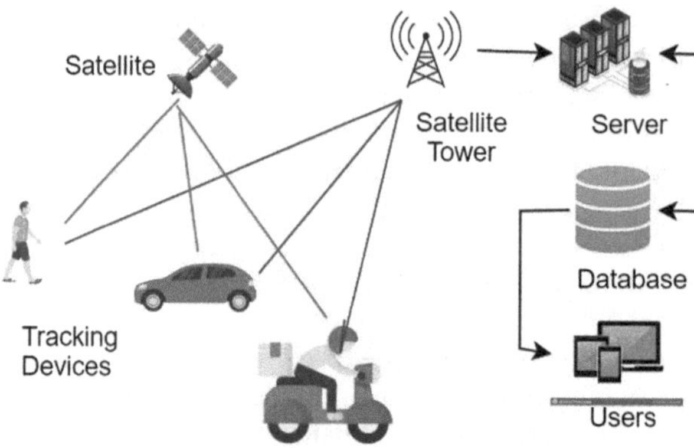

Figure 12.1 Vehicle tracking system.

location on a digital map. Several vehicle monitoring systems allow remote control of a vehicle's functions, such as the ability to disable an engine or close or open the door in case of emergency. The insurance cost can also be reduced by using vehicle monitoring devices [10].

12.3.2 Camera processing system

In many countries, modern tech called picture processing has been implemented. The majority of countries utilize this technology to reduce traffic. Vehicles can also be tracked using several algorithms, and the first camera can record their movement too [11].

12.3.3 Location detector and immediate recovery of accident

As is already widely known, it is possible to locate where a car incident happened in a number of ways. When accidents occur, it is challenging to bring help to a victim since neither the hospitals nor the police, nor has the victim's family received notice of the accident. Therefore, there has been a significant loss of human life. To get rid from this, SMS can be automatically send to the system's designated numbers. The sensors trigger the GPS tracker using Bluetooth Technology as a medium. It stands in the middle of the sensors and the GPS and works like a bridge. However, MESA technology may now activate GPS and broadcast the position coordinates to designated numbers in addition to Bluetooth technology [12].

12.3.4 VANET system

According to Kandekar [5] a wireless technology known as a VANET uses wireless connections placed upon every node (vehicle) to communicate information. As the nodes in VANET connect throughout a middle node that is located within their wireless range, every node serves as both the network's candidate and router. An autonomous network is VANET. It is independent of any kind of permanent communications infrastructure. Although some permanent nodes serve as gateways to the web or roadside devices to facilitate vehicle networking for providing geographic information. Figure 12.2 clearly shows that how VANET get information of accident. The major features of VANET are a set of perceptions of mobility, speed, and rapid pattern movement. According to Patel [13], this also results in quick changes in network structure. Vehicles perform as nodes in a specific sort of MANET that is also known as VANET. In distinction to MANET, cars travel along appointed roads, and their rate of speed is specified by speed signs. In addition, all automobiles must tolerate traffic signs and signals [14].

This study has been categorized in 3 parts for this accident identification and prevention techniques as shown in Figure 12.3 as detection techniques, prevention techniques and hybrid techniques. These are further subcategorized in controller based and AI based techniques.

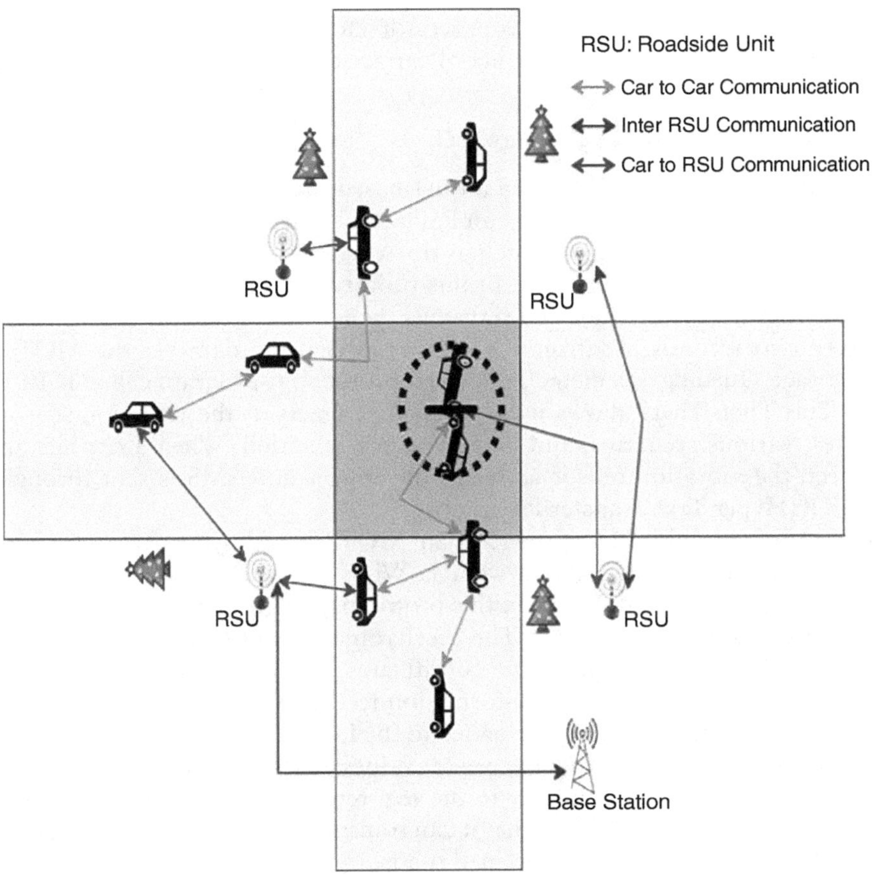

Figure 12.2 VANET sample of an accident report.

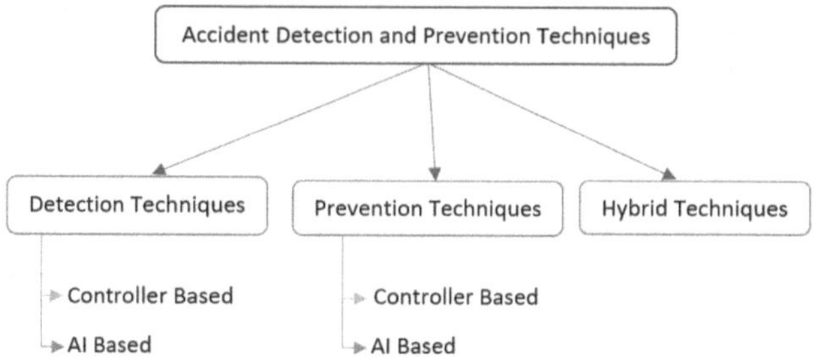

Figure 12.3 Classification of several methods for preventing and detecting accidents.

12.4 ACCIDENT DETECTION TECHNIQUES

12.4.1 Controller based accident detection techniques

Several researchers have carried out by using various methods like NodeMCU, Arduino, Microcontroller microprocessor etc., and their research on the detection, monitoring, and alerting of car accidents as discussed below.

12.4.1.1 Using ESP8266 NodeMCU

Bansal and Garg [15] developed an IoT-based prototype system for detecting and notifying accident using an ESP8266 NodeMCU as the brain and a simple and direct vibration switch as the sensor system. This prototype system employs Wi-Fi technology. In this prototype vibration sensor continually senses the disturbance and transmits them to the NodeMCU controller. Every two seconds, Adafruit IO receives all sensing data via the MQTT (Message Queuing Telemetry Transport) protocol. A program called IFTTT (IF This Then That) allows users to create triggers at the vibration sensor to test various scenarios, and take appropriate action. When a car hits an object, the vibration sensor activates the trigger and SMS is sent through HTTP (Hyper Text Transfer Protocol).

Kodali and Sahu [16] proposed an MQTT-based system for accident detection. The basis of this system is Wi-Fi technology. The controlling mechanism is the NodeMCU. In this prototype, an accelerometer module is employed as a sensing device. The accelerometer continually measures the vehicles' X, Y, and Z orientation coordinates, and the NodeMCU uses the MQTT protocol to send that information to the LOSANT cloud platform. These values are continually uploaded to the LOSANT dashboard, and once it notices any deviation from the pre-set values, it sends accident coordinates through an email automatically to the registered email account. The major drawback of this prototype is that it can transmit false alert messages due to speed bumps and poorly maintained roads.

12.4.1.2 Using Arduino

Basheer *et al.* [2] proposed an IoT prototype for detecting and notifying about an accident. This prototype system creates technologies based on the GSM (Global System for Mobile) protocol, vibration sensor, which is powered by the Arduino Uno R3. If the received vibration sensor value exceeds the pre-set value, a message about accident coordinates is sent to the registered mobile numbers. A GPS (Global Positioning System) module has also been used to locate the exact coordinates of the place of the accident. The main disadvantages of this system are that it can send alarms even if the accident is not occurred due to ill-maintained roads and speed breakers and also sending of information isn't secure. A sample of accident detection and notification techniques is shown in Figure 12.4.

Singh *et al.* [17] described techniques for accident detection by improving the response time in care of emergency. They have used Arduino mega as a main controller and for accident detection, vibrator sensor and a gyro sensor to determine accident coordinates. Whenever the measured value will be greater than the pre-set value the system will detect an accident then GPS will send location to registered user's mobile number. They have provided an additional switch to stop sending message when accident will not major. Bari *et al.* [18] proposed accident detection system using Arduino UNO. They have installed two infrared sensors on the roadside to determine the

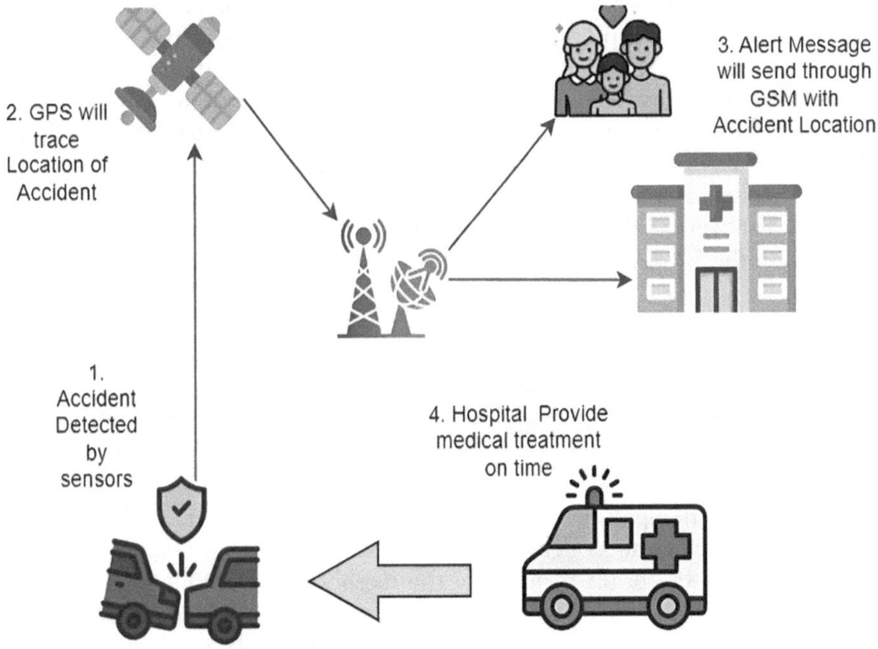

Figure 12.4 Accident detection using GSM and GPS.

location of vehicle. Then, LEDs mounted on either side alert drivers, monitoring the vehicle movement at the turn. Here Vibration sensor is used to detect accidents and GPS module to get the coordinates of that place and a GSM module to send alert notifications to the police stations, rescue team, and family members.

12.4.1.3 Using Raspberry Pi

Rani *et al.* [8] proposed an IoT-based prototype for accident detection and tracking using Raspberry Pi and vibration sensor. If the vibration sensor detects a collision or any barrier in the route of the vehicle, it sends an alert to the controller or R-pi3. The location of the accident are detected and sent to the GPS receiver. These location details are sent through WhatsApp messages to the registered mobile numbers. The user would access these coordinates online using their mobile phone. Python programming was used to create this prototype raspberry pi. Karande *et al.* [19] also developed a raspberry pi-based system. The alarm message is sent using both GSM and Wi-Fi; if one stops working, the other could still operate. The main benefits of this prototype are the camera module, which provides images of the surrounding area and may be highly useful in locating the location and in police investigations. An accelerometer is used by this system to identify accidents.

Kumar and Deepak [20] proposed a system to detect accidents and transmit an alert message which does not require any human interference before or after the accident is pinpointed. GPS, a Raspberry Pi, an Android Apps, and a gyroscope are used as major modules. The system also gathers data on the driver's blood groups. The gyro, that is fitted in the automobiles, monitors the rotation angle and as soon as an accident takes place, or crossed the threshold limit, it sends a notification to the server. The application, which is installed in ambulance will constantly get details from the server and assists the ambulance in discovering the accident spot by directing it along a less-crowded route using the Google Maps API (Application Programming Interface).

12.4.1.4 Using ARM microprocessor

Jose *et al.* [21] developed an embedded system for automatic crash prediction and notification technology as shown in Figure 12.5. Here, they have given an example of a concept for a combined system that could be effectively used in automobiles. Three apps are being combined into a single module that is being run on an ARM11 CPU. They have used ARM 11 and sonar detectors to collect distance between two vehicles, while ARM 11 functions as the brain to identify the likelihood of a collision. In order to provide assistance to accident victims as soon as possible, the GSM/GPS module will send an alert notification to the closest police station or hospital. The location of the vehicles is included in this alert message, ensuring that the rescuers will be immediately reached.

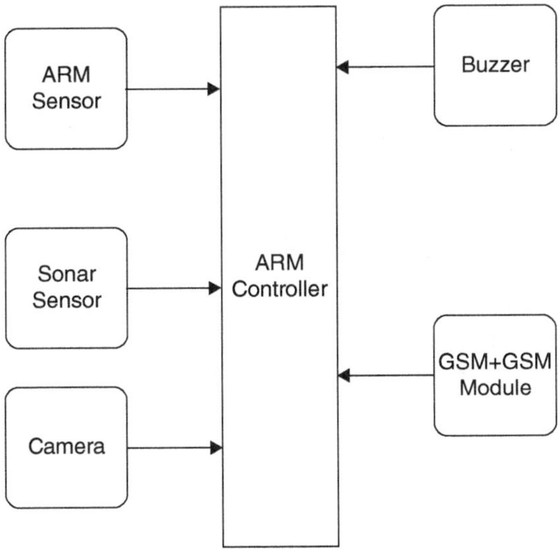

Figure 12.5 Schematic model for crash predication and alerting module.

Wen and Meng [22] proposed a design to detect accident vehicles using ARM microprocessor LPC2129 as the main controller of the system. GPS module and GMS module have been used in this system for tracking the location of vehicles and sending notifications to authorized persons respectively. The specified individual sends an SMS request to receive the exact site of the vehicles. The GSM module gets the request and approves it. Then the Spartan processor processes the request. The CPU delivers the message to the GPS module. The design has a weakness since external environmental conditions might have a huge impact on a system's performance.

12.4.1.5 Using other microcontrollers

Tushara and vardhini [23] developed a system to identify accidents using Atmel microcontroller AT89S52 to handle functions like detecting and notifying. The design tries to lower the time needed to take action whenever an accident is detected. This device pre-fetches a contact number to which an alarm is delivered on the occurrence of an accident. A vibration sensor has been employed to locate the accident. Because it just uses one sensor, the system faces a significant chance of delivering a faulty result. The system's fault is that, although it transmits an attentive message when an accident occurs, the accident place has not been disclosed. Patil *et al.* [24] proposed a device to get an accident details. This device will identify accidents and will notify the closest police station and emergency teams. The emergency squads receive messages via GSM. In the case of accidents, the device can instantly notify emergency services because it regularly monitors the car.

Renessa's microcontroller has been used here to connect the GSM and GPS system. The site specified by GPS is mentioned in an alert message transmitted through GSM for communications. GPS, GSM, and piezoelectric sensors make up its primary modules. The technology rapidly monitors the location to ensure that in the situation of an accident, the place may be rapidly reported. The biggest disadvantage of this technique is that there is no switch present to quit sending warning messages in the matter of no major injury, and also, because the site is regularly recorded, much energy may be wasted. Megalingam *et al.* [25] developed a prototype to detect accidents and inform as soon as possible to rescue team by processing and computing the accelerometer readings using PIC16F877A microcontroller. This system can be installed easily in any vehicle. An RF (Radio Frequency) module has been used to transmit information to the rescue team. The only disadvantage in this device is that the RF module has a short range. Amin *et al.* [26] proposed a system that considers speed as the main reason for accidents. They have used a GPS receiver as a speed monitoring system. It constantly checks the speed of the vehicle and whenever the speed will be higher than pre-set value then it will consider the accident has occurred and it will transmit an alert notification using the GSM module to emergency services. They used a PIC18F4550 microcontroller as the brain of the device. The main drawback of this design is that it can transmit false messages as it has used only one sensor for accident detection. Chandran *et al.* [27] developed a smart helmet and they named that helmet "Konnect". In this, they have used Wi-Fi module to connect with the cloud. An ABMA222 tri-axial accelerometer is used to detect x, y, and z coordinates, and a GPS module to get the location of the person. A microcontroller has been used for processing purposes and a threshold value is pre-set in it. Whenever a sudden change crosses the pre-set value, the controller will detect an accident and immediately send alert information with the help of the cloud to the authorized person without any delay.

12.4.2 AI Based Accident Detection Techniques

12.4.2.1 Using Fuzzy Logic

Alkandari *et al.* [28] represented a system to get accident details in the field of traffic lights with the use of a fuzzy logic method that notices accidents. The method is composed of two subunits: the identification technique and the notification system. The technique is established on the Webster Method, with so many improvements. The systems gather information regarding different-different sectors, such as the growth in the number of vehicles in a lane, the speeds of automobiles in certain lanes, and so on. The system's main components are crisp input and output, membership processes, fuzzy method, and linguistic variables. The linguistic factors cross proportion, zone condition, accident position, and section speed assist in specifying an accident. The system's response is constructed by utilizing fuzzy logic to linguistic inputs; regulated by this input and logic, proper steps are executed to enhance traffic flow.

12.4.2.2 Using SVM (Support Vector Machine)

Pan and Wu [29] presented a technique to detect accidents with the help of mobile sensors and SVM. Their procedure was especially concentrated on identifying accidents on urban roads because these roads are more likely to cause an accident because of the existence of numerous flow-disturbing entities e.g., bus stops, traffic control signals, and so on, as opposed to freeways where road traffic is not interrupted. This method utilizes VANETs, where all vehicles might capture traffic patterns such as location, identification, speed, and lane situation using vehicle-mounted sensors. The traffic information is recorded by RSU (Roadside Unit) and transmitted via On Board Units for some more processing. To identify an accident, three traffic features are considered: vehicle speed, acceleration, and route status. On-board mobile sensors can provide a wider range of observation.

12.4.2.3 Using Artificial Neural Network

Ghosh *et al.* [30] designed a system that gathered video footage from CCTV posted on streets to detect an accident. Every frame of footage is supplied to the Convolutional Neural Networks (CNN) model, which can differentiate between accident and non-accident frames of footage. The Raspberry Pi B+ Model is utilized in this as a remote computer to connect CCTV. The Inception v3 model is utilized to identify accidents after being instructed on different pairs of video data that contain accident and non-accident frames. Individually video frame is screened by the model, and a decision is made based on whether the frame is an accident frame or not. A prophecy boundary is pre-set, and if it is crossed, the Raspberry Pi detects an accident and it activated the GSM module, which transmits an alert notice and accident site coordinate to the nearest emergency services. The weakness of this design is that the system was based on data from severe accident scenarios, therefore it might not capture all accident cases if the accident is minor or small.

12.5 ACCIDENT PREVENTION TECHNIQUES

12.5.1 Controller based accident prevention systems

12.5.1.1 Using arduino

Eduku *et al.* [31] developed a driver drowsiness check system that uses an eye blink system sensor and auto brake system to improve traffic -hazards by decreasing the car speed and stopping the vehicle whenever drowsiness is found in the driver. In this technique, an RF (Radio Frequency) module is used to get information about nearby vehicles and an IR sensor is used to monitor the eye blink. Arduino is used as a controller in this system. Its key elements include an indicator that notifies the driver if indications of drowsiness are found, auto braking systems to decrease the speed of the vehicle and bring it to a stop, and an RF module to transmit warnings to enclosing moving vehicles. The main downside of the technology is the low RF module range. Similarly, Rao and Yellu [32] proposed an Arduino-based accident prevention system using an alcohol sensor to prevent drunken driving as depicted in Figure 12.6. This sensor is used to get information of alcohol in the blood. In the controller, a threshold value is set and whenever the detected alcohol reading is higher than the pre-set value the vehicle will stop. Here GPS is used to detect the location and GMS is used to send alert messages to registered cell phone numbers. This system concentrates on protection to avoid any unsafe conditions. Since it restricts drunk drivers from operating vehicles, the system will also protect the broader public.

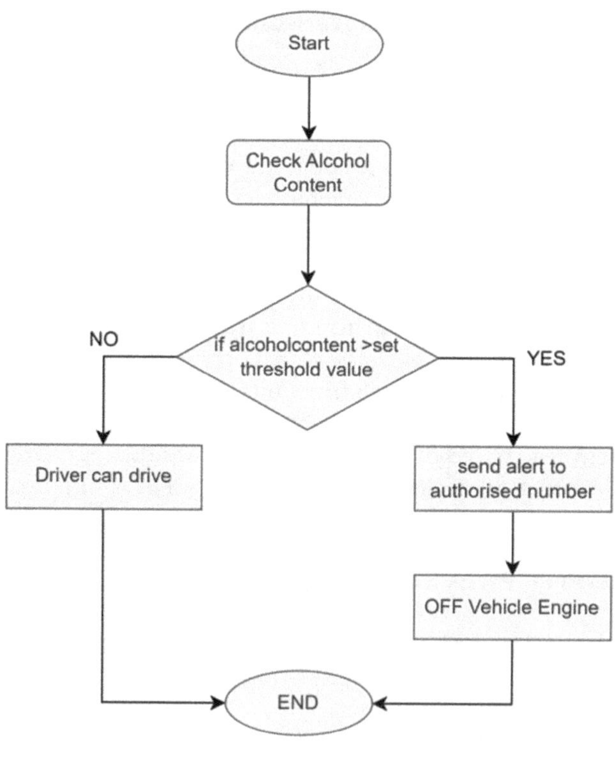

Figure 12.6 Flowchart of alcohol sensor-based system.

12.5.2 AI based accident prevention systems

12.5.2.1 Driver destruction identifier using CNN

Distracted driving is the main reason of road accident. Basubeit *et al.* [33] utilized CNN approach to recognize inattentive driving. Driver destruction can be divided into three groups: visual, manual, and cognitive. The monitoring of careless drivers may be carried out with a variety of techniques, but this methodology focuses primarily on collecting visual data. To monitor the driver's actions, a webcam has been attached inside the vehicle. A deep learning algorithm is applied to pinpoint the difference between inattentive and normal driving. This offers a review of deep CNNs that have been trained. The collection of data is classified into 10 unique classifications to recognize various forms of distraction, such as texting, drinking, conversing with other passengers, etc. The models were initially trained with the Picture net data set before being retrained on the State Farm data set to check distracted driving action.

12.5.2.2 Driver drowsy identifier using representation learning

Dwivedi *et al.* [34] proposed a technique to notify the drowsy condition of driver with the help of representation learning to control accidents happened due to sleep. The features are trained using CNN, and the system collects the visual characteristics from the data. A soft-max classifier layer employs these characteristics to define frames as sleepy or non-sleepy. The face properties are produced using a convolutional neural representation learning method, which combines characteristics of features to represent the difficult interconnections of raw data. The algorithm is divided into two complication layers and a hidden layer with a nonlinear activation function that is connected to the degeneration layer for identification. The video frames are retrieved and sent into a face identifier. The pinpointed faces are resized to square pictures, which are then regularized before being input into a convolutional neural network. The classifier operates well over a broad range of data sets. The methodology has the key advantage that it can retrieve more smart characteristics from raw data input than traditional techniques.

12.5.2.3 Pedestrian identifier using CNN

Szarvas *et al.* [35] presented a CNN-based technique for accident detection. The software is checked on walkers in an urban area. The SVM classifier and the CNN classifier are compared, and the results demonstrate that CNN is much more efficient than SVM when utilizing conventional picture data. The photographs of the applicants are transmitted to the classifier. The CNN provides a classification score for every candidate and the candidates with scores greater than the criteria are kept in the raw identification list.

That list is also arranged as per classifier scores. To get rid of multiple outcomes for only one pedestrian, a multiple detection merging method is used. When the merging process has complete, the final classification list provides the detection system's results. This method obtained high precision through automated feature optimization and regularization of neural networks.

12.6 HYBRID TECHNIQUES

Road accidents are very common these days. In almost all cases, accidents are minor, and if patients are retrieved quickly, their life could be saved. Several hybrid techniques utilize both accident identification and protection techniques as shown in Figure 12.7.

Berade *et al.* [36] proposed a technique for vehicle accident detection as well as checks for seat belt and condition of driver that if he/she is drunk or not based on an alcohol sensor and a seat belt detector. When the sensor notices an accident, microcontrollers transmit data to GPS, which monitors the site and delivers the alert notification through the GPS module. An accident is identified by limit switches installed in the front and at the rear of the vehicle. The system not only identifies accidents but also tries to avoid them. Nanda *et al.* [37] introduced a method that identifies accidents with the use of vibration sensors. This system is designed for bike drivers because they are more prone to major damage because motorcycles lack protective mechanisms as compete with other automobiles. GPS monitors the accident site, and GSM is deployed for transmitting the information. The system also includes a method for figuring out whether the driver is sleeping or not. The technology also checks to see if the driver has drunk alcohol. The camera installed in front of the bike detects traffic lights and automatically reduces speed. To prevent accidents, it utilizes a variety of techniques such as traffic signal identification, and driver condition monitoring. The system concentrates on avoiding accidents and not just reporting them, which assists to save lives. Murshed and Chowdhury [38] proposed a technique that consistently analyses and keeps a safe space between vehicles and objects via distance sensors as depicted in Figure 12.8. When a certain distance is crossed, the suggested method warns the driver to reduce the speed and slows down on its own. When an accident occurs due to an unknown scenario, an email notification with vehicle information is delivered to the responsible individual. The ultrasonic sensor monitors the vehicle's distance. The Raspberry Pi, LED Buzzer, Ultrasonic Sensor, and Servo Motor are all present in the system. It reduces the risk of injury by slowing down the vehicle when a threshold point is crossed.

Ahmed and Jawarkar [39] utilized a mobile phone, a microcontroller, and an accelerometer to alert the user in case of risky driving. It also has the capability to take images when an accident occurs and transmit multimedia to registered contacts. It is environmentally safe since used mobile phones

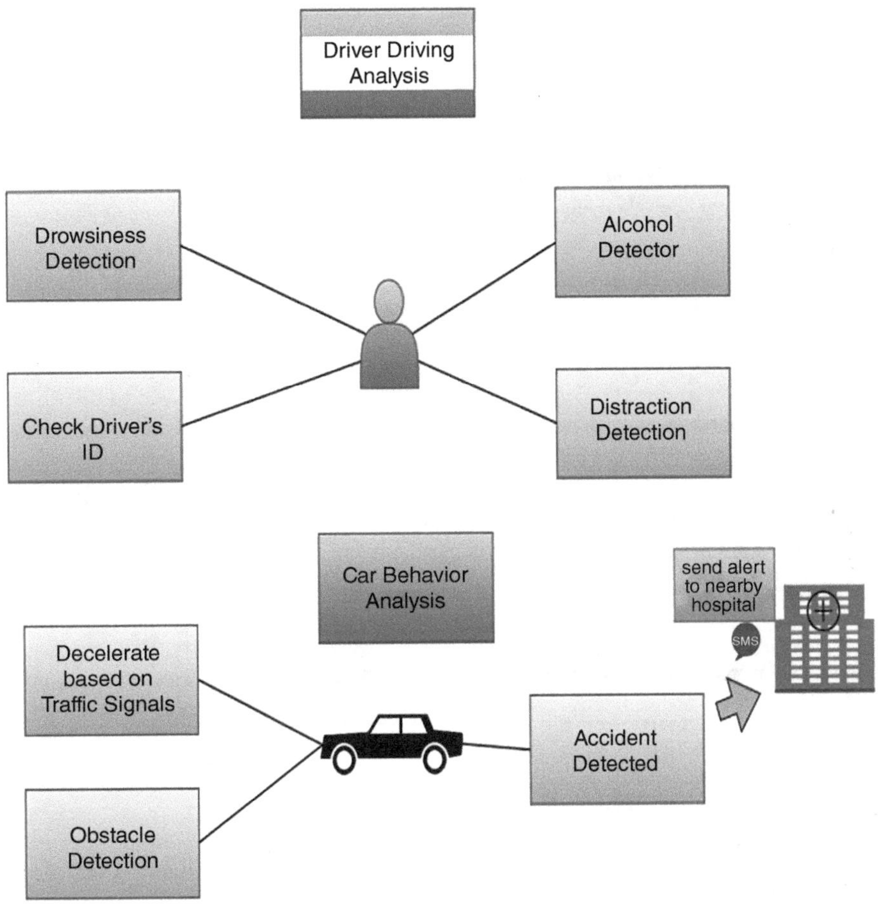

Figure 12.7 Accident identification and protection based Hybrid technique.

are recovered. It is simple to install on present vehicles. The major drawback of this technique is that when incidents happen, alarm messages must be delivered personally by the driver. Priyanka *et al.* [40] used smart helmets as a device for accident identification and prevention. The technology will not authorize the person to drive unless he wears a helmet. The pressure sensor and accelerometer are utilized to identify whether the rider is wearing a helmet or not. The drivers' breathing is analysed for alcohol level by alcohol detectors installed in their helmets. If the levels exceed the pre-set limit value, the bike will not start. In case of accident, as the driver's helmet hits the ground, pressure and accelerometer sensor identifies the movement and instantaneously transmit accident coordinates to the emergency phone numbers and emergency crews. The device could be incredibly beneficial to both the rider and the public, as drunken driving threatens the life of the motorist and other individuals moving on the roads.

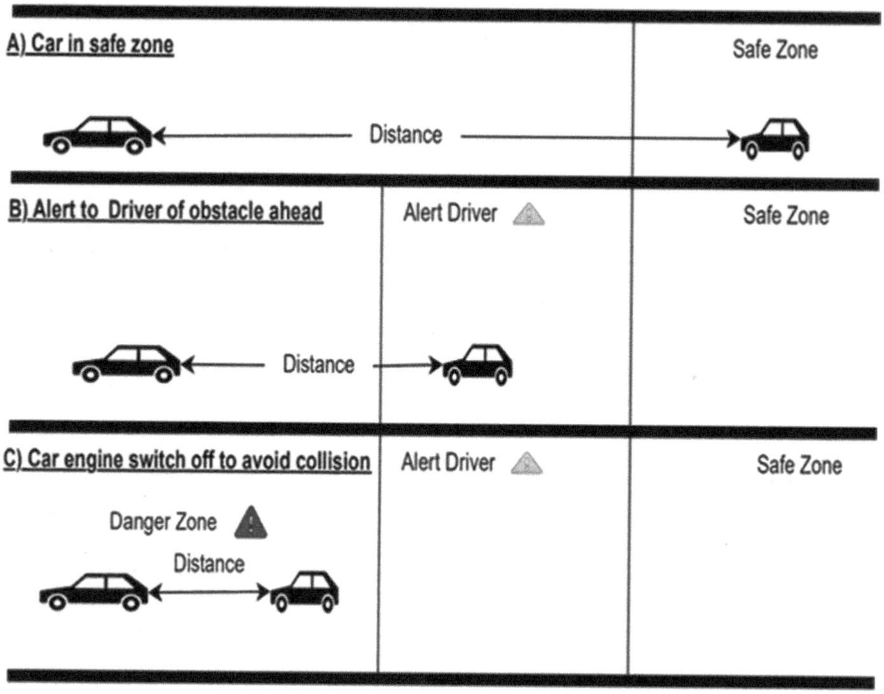

Figure 12.8 Object distance-based accident prevention system.

12.7 CONCLUSIONS

Different accident detection and notification systems have been researched and developed over time. This chapter aimed to pinpoint their benefits and drawbacks using a comparative analysis. For isolated places, a more dependable system might be created to provide speedier medical care. For that combination of GPS and Wi-Fi should be used to alert the local police station and hospitals. To further enhance the driver's security, the device can be interfaced with the vehicle's airbag system and a camera module.

REFERENCE

1. Mohan, D., Tsimhoni, O., Sivak, M., & Flannagan, M. J. (2009). "Road safety in India: Challenges and opportunities".
2. Basheer, F. B., Alias, J. J., Favas, C. M., Navas, V., Farhan, N. K., & Raghu, C. V. (2013, August). "Design of accident detection and alert system for motor cycles". In *2013 IEEE Global Humanitarian Technology Conference: South Asia Satellite (GHTC-SAS)* (pp. 85–89). IEEE.
3. Foroughi, P., Beyranvand, H., Gagnaire, M., & Al Zahr, S. (2020, July). "User association in hybrid UAV-cellular networks for massive real-time IoT

applications". In *IEEE INFOCOM 2020-IEEE Conference on Computer Communications Workshops (INFOCOM WKSHPS)* (pp. 243–248). IEEE.
4. Manuja M et al., (2019). "IoT based on automatic accident detection and rescue management in Vanet". *SSRG International Journal of Computer Science and Engineering (SSRG – IJCSE)*, ISSN: 2348 – 8387. 36–41.
5. Kandekar, S. N. (2012). "Issues and future developments in Ad hoc Routing Protocols in VENET". *International Multidisciplinary e–Journal.*
6. Lee, S., Tewolde, G., & Kwon, J. (2014, March). "Design and implementation of a vehicle tracking system using GPS/GSM/GPRS technology and smartphone application". In *2014 IEEE world forum on internet of things (WF-IoT)* (pp. 353–358). IEEE.
7. Nandaniya, K., Choksi, V., Patel, A., & Potdar, M. B. (2014). "Automatic accident alert and safety system using embedded GSM interface". *International Journal of Computer Applications*, 85(6).
8. Rani, B., Sam, R. P., & Kamatam, G. R. (2018). "A review on vehicle tracking and accident detection system using accelerometer". *International Journal of Applied Engineering Research*, 13(11), 9215–9217.
9. Bansal, B. N., & Garg, V. (2021). "Vehicle Accident Detection, Tracking and Notification Systems-A comparative study".
10. Harshadbhai, P. K. (2013). "Design of GPS and GSM-based vehicle location and tracking system". *International Journal of Science and Research*, 2(1), 165–168.
11. Hendry, M. (2015). *Near Field Communications Technology and Applications.* Cambridge University Press.
12. Verma, P., & Bhatia, J. S. (2013). "Design and development of GPS-GSM based tracking system with Google map-based monitoring". *International Journal of Computer Science, Engineering and Applications (IJCSEA)*, 3(3), 33–40.
13. Patel, M. K. (2012). "Comparative study of vehicular ad-hoc network mobility models and simulators". *International Journal of Computer Applications*, 47(6), 38–43.
14. Grilli, G. (2010). "Data dissemination in vehicular networks". Philosophiæ Doctor (PhD) dissertation in Computer Science and Automation Engineering/ University of Rome Tor Vergata.
15. Bansal, B. N., & Garg, V. (2019). "Development of message queuing telemetry transport (MQTT) based vehicle accident notification system". *International Journal of Engineering and Advanced Technology*, 9(2), 268–273.
16. Kodali, R. K., & Sahu, S. (2017, December). "MQTT based vehicle accident detection and alert system". In *2017 3rd International Conference on Applied and Theoretical Computing and Communication Technology (iCATccT)* (pp. 186–189). IEEE.
17. Singh, J., Velu, V., & Nirmal, U. (2021). "Vehicle Accident Detection System using Internet of Things (VADS -IoT)". *2021 IEEE 11th IEEE Symposium on Computer Applications & Industrial Electronics (ISCAIE)* (pp. 353–359).
18. Bari, A. S., Falalu, M. A., Umar, M. A., Sulaiman, Y. Y., Gamble, A. M., & Baballe, M. A.. (2022). "Accident detection and alerting systems: A review". *Global Journal of Research in Engineering & Computer Sciences*, 2(4), 24–29. https://doi.org/10.5281/zenodo.7063008
19. Karande, I., Deshpande, G., Kumbhar, S., & Deshmukh, A. V. (2016). "Intelligent anti-theft tracking and accident detection system for automobiles

based on the internet of things". *International Journal of Innovative Research in Computer and Communication Engineering*, 4(3), 4142–4149.
20. Kumar, H. N., & Deepak, G. (2017). "Accident detection and intelligent navigation system for emergency vehicles in urban areas using IoT". *International Journal of Engineering and Technology*, 3(6), 330–334.
21. Jose, S. K., Mary, X. A., & Mathew, N. (2013). "Arm 11-based accident alert and vehicle tracking system". *International Journal of Innovative Technology and Exploring Engineering (IJITEE)*, 2(4).
22. Wen, Z., & Meng, J. (2011, May). "Design of vehicle positioning system based on ARM". In *2011 International Conference on Business Management and Electronic Information* (Vol. 4, pp. 395–397). IEEE.
23. Tushara, D. B., & Vardhini, P. H. (2016, March). "Wireless vehicle alert and collision prevention system design using Atmel microcontroller". *In 2016 International Conference on Electrical, Electronics, and Optimization Techniques (ICEEOT)* (pp. 2784–2787). IEEE.
24. Patil, U., More, P., Pandey, R., & Patkar, U. (2017). "Tracking and recovery of the vehicle using GPS and GSM". *International Research Journal of Engineering and Technology*, 4(3), 2074–2077.
25. Megalingam, R. K., Nair, R. N., & Prakhya, S. M. (2010, September). "Wireless vehicular accident detection and reporting system". In *2010 International Conference on Mechanical and Electrical Technology* (pp. 636–640). IEEE.
26. Amin, M. S., Jalil, J., & Reaz, M. B. I. (2012, May). "Accident detection and reporting system using GPS, GPRS, and GSM technology". In *2012 International Conference on Informatics, Electronics & Vision (ICIEV)* (pp. 640–643). IEEE.
27. Chandran, S., Chandrasekar, S. and Elizabeth, N. E.. (2016, December) "Konnect: An Internet of Things(IoT) based smart helmet for accident detection and notification. In *Proceedings of the IEEE Annual India Conference (INDICON)* (pp. 1–4).
28. Alkandari, A., Al-Shaikhli, I. F., Najaa, A., & Aljandal, M. (2013, December). "Accident detection and action system using fuzzy logic theory". In *2013 International Conference on Fuzzy Theory and Its Applications (iFUZZY)* (pp. 385–390). IEEE.
29. Pan, B., & Wu, H. (2017, August). "Urban traffic incident detection with mobile sensors based on SVM". In *2017 XXXIInd General Assembly and Scientific Symposium of the International Union of Radio Science (URSI GASS)* (pp. 1–4). IEEE.
30. Ghosh, S., Sunny, S. J., & Roney, R. (2019, March). "Accident detection using convolutional neural networks". In *2019 International Conference on Data Science and Communication (IconDSC)* (pp. 1–6). IEEE.
31. Eduku, S., Alhassan, M. O., & Sekyi-Ansah, J. (2016). "Design of vehicle accident prevention system using wireless technology". *International Journal of Scientific and Research Publications*, 7(10), 397–403.
32. Narayana Rao, T. V., & Yellu, K. R. (2017). "Preventing drunken driving accidents using IoT". *International Journal of Advanced Research in Computer Science*, 8(3).
33. Basubeit, O. G., How, D. N. T., Hou, Y. C., & Sahari, K. S. M. (2019). "Distracted driver detection with deep convolutional neural network". *International Journal of Recent Technology and Engineering*, 8(4), 6159–6163.

34. Dwivedi, K., Biswaranjan, K., & Sethi, A. (2014, February). "Drowsy driver detection using representation learning". In *2014 IEEE international advance computing conference (IACC)* (pp. 995–999). IEEE.
35. Szarvas, M., Yoshizawa, A., Yamamoto, M., & Ogata, J. (2005, June). "Pedestrian detection with convolutional neural networks". In *IEEE Proceedings. Intelligent Vehicles Symposium, 2005* (pp. 224–229). IEEE.
36. Berade, P., Patil, K., Tawate, P., & Ghewari, M. U. (2018). "Intelligent accident identification and prevention system using GPS and GSM modem". *International Research Journal of Engineering and Technology*, 5(2), 1915–1919.
37. Nanda, S., Joshi, H., & Khairnar, S. (2018, August). "An IOT based smart system for accident prevention and detection". In *2018 Fourth International Conference on Computing Communication Control and Automation (ICCUBEA)* (pp. 1–6). IEEE.
38. Murshed, M., & Chowdhury, M. S. (2019, January). "An IoT based car accident prevention and detection system with smart brake control". In *Proceeding of International Conference on Applications and Techniques in Information Science (iCATIS)* (Vol. 23).
39. Ahmed, V., & Jawarkar, N. P. (2013, December). "Design of low cost versatile microcontroller based system using cell phone for accident detection and prevention". In *2013 6th International Conference on Emerging Trends in Engineering and Technology* (pp. 73–77). IEEE.
40. Priyanka, C., Darshini, C. P., Shavi, V. G., & Begum, S. (2018). "Two-wheeler safety system for accident prevention detection and reporting". *International Journal of Engineering and Computer Science*, 7(03), 23680–23682.

Chapter 13

HEMT biosensors integrated with AI and ML for IoT devices

Yogesh Kumar Verma

13.1 INTRODUCTION

The HEMT device has been used recently in different potential applications, such as biosensors and low-noise amplifiers, due to its interesting properties, including higher breakdown voltage and higher mobility of electrons. The higher mobility of electrons is possible due to the reduced ionized impurity scattering mechanism that is prominent in conventional MOSFETs. There is a wide bandgap material introduced over the narrow bandgap material that results in the diffusion of electrons at the hetero-interface [1–21]. There are several potential features of HEMT biosensors, such as high sensitivity and specificity, fast detection and response time, compact and portable design, and label-free detection. The integration of AI and ML provides data analysis and pattern recognition, real-time signal processing, predictive modeling and classification, and automated decision-making. The other applications include point-of-care diagnostics, disease biomarker detection, personalized medicine, environmental monitoring, and food safety testing. The HEMT sensor detects the biomolecular interactions, analyzes and interprets the sensor data, handles sample preparation and delivery, and also enables the remote monitoring and data transmission. The research fields in this aspect include graphene-based HEMT biosensors, nanoparticle-enhanced sensing, deep learning-based signal processing, wearable and implantable devices, and point-of-care diagnostics for infectious diseases [22–40]. There are several organizations working in the field of HEMT biosensors with AI and BL, such as Roche Diagnostics, Abbott Laboratories, Siemens Healthineers, IBM Watson Health, and Google Life Science.

13.2 DESCRIPTION OF THE STRUCTURE

The major components of the structure are substrate, channel layer, gate electrode, source/drain electrodes, sensing layer, and dielectric layer. There are two possible architectures of HEMT, i.e., planar structure and vertical

 DOI: 10.1201/9781003643173-13

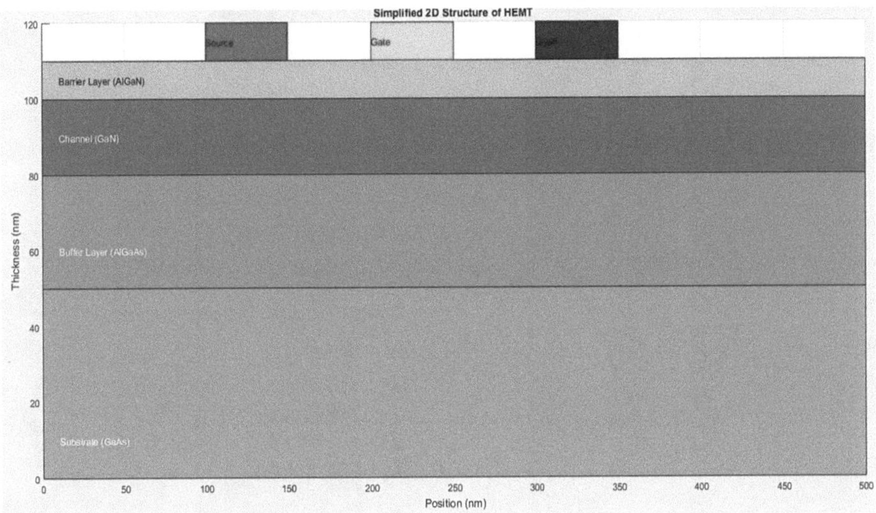

Figure 13.1 Device structure.

structure. The sensing mechanism is basically classified into three different categories, such as biomolecular binding, charge transfer, and current modulation. There are two different types of HEMT biosensors, i.e., depletion-mode and enhancement-mode HEMT. The higher mobility of electrons at the hetero-interface enables high sensitivity and fast response. Figure 13.1 represents the device structure.

13.3 RESULTS AND DISCUSSIONS

The gate-source capacitance is a critical parameter in HEMT-based biosensors, influencing sensing performance and sensitivity. This capacitance affects the sensing sensitivity and detection limits. It also influences the device noise and signal-to-noise ratio. It also impacts the device's stability and reliability. Figure 13.2 represents the calculation of drain current and output conductance. Figure 13.3 represents the calculation of gate-source capacitance. There are several factors affecting the gate-source capacitance, such as gate length and width, gate-source spacing, channel material and thickness, dielectric layer thickness and material, and operating frequency and voltage. Figure 13.4 represents the calculation of the cutoff frequency.

The higher mobility of electrons and low-noise levels make HEMT a potential contender for biosensing applications.

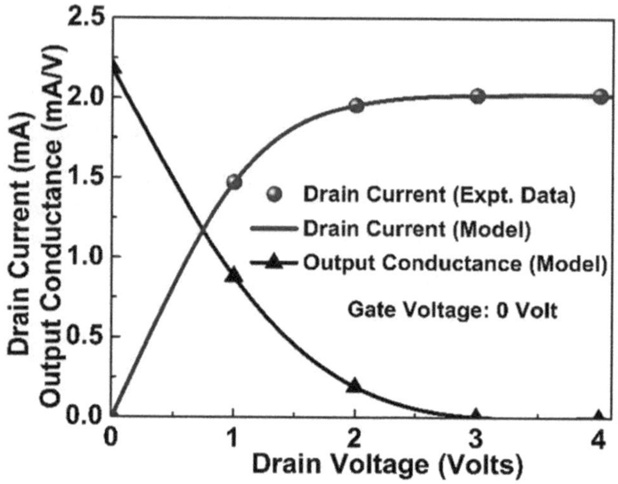

Figure 13.2 Calculation of I_D and g_d.

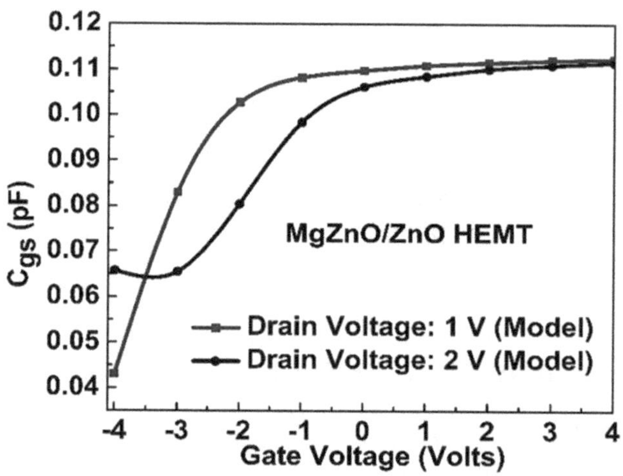

Figure 13.3 Calculation of C_{gs}.

13.4 CONCLUSION

The biosensors based on HEMT have emerged as a promising candidate in detecting biomolecules with high sensitivity and speed. The potential applications of this study are in the fields of healthcare and environmental monitoring. The future research includes the development of HEMT materials and structures, investigation of new sensing mechanisms and modalities, integration of HEMT biosensors with microfluids and lab-on-a-chip systems, exploring applications in personalized medicine and precision health, and development of wearable and implantable HEMT-based biosensors.

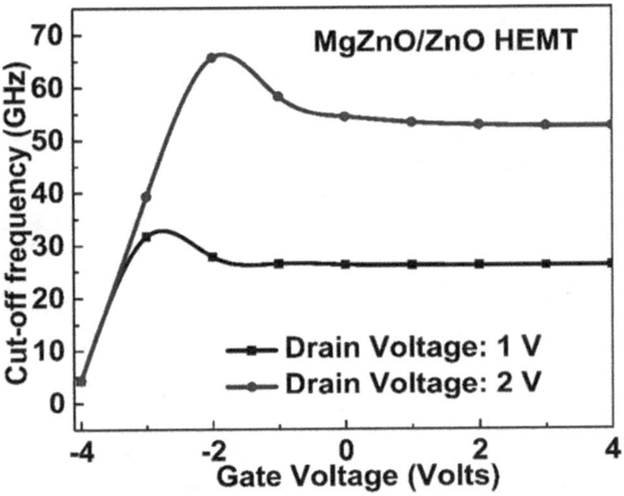

Figure 13.4 Calculation of f_T.

REFERENCES

[1] O. Ambacher, J. Smart, J. R. Shealy, N. G. Weimann, K. Chu, M. Murphy, W. J. Schaff, and L. F. Eastman, "Two-dimensional electron gases induced by spontaneous and piezoelectric polarization charges in N- and Ga-face AlGaN/GaN heterostructures," *J. Appl. Phys.*, Vol. 85, No. 6, pp. 3222–3233, 1999.

[2] S. Ghosh, A. Dasgupta, S. Khandelwal, S. Agnihotri, and Y. S. Chauhan, "Surface-potential-based compact modeling of gate current in AlGaN/GaN HEMTs," *IEEE Trans. Electron Devices*, Vol. 62, No. 2, pp. 443–448, 2015.

[3] S. A. Ahsan, S. Ghosh, K. Sharma, A. Dasgupta, S. Khandelwal, and Y. S. Chauhan, "Capacitance modeling in dual field-plate power GaN HEMT for accurate switching behavior," *IEEE Trans. Electron Devices*, Vol. 63, No. 2, pp. 565–572, 2016.

[4] M. D. Ganeriwala, C. Yadav, N. R. Mohapatra, S. Khandelwal, C. Hu, and Y. S. Chauhan, "Modeling of charge and quantum capacitance in low effective mass III-V FinFETs," *IEEE J. Electron Devices Soc.*, Vol. 4, No. 6, pp. 396–401, 2016.

[5] C. Yadav, J. P. Duarte, S. Khandelwal, A. Agarwal, C. Hu, and Y. S. Chauhan, "Capacitance modeling in III–V FinFETs," *IEEE Trans. Electron Devices*, Vol. 62, No. 11, pp. 3892–3897, 2015.

[6] C. Yadav, P. Kushwaha, S. Khandelwal, J. P. Duarte, Y. S. Chauhan, and C. Hu, "Modeling of GaN-based normally-off FinFET," *IEEE Electron Device Lett.*, Vol. 35, No. 6, pp. 612–614, 2014.

[7] S. Khandelwal and T. A. Fjeldly, "Analysis of drain-current nonlinearity using surface-potential-based model in GaAs pHEMTs," *IEEE Trans. Microw. Theory Tech.*, Vol. 61, No. 9, pp. 3265–3270, 2013.

[8] S. Khandelwal, Y. S. Chauhan, and T. A. Fjeldly, "Analytical modeling of surface-potential and intrinsic charges in AlGaN/GaN HEMT devices," *IEEE Trans. Electron Devices*, Vol. 59, No. 10, pp. 2856–2860, 2012.

[9] J. D. Ye, S. Pannirselvam, S. T. Lim, J. F. Bi, X. W. Sun, G. Q. Lo, and K. L. Teo, "Two-dimensional electron gas in Zn-polar ZnMgO/ZnO heterostructure grown by metal-organic vapor phase epitaxy," *Appl. Phys. Lett.*, Vol. 97, No. 11, pp. 2010–2012, 2010.
[10] M. A. Khan, R. Singh, S. Mukherjee, and A. Kranti, "Buffer layer engineering for high (≥1013 cm-2) 2-DEG density in ZnO-based heterostructures," *IEEE Trans. Electron Devices*, Vol. 64, No. 3, pp. 1015–1019, 2017.
[11] S. Khandelwal, N. Goyal, and T. A. Fjeldly, "A physics-based analytical model for 2DEG charge density in AlGaN/GaN HEMT devices," *IEEE Trans. Electron Devices*, Vol. 58, No. 10, pp. 3622–3625, 2011.
[12] H. Tampo, H. Shibata, K. Maejima, A. Yamada, K. Matsubara, P. Fons, S. Kashiwaya, S. Niki, Y. Chiba, T. Wakamatsu, and H. Kanie, "Polarization-induced two-dimensional electron gases in ZnMgO/ZnO heterostructures," *Appl. Phys. Lett.*, Vol. 93, No. 20, pp. 11–14, 2008.
[13] Kelin J. Kuhn, Martin D. Giles, David Becher, Pramod Kolar, Avner Kornfeld, Roza Kotlyar, Sean T. Ma, Atul Maheshwari, and Sivakumar Mudanai, "Process technology variation," *IEEE Trans. Electron Devices*, Vol. 58, No. 8, pp. 2197–2208, 2011.
[14] M. S. Shur, R. Gaska, A. Khan, and G. Simin, "Wide band gap electronic devices," in *Proceedings of the Fourth IEEE International Caracas Conference on Devices, Circuits and Systems* (Cat. No. 02TH8611) at Oranjestad Aruba Netherlands, pp. D051-D051, 2002.
[15] R. L. Anderson, "Germanium-gallium arsenide heterojunctions," *IBM J. Res. Dev.*, Vol. 4, No. 3, pp. 283–287, 1960.
[16] R. Dingle, H. L. Störmer, A. C. Gossard, and W. Wiegmann, "Electron mobilities in modulation-doped semiconductor heterojunction superlattices," *Appl. Phys. Lett.*, Vol. 33, No. 7, pp. 665–667, 1978.
[17] L. C. Witkowski, T. J. Drummond, C. M. Stanchak, and H. Morkoc, "High mobilities in AlxGa1-xAs-GaAs heterojuntions," *Appl. Phys. Lett.*, Vol. 37, No. 11, pp. 1033–1035, 1980.
[18] W. I. Wang, C. E. C. Wood, and L. F. Eastman, "Extremely high electron mobilities in modulation-doped GaAs-AlxGa1-xAs heterojunction superlattices," *Electron. Lett.*, Vol. 17, No. 1, pp. 36–37, 1981.
[19] D. Delagebeaudeuf and N. T. Linh, "Metal-(n) AlGaAs-GaAs two-dimensional electron gas FET," *IEEE Trans. Electron Devices*, Vol. 29, No. 6, pp. 955–960, 1982.
[20] O. Ambacher, B. Foutz, J. Smart, J. R. Shealy, N. G. Weimann, K. Chu, M. Murphy, A. J. Sierakowski, W. J. Schaff, and L. F. Eastman, "Two dimensional electron gases induced by spontaneous and piezoelectric polarization in undoped and doped AlGaN/GaN heterostructures," *J. Appl. Phys.*, Vol. 87, No. 1, pp. 334–344, 2000.
[21] M. S. Shur, "GaN based transistors for high power applications," *Solid. State. Electron.*, Vol. 42, No. 12, pp. 2131–2138, 1998.
[22] D. C. Look, "Recent advances in ZnO materials and devices," *Mater. Sci. Eng. B*, Vol. 80, No. 1–3, pp. 383–387, 2001.
[23] Ü. Özgür, Ya. I. Alivov, C. Liu, A. Teke, M. A. Reshchikov, S. Dogan, V. Avrutin, S. -J. Cho, and H. Morkoc, "A comprehensive review of ZnO materials and devices," *Appl. Phys.*, Vol. 98, No. 4, pp. 041301, 2005.
[24] B. K. Meyer, H. Alves, D. M. Hofmann, W. Kriegseis, D. Forster, F. Bertram, J. Christen, A. Hoffmann, M. StraBburg, M. Dworzak, U. Haboeck, and A. V.

Rodina, "Bound exciton and donor-acceptor pair recombinations in ZnO," *Phys. Status Solidi Basic Res.*, Vol. 241, No. 2, pp. 231–260, 2004.
[25] R. Singh, M. A. Khan, S. Mukherjee, and A. Kranti, "Analytical model for 2DEG density in graded MgZnO/ZnO heterostructures with cap layer," *IEEE Trans. Electron Devices*, Vol. 64, No. 9, pp. 3661–3667, 2017.
[26] R. Singh, M. A. Khan, S. Mukherjee, and A. Kranti, "Role of surface states and interface charges in 2DEG in sputtered ZnO heterostructures," *IEEE Trans. Electron Devices*, Vol. 65, No. 7, pp. 2850–2854, 2018.
[27] C. Yadav, M. D. Ganeriwala, N. R. Mohapatra, A. Agarwal, and Y. S. Chauhan, "Compact modeling of gate capacitance in III-V channel Quadruple-Gate FETs," *IEEE Trans. Nanotechnol.*, Vol. 16, No. 4, pp. 703–710, 2017.
[28] I. Ferain, C. A. Colinge, and J.-P. Colinge, "Multigate transistors as the future of classical metal-oxide-semiconductor field-effect transistors," *Nature*, Vol. 479, No. 7373, pp. 310–316, 2011.
[29] L. He, T. K. Chiang, J. J. Liou, W. Zheng, and Z. Liu, "A new analytical subthreshold potential/current model for quadruple-gate junctionless MOSFETs," *IEEE Trans. Electron Devices*, Vol. 61, No. 6, pp. 1972–1978, 2014.
[30] C.-T. Lee, W.-S. Chen, and H.-Y. Lee, "Quadruple gate-embedded T structured GaN-based Metal-Oxide-semiconductor high-electron mobility transistors," *IEEE J. Electron Devices Soc.*, Vol. 1, No. 6, pp. 63–67, 2018.
[31] D. Sharma and S. K. Vishvakarma, "Analyses of DC and analog/RF performances for short channel quadruple-gate gate-all-around MOSFET," *Microelectronics J.*, Vol. 46, No. 8, pp. 731–739, 2015.
[32] M. Li and Y. Wang, "2-D analytical model for current–voltage characteristics and transconductance," *IEEE Trans. Electron Devices*, Vol. 55, No. 1, pp. 261–267, 2008.
[33] A. U. Manual, *3-D device simulator*, Silvaco Inc., St. Clara, CA, 2012.
[34] T. Mimura, S. Hiyamizu, T. Fujii, and K. Nanbu, "A new field-effect transistor with selectively doped GaAs/n-AlxGa1-xAs heterojunctions," *Jpn. J. Appl. Phys.*, Vol. 19, No. 5, pp. L225, 1980.
[35] Y.-F. Wu et al., "Very high breakdown voltage and large transconductance realized on GaN heterojunction field effect transistors," *Appl. Phys. Lett.*, vol. 69, no. 10, pp. 1438–1440, 1996.
[36] M. A. Khan et al., "AlGaN/GaN metal oxide semiconductor heterostructure field effect transistor," *IEEE Electron Device Lett.*, vol. 21, no. 2, pp. 63–65, 2000.
[37] P. M. Solomon and H. Morkoc, "Modulation-doped GaAs/AlGaAs heterojunction field-effect transistors (MODFET's), ultrahigh-speed device for supercomputers," *IEEE Trans. Electron Devices*, vol. 31, no. 8, pp. 1015–1027, 1984.
[38] K. Lee, M. S. Shur, T. J. Drummond, and H. Morkoc, "Parasitic MESFET in (Al, Ga) As/GaAs modulation doped FET's and MODFET characterization," *IEEE Trans. Electron Devices*, vol. 31, no. 1, pp. 29–35, 1984.
[39] K. B. Madhavi, S. L. Tripathi, S. Saxena et al. "Design and analysis of 18 nm multichannel FinFET as biosensor for detection of biological species," *Silicon*, Vol. 15, pp. 6313–6322, 2023. https://doi.org/10.1007/s12633-023-02519-y
[40] S. L. Tripathi, L. P. Dasari, I. Wijayanto, D. Ghai, & M. Mahmud, "Assistive technology for neuro-rehabilitation applications using machine learning techniques," in *Machine Learning Models and Architectures for Biomedical Signal Processing* (pp. 179–191). Academic Press, Elsevier, 2025. https://doi.org/10.1016/B978-0-443-22158-3.00008-9

Chapter 14

Beamforming and its significance

Shahid Hamid and Shakti Raj Chopra

14.1 INTRODUCTION

Before delving into beamforming, it's important to understand beam steering, which involves directing the main portion of the probe in one of several predetermined directions. There are several methods used to determine the appropriate direction at the transmitter, such as transmitting orthogonal vector streams along each of the intended paths. Beamforming improves information strength levels in omnidirectional antennas, as signal strength tends to decrease from the cell core to the cell circumference. By channelling all wattage in a single orientation, SINR is significantly enhanced at the cell boundaries, and multiple sovereign cells can function in the same area without coordination or cooperation, and can be located close to each other. In this scenario, all broadcast antennas are co-located and use beamforming to guide energy in a particular direction [1].

By raising the quantity of antennas, energy is more effectively concentrated in a narrower beam, enhancing the signal strength for any user on the receiving side but degrading the signal for other users. In the literature and throughout this chapter, the words "BF" and "precoding/combining" are frequently used interchangeably. Ground stations are concurrently trained as they quantify information streams from various user terminals, and the number of vectors that these antennas obtain is precisely correlated to the total amount of UEs. Additionally, the number of user terminals that can be accommodated in a cell increases as coherence time increases. A CSI factor is exploited in the transmission between the user terminal and the mesh during channel training in a TDD system to characterise the channel condition and choose the genuine precoding matrix [2]. In a TDD system, the CSI of the downlink is determined by measuring the uplink channel. This approach takes advantage of channel reciprocity, which is a property of some wireless communication systems where the uplink and downlink channels have similar characteristics. By leveraging this property, the system can efficiently estimate the downlink channel quality without having to directly measure it [3]. In FDD systems, the uplink and downlink channels are trained separately. This means that the training time for the downlink channel is dependent on

DOI: 10.1201/9781003643173-14

the number of antennas, making it difficult to operate large antenna arrays effectively. Consequently, alternative strategies are needed to obtain accurate CSI estimates in FDD systems [4].

14.2 ANALOGUE BF

The wireless signal is broken down using a power divider and then routed via a beamformer, which can adjust the magnitude (a_k) and period (θ_k) of the transmissions in each of the routes leading to the antenna stack; number of antenna elements decides the power divider. At the receiver end, the signal from each antenna in the array is given a complicated weight (Figure 14.1). The amplitude and phase of a complex weight are both present. After that, the signals are merged into a single output.

$$w_k = a_k e^{j\sin(\theta_k)}$$

$$w_k = a_k \cos(\theta_k) + j a_k \sin(\theta_k)$$

where w_k is the complex weight for kth antenna.

14.3 DIGITAL BF AND ITS PARAMETERS

The diagram below depicts a simplified digital BF architecture. In this case, each antenna element has its own specialised RF chain, DACs, and ADCs. Using the sampling analogy, this means that the gain and phase of each spatial sample are individually changed together with baseband processing before up-conversion at Tx or after down-conversion at Rx. This enables the genuine implementation of mathematical algorithms with maximal freedom, the majority of which regard each antenna output as an

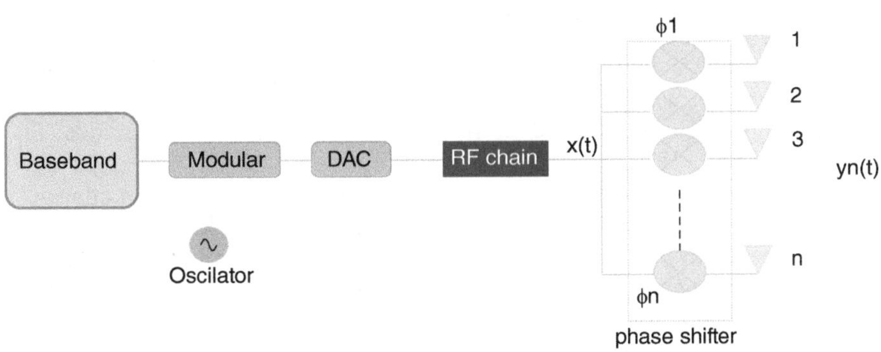

Figure 14.1 Analogue beamforming.

accessible sample. In addition to maximising signal intensity in a desired direction, nulls can be formed in unwanted directions to suppress interference. These techniques are designed to enhance the signal quality by reducing interference and noise and to optimise the transmission of data to the intended recipient. Non-linear precoding techniques are used to augment execution [5].

14.3.1 Zero-forcing

It is a spatial data processing algorithm that is commonly used in wireless communication systems. It utilises multiple broadcasting antennas to filter out interference signals from multiple users and to compensate for IUI at each user. In massive MIMO systems, the precoding matrix of zero-forcing (ZF) is designed to minimise interference among user streams, making it an effective linear precoding approach. [6]. ZF works well in low-noise or high-power applications. The power of the pre-coded signal is considerably boosted when the channel is poorly established, which causes a decreased SNR at the demand end. IUI is eliminated at the disadvantage of increased broadcast wattage. According to [7], interference is reduced by delivering information bits in the projected user's orientation together with nulls in the other users' directions [8]. The false-inverse of the H is used in conjunction with the H^{-1} as a broadcast screen to create the precoding matrix. The data stream that is primarily routed towards the terminal is premultiplied using a precoding matrix to perform the IUI suppression. Consequently, the BS's ZF precoding matrix is given as:

$$w_{zf} = H^H \left(HH^H \right)^{-H}$$

better results when employed in noise-free environment. In the channel with additive noise, it will amplify noise also [9].

14.3.2 Maximum-ratio transmission

It aims to maximise the SINR at the anticipated objective. This technique is widely used in wireless communication systems to improve the quality of the transmitted signal and reduce the effects of noise and interference. The matched filter (MF) linear precoding technique's fundamental tenet is to pre-code the transmitted signal in a way that it is matched to the channel's properties. The transmit signal is multiplied by a complex-valued weight vector to perform the precoding. By utilising channel knowledge, this weight vector is intended to maximise the signal-to-noise ratio at the receiver. One benefit of the MF linear precoding technique is that it may be applied to a number of communication systems, such as BF, space-time coding, and

MIMO systems. To lessen the impacts of ISI and ICI in MIMO systems, MF linear precoding can be utilised. Matching filter linear precoding can be employed in BF systems to increase the antenna array's directivity and raise the SINR. A MF linear precoding technique can enhance diversity gain and lower error rates for space-time coding schemes.

$$w_{MF} = H^H$$

MRT, as an enormous array M's antenna count rise, making MRT the best precoder. In contrast to ZF, MRT increases the power of the symbol rather than attempting to cancel IUI. MRT precoding, for instance, works superior to ZF alone provided the quantity of simultaneous clients is very elevated, users are very mobile, or the base station hardware capability is very limited.

14.3.3 Regularised-zero forcing

Regularised zero-forcing (RZF) precoding uses regularised channel inversion rather than a conventional channel reversal [10]. It is a type of linear precoding used in MU-MIMO communication systems to lessen user interference. When numerous users share the same frequency band, it results in interference between the users. With RZF precoding, users' interference is eliminated while the received signal power is increased. The pseudo-inverse of the channel matrix with regularisation is used in RZF precoding to produce a linear transformation of the broadcast signal. The complexity of the RZF precoding matrix is managed by the regularisation parameter in order to prevent overfitting to the training set of data. The MSE of the received signal is desired to be reduced to the absolute minimum using the RZF precoding matrix, thus also known as MMSE. RZF precoding has been demonstrated to outperform conventional ZF precoding in terms of overall system performance, particularly in situations with high SNR. The regularisation term, which reduces system noise and interference, is primarily responsible for the performance increase. For MIMO wireless communications systems, it is considered an extremely sophisticated straight precoder since it can trade off the advantages of both MRT and ZF precoders [11]. The RZF precoding matrix is provided by

$$w_{RZF} = \beta\left(HH^H + \alpha I_k\right)^{-1} H^H$$

The parameter for power normalisation is, β and k/Pd = α is the regularisation factor to be optimised. RZF precoder tends to ZF precoder at small α while it tends to MRT precoder at lower α. I_k is the identity matrix, P_d is the downlink power assigned to each user. The system complexity increases as the number of antenna elements rises (Table 14.1).

Table 14.1 Advantages of beamforming techniques

Linear precoding	*Advantages*	*Disadvantages*
ZF	• Zero forcing can help reduce the effects of channel distortion. • By equalising the signal, zero forcing can improve the overall signal quality and reduce errors in the received signal. • Zero forcing is a relatively simple equalisation technique, which makes it easy to implement in digital signal processing systems. • Zero forcing can be highly effective in simple communication channels where the signal distortion is mainly caused by a delay and a linear attenuation. • Zero forcing can provide low latency in digital signal processing systems. Also Zero forcing is compatible with different modulation schemes, including QPSK, BPSK, and other digital modulation schemes.	• In noisy case it works badly. • Power consumption is more. • The implementation complexity is more. • Performs badly when number of users grow. • In contaminated channels, it amplifies both signal and noise power.
MRT	• When there are more BS antennae than users, performance is close to ideal. • Considering that channel matrix inversion is not present, computational complexity is lower. • It is preferred in high-mobility cases. • Presents high performance at lower SNR.	• MRT shows lower achievable rate when number of base station antennas is low. • In complex signal processing cases, power consumption is high. • Small errors in channel estimation can significantly degrade MRT performance.
RZF	• It is ideal if all users experience the same SINR and average channel attenuation ratio. • Compared to ZF and MRT, RZF shows better sum rate performance.	• When dealing with large and complex communication channels, processing will also be complex and high-power-consuming. • Sensitivity to channel estimation errors. • Suboptimal performance in highly correlated channels. • Limited performance in multiuser scenarios.

14.3.4 Bit error rate

The amount of inaccuracies in bits obtained across a channel is measured by the BER. When bits are altered as a result of interference, noise, distortion, or synchronisation issues, this happens. It can alternatively be described as the proportion of incorrect bits to all bits communicated in a particular amount of time [12]. The transmission modulation technique employed in a download single-cell massive MIMO communication scheme determines the BER. In a huge MIMO structure with ZF precoding and grey-coded square QAM modulation, the mean BER for the k_{th} user is calculated as [13]:

$$P_e(\gamma k) \approx \frac{c_N}{2d_N} \frac{\Gamma\left(\tau + \frac{1}{2}\right) \Big/ \Gamma(\tau + 1)}{\left(\gamma k d_N^2 + 1\right)^{\left(\tau + \frac{1}{2}\right)} \sqrt{\pi \gamma k}}$$

Where the degree of freedom is given as τ = M − K, M is quantity of base station antennas, and k is amount of users. Transmission SNR of the end terminal is given by, $\gamma k = P_T / K\sigma^2$. P_T is the overall broadcast power at the base station shared out equally among users. C_N and d_N are constants.

14.3.5 Achievable Sum rate

The achievable rate is another parameter used to evaluate the effectiveness of huge MIMO systems. The lowest spectral efficiency over a fading channel that a MIMO system is capable of achieving is specified, or achievable sum rate in linear precoding techniques can be expressed mathematically as the sum of the achievable data rates of all the users in the system. The achievable data rate for each user depends on the channel conditions and the linear precoding scheme used. In this regard, the possible rate per user in a single-cell downlink massive MIMO network is governed by the Shannon channel capacity theory and is given as [14].

$$R_k = B \log_2 (1 + SINR_k)$$

And the achievable sum rate for k users will be given as:

$$R_k = B \sum_{k=1}^{K} \log_2 (1 + SINR_k)$$

Where B denotes the bandwidth of the system.

14.3.6 Power expenditure

The downlink massive MIMO system's overall power consumption is determined by the quantity of power broadcast and the voltage used by the circuit [15].

$$P_t = \mu \sum_{k=1}^{K} p_k + MP_c + P_f$$

Where pf is the voltage expenditure at the BS regardless of the quantity of broadcast antennas, pk is the downstream broadcast power allotted to each user, pc is the fixed device voltage expenditure per antenna, and μ is the reverse of the voltage amplifier performance at the base station.

14.3.7 Energy efficiency

The energy efficiency of a linear precoding technique can be measured in terms of the amount of bits that can be broadcast per unit of energy expended. This metric is known as the energy efficiency or spectral efficiency per unit energy. Generally, the higher the energy efficiency, the better the performance of the system.

$$EE = \frac{Achievable\ Sum\ Rate}{Total\ Power\ Consumption}$$

The ideal precoding technique for huge MIMO systems—one that achieves a desired throughput performance while minimising complexity—has been extensively researched. But even though linear precoders have the benefit of being simple, their lack of precoding accuracy cannot be ignored. Significantly, linear precoding performs well when the channel correlation between the users is minimal, but when this correlation is weighty, the performance suffers greatly. On the other hand, non-linear precoding methods have the prospective to considerably improve the massive MIMO functioning in 5G because they are often more resilient to channel correlation between UEs (Figure 14.2).

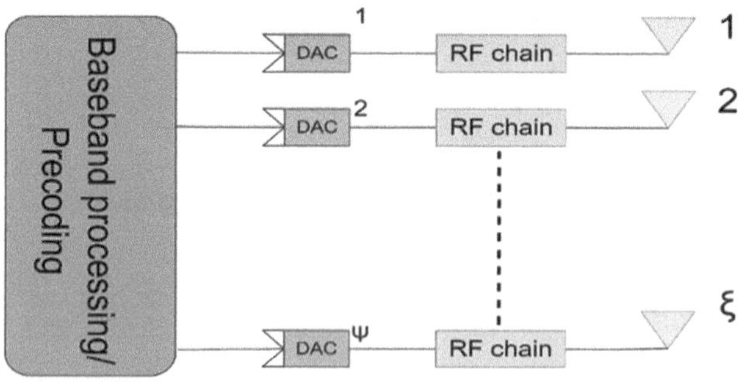

Figure 14.2 Digital beamforming.

To achieve good flexibility, control over amplitude and power consumption, and cost, hybrid BF is introduced [16], The design of hybrid BF techniques is to jointly tune analogue and digital beamformers to increase the achievable rate. The detailed discussion on hybrid BF will be presented next.

14.4 HYBRID BEAMFORMING

Nearly 20 years ago, hybrid BF was developed, and as a specific instance, peer-to-peer sole-stream broadcasting in sub-6 GHz facilities was researched. Over a decade later, in mm-wave structures, hybrid BF was investigated and attracted significant interest from the academic and industry worlds [9]. In this instance, compared to analogue BF, hybrid BF provides spatial division multiple access and multi-stream broadcasting with spatial multiplexing. With significantly less complicated technology, it achieves spectral efficiency comparable to completely digital BF. It is considered to be a good option for transmission architecture in mm-wave technology as a result.

$$y = \begin{bmatrix} H_1 \\ \vdots \\ H_N \end{bmatrix} \begin{bmatrix} A_1 & \cdots & 0 \\ \vdots & \ddots & \vdots \\ 0 & \cdots & A_N \end{bmatrix} D \begin{bmatrix} s_1 \\ \vdots \\ s_N \end{bmatrix} + n$$

H represents the downlink channel of the ith user. Ai is analogue precoder on the ith transceiver, and D is digital BF matrix. n is the noise vector, and Si is the broadcast information for the ith user. The analogue BF Ai and the digital BF D can be jointly optimised under specific channel availability assumptions. In general, uplink sounding in a TDD system or downlink address signals are the two ways to collect downlink CSI. The DL/UL channel mutuality tuning, using conventional methods at the base station, may not perform effectively due to the high complexity of the tuning circuits, therefore, the first way is not simple. Furthermore, measuring the CSI per antenna is virtually impossible due to the N*M structure (N = quantity of transceivers and M = quantity of active antennas per transceivers). The second approach is likewise useless because each transceiver's antennas actually function as a single logical antenna, and it is still unknown how the guide signals should be delivered. One practical assumption about channel information is that the BS is aware of partial channel information, such as AoA and AoD, If H and A are known, it is possible to determine the effective channel matrix HA. The total rate can then be maximised by designing the digital BF matrix D.

Consider the BF structure N by M. With M antennas, each transceiver pointing at one user (there are N users in total), perfect analogue BF is assumed. In the case of a sufficiently big M, there is no inter-user interference. It is possible to derive the N user total capacity of this structure as

$$C = W * N * \log\left(1 + \frac{M\eta_{PA}P}{wN_0}\right)$$

Relationship between EE and SE: In the formula mentioned above, W stands for bandwidth, P corresponds to each transceiver's broadcast power, η_0 for PA productivity, and N0 for thermal noise level. [17]. Channel gain is assumed to be unit. Then, this structure's SE is expressed as,

$$\eta_{SE} = \frac{C}{W} = N * \log\left(1 + \frac{M\eta_{PA}P}{wN_0}\right)$$

$$P_{total} = NP + NP_0 + P_{common} + NMP_{rf-circuit}$$

$$\eta_{EE} = \frac{C}{P_{total}}$$

Hybrid BF is still dealing with a number of serious problems, which could limit its potential to be used in real-world applications. Hardware complexity has been greatly decreased when compared to fully digital BF, but it is still a major challenge, specifically given the price and energy requirements of mm-wave tools [18]. Therefore, hardware-efficient hybrid BF architectures should be created (Figure 14.3).

Hybrid BF issues are inherently non-convex and difficult to address. A comprehensive strategy should be used to solve these implementation issues for hybrid BF. We particularly require a thorough analysis that takes into account the three crucial factors listed below: Spectral efficiency, hardware efficiency, and computational efficiency. There are two parts to the hybrid beamformer: a digital part and an analogue part. The RF chains that make up the digital portion share a common structure with the many plans under discussion. The digital component of hybrid BF, denoted as, $F_{BBk,f} \in \mathbb{C}^{N_{RF}^{t} * N_s}$ can be used for each user on each sub-carrier, similar to the conventional

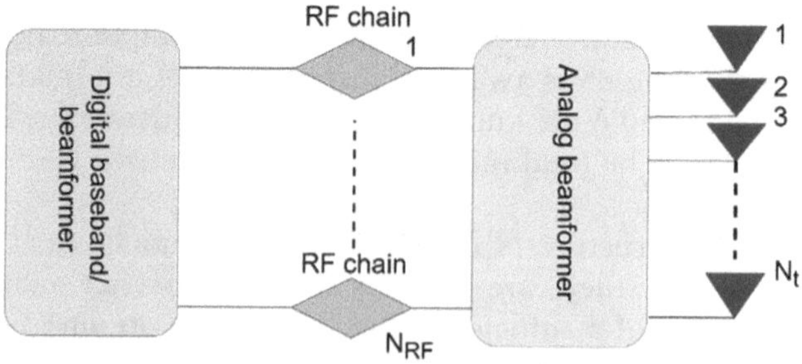

Figure 14.3 Hybrid beamforming.

fully digital BF. In contrast, this is not true of the hybrid BF analogue network or component. The analogue network $\left[F_{RF} \in \mathbb{C}^{N_{RF}^{t} * N_t}\right]$ is a part that all users and subcarriers share because digital beamformers combine what is sent for all users, while analogue RF BF is a post-IFFT procedure. In addition, the analogue network is the primary element that distinguishes various hybrid BF topologies. In particular, the spectral efficiency that may be achieved and the design of the algorithms are both significantly impacted by the analogue network's structure.

Power amplifiers, phase shifters, and switches are important hardware elements in the analogue RF domain. Each antenna element requires a power amplifier as a basic component in conventional completely digital BF, and IC design has focused heavily on realising low-power amplifiers. of contrast, phase shifters are the recently developed hardware components of hybrid BF systems. They were first used in military radar systems. Phase shifters are currently quite expensive, costing up to $100 USD even for low resolution[2], and hardware manufacturers are not yet prepared to offer them for commercial usage. It encourages alternative structures to reduce the number of phase shifters or replace them with other components. To simplify the circuitry, Roi et al. [19] suggested replacing phase shifters with switches. Phase shifters and/or switches play a major role in the hardware effectiveness of the analogue network since power amplifiers are essential and difficult to replace. In fact, switches only require binary states and perform better than phase shifters in terms of cost, power consumption, and implementation difficulty. But being limited to the on-off state will unavoidably result in spectral efficiency performance degradation. The key way hybrid BF structures vary from conventional BF structures is in how they construct the analogue network using the aforementioned technology. In particular, the mapping approach and hardware implementation play a major role in determining the analogue network structure. In this work presented, we will discuss mapping approach in detail only. The mapping strategy determines the connections between the antenna components and RF chains. There are two fundamental mapping procedures, fully-connected mapping and partially-connected mapping.

Fully connected and partially connected Hybrid BF:

This suboptimal approach to BS EE maximisation attempts to lower baseband processing and RF chain energy usage. Even with less-than-ideal solutions, a boost of more than 170% is seen when BS antennas and a sufficient amount of UEs are used to jointly optimise power and budget efficiency. In the work [6], for the downstream link mmWave huge MIMO systems, a multiuser hybrid BF is devised. Analogue BF is used to increase connection gain and select the optimal beams that will maximise the output of the intended recipient and minimise disturbance from any additional users. Nevertheless, digital BF maximises the EE of its target user while increasing the spatial multiplexing gain via a zero-gradient approach. Due to the irregular nature of the optimisation challenge and to prevent complexities, the

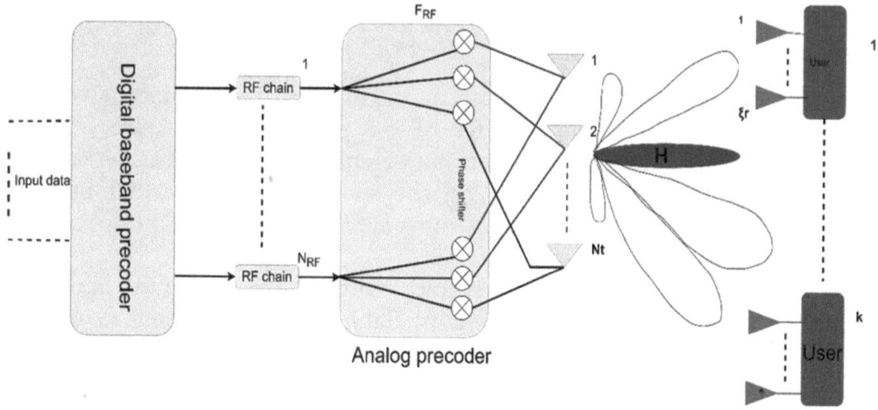

Figure 14.4 Fully connected hybrid-beamforming.

authors use the fixed magnitude hypothesis on the DB and numerically solve the AB difficulties with a per antenna energy limiting (Figure 14.4).

This is based on fully connected hybrid BF with hybrid digital and analogue inputs at both the user terminals and the BS. [15] presents a useful hybrid BF for multiuser large MIMO systems that combines analogue BF with digital BF with ZF precoding. The suggested techniques demonstrate that hybrid BF with additional RF chains can perform better than traditional digital precoding.

It is anticipated that the amalgamation of large MIMO systems with a hybrid BF setup that includes a substantially decreased RF chain number will significantly improve EE and SE for 5G [20]. At high energy bands, such as mmWave, the transmission signal can be adaptively altered in accordance with the channel while overpowering the unfavourable channel characteristics [21, 22] (Figure 14.5).

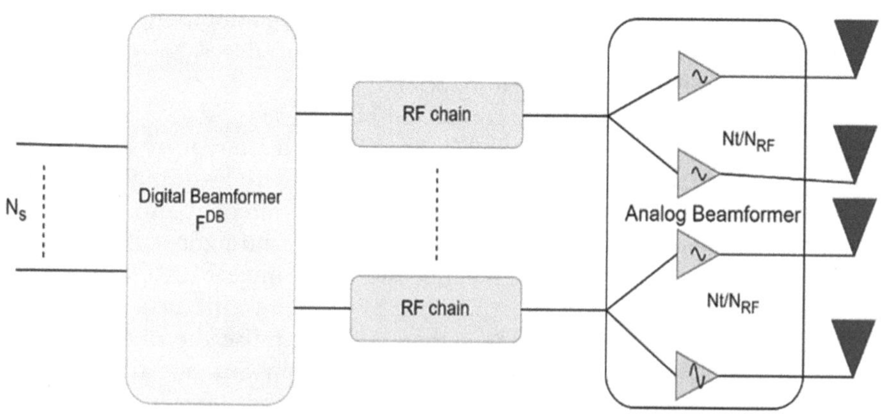

Figure 14.5 Partially connected hybrid-beamforming.

For employing BF to enhance the efficiency, different algorithms are applied, LMS algorithm, Sample Matrix inversion algorithm, and Recursive Least Square algorithm, while for hybrid BF the optimal algorithm can be one of the adaptive algorithms, such as MVDR or the combination of both.

14.5 RESULTS

In this section, the results of some beamforming techniques are discussed, so that you can imagine how important the variable changes are in the technology development. In Figure 14.6 it has been shown by Jun Zhang how digital beamforming shows dominance in spectral efficiency as compared to fixed phase shifters with different numbers of RF chains (η).

Figure 14.7 provides a brief discussion and displays the results of the hybrid BF algorithm in combination with digital linear precoding techniques such as matched-filter and regularised zero forcing in the MIMO and massive-MIMO systems in the rapid fading channel with four different receivers—MF, ZF, RZF, and regularised hybrid zero-forcing (HRZF). The results are evident that the performance of the hybrid regularised zero forcing technique dominated all other presented techniques like MF, ZF, etc.

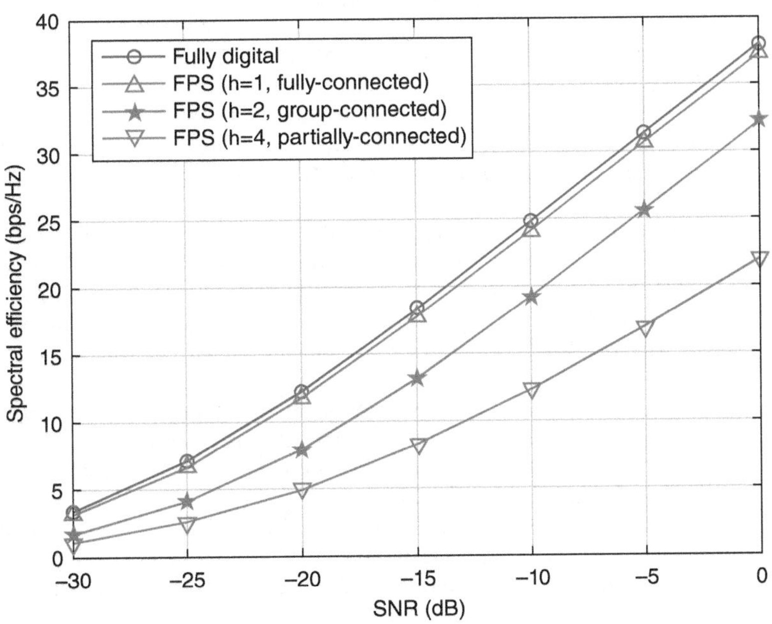

Figure 14.6 Spectral efficiency vs signal-to-noise ratio.

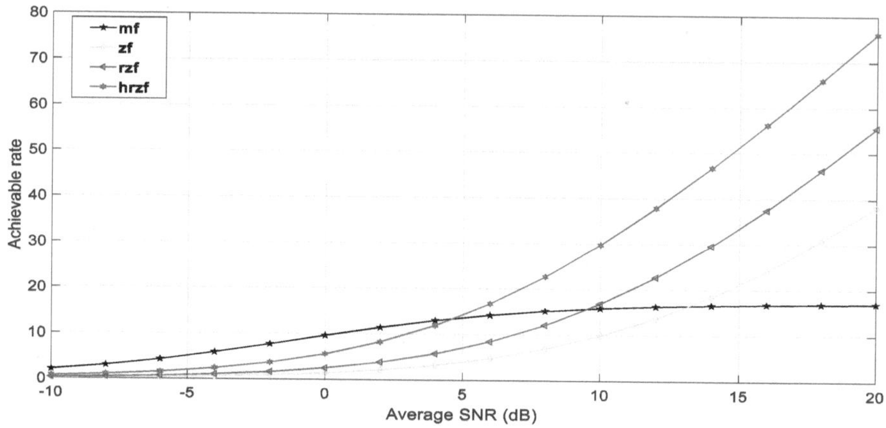

Figure 14.7 Achievable rate vs signal-to-noise ratio.

REFERENCES

[1] A. Boufrioua, *Smart Miniaturized Wideband/Multiband and Reconfigurable Antenna for Modern Applications*, Nova Science Publishers, Incorporated, 2018.

[2] J. P. Cosmas, *Massive MIMO Systems for 5G Communications*, Springer, 2021.

[3] A. N. U. A. N. M. Mahyuddin, "A comprehensive survey on millimeter wave communication for fifth generation wireless networks: Feasibility and challanges," *IEEE*, vol. 8, pp. 62367–62414, 2020.

[4] Q. Li, A. Zhang, P. Liu, J. Li, & C. Li, "A novel CSI feedback approach for massive MIMO using LSTM-attention CNN," *IEEE Access*, vol. 8, pp. 7295–7302, 2020.

[5] O. Raeesi, A. Gokceoglu, Y. Zou, E. Björnson, & M. Valkama, "Performance analysis of multi-user massive MIMO downlink under channel non-reciprocity and imperfect CSI," *IEEE Transactions on Communications*, vol. 66(6), 2456–2471, (2018).

[6] S. R. Chopra, A. Gupta, S. Tanwar, C. O. Safirescu, T. C. Mihaltan, & R. Sharma. "Multi-User Massive MIMO System with Adaptive Antenna Grouping for Beyond 5G Communication Network," *Mathematics* 10, no. 19, p. 3621, 2022.

[7] Y. G. A. Kaifeng Guo, "Distributed antennas aided secure communication in MU-massive-MIMO with QoS guarantee," in *2015 IEEE 82nd Vehicular Technology Conference (VTC2015-Fall)*, Boston, MA, USA, 2015.

[8] Shakti Raj, Akhil Gupta, & Himanshu Monga, "Performance analysis of space-time Trellis codes in Rayleigh fading channel," in *International Conference on Harmony Search and Nature Inspired Optimization Algorithms (ICHSA)*, pp. 957–967, Springer, Singapore, 2019.

[9] Shakti Raj Chopra, Akhil Gupta, & Rakesh Kumar Jha, "Performance analysis of grouped multilevel space-time Trellis coding technique using cognitive radio in different deployment models", *Wireless Communications and Mobile Computing*, vol. 2019, p. 5280615, 2019.

[10] L. Liang, W. Xu, & X. Dong, "Low-complexity hybrid precoding in massive multiuser MIMO systems," *IEEE Wireless Communi*cations, vol. 3, no. 6, pp. 653–656, 2014.
[11] S. Raj, & A. Gupta, "Designing and testing of quadrifilar helix antenna for NOAA weather satellite", in *2019 6th International Conference on Signal Processing and Integrated Networks (SPIN)*, ©2019 IEEE.
[12] P. SinghParihar, R. Saraswat, & S. Maheshwari, "Energy and spectral efficiency of very large multiuser MIMO systems," *International Journal of Computer Applications*, vol. 111, no. 5, pp. 4–7, 2015.
[13] M. A. Albreem, A. H. Al Habbash, A. M. Abu-Hudrouss, and S. S. Ikki, "Overview of precoding techniques for massive MIMO," *IEEE Access*, vol. 9, pp. 60764–60801, 2021.
[14] S. Dutta, C. N. Barati, D. Ramirez, A. Dhananjay, J. F. Buckwalter, & S. Rangan, "A case for digital beamforming at mmWave," *IEEE Transactions on Wireless Communications*, vol. 19, no. 2, pp. 756–770, 2020, doi: 10.1109/TWC.2019.2948329
[15] Shahid Hamid, Shakti Raj Chopra, Akhil Gupta, Sudeep Tanwar, Bogdan Cristian Florea, Dragos Daniel Taralunga, Osama Alfarraj, & Ahmed M. Shehata. "Hybrid beamforming in massive MIMO for next-generation communication technology," *Sensors*, vol. 23, no. 16, p. 7294, 2023.
[16] O. E. Ayach, S. Rajagopal, S. Abu-Surra, Z. Pi, & R. W. Heath, "Spatially sparse precoding in millimeter wave MIMO systems," *IEEE Transactions on Wireless Communications*, vol. 13, no. 3, pp. 14991513, March 2014.
[17] Y. Ren, Y. Wang, C. Qi, & Y. Liu, "Multiple-beam selection with limited feedback for hybrid beamforming in massive MIMO systems," *IEEE Access*, vol. 5, pp. 13327–13335, 2017.
[18] S. R. Chopra, & A. Gupta, "Error analysis of grouped multilevel space-time Trellis coding with the combined application of massive MIMO and cognitive radio", *Wireless Personal Communication*, vol. 117, pp. 461–482, 2021.
[19] D. S. Shajahan Kutty, "Beamforming for millimeter wave communications: an inclusive survey," *IEEE*, vol. 18, no. 2, pp. 949–973, 2015.
[20] Rahul Sharma, Shakti Raj Chopra, & Akhil Gupta. "Power optimization of unmanned aerial vehicle-assisted future wireless communication using hybrid beamforming technique in disaster management," *IOP Conference Series: Earth and Environmental Science*, vol. 1285, no. 1, p. 012025, 2024.
[21] Rahul Sharma, Shakti Raj Chopra, & Akhil Gupta. "UAV assisted next generation wireless communication network," in *2024 International Conference on Emerging Smart Computing and Informatics (ESCI)*, pp. 1–5. IEEE, 2024.
[22] S. Pragati L. Tripathi, S. R. Patre, S. Singh, & S. P. Singh, "Triple-band microstrip patch antenna with improved gain," in *2016 International Conference on Emerging Trends in Electrical Electronics & Sustainable Energy Systems (ICETEESES)*, pp. 106–110, 2016, doi: 10.1109/ICETEESES.2016.7581362

Chapter 15

Exploring the evolution of Artificial Intelligence methods

Tracing the path from traditional to contemporary approaches

M. Nagabhushana Rao, S. K. HimaBindu, K. Aruna Kumari, M. Nagalakshmi and A. Leela Sravanthi

15.1 INTRODUCTION

Artificial intelligence (AI) has now become a really powerful tool that has spread into practically every field, such as healthcare, finance, transport, and others. In this section, some of the modern AI features concerning its implications for these areas will be elaborated in more detail.

15.2 BACKGROUND AND CONTEXT OF ARTIFICIAL INTELLIGENCE

This subsection provides an overview of the history and evolution of AI. It traces the key milestones that have shaped its journey, offering insights into how AI has developed over time. It therefore assists the readers to build up comprehensive knowledge and perspective of history of AI to understand the basis and background for advancement in this area in the present world.

15.3 SIGNIFICANCE OF ARTIFICIAL INTELLIGENCE ACROSS DIFFERENT DOMAINS

The section of this paper aims at discussing how AI is effecting change in different fields such as the health sector, commerce, and traffic, among others. It focuses on the powerful potential of AI to increase effectiveness, raise output, and correctness across numerous uses.

15.4 MOTIVATION BEHIND THE STUDY

This subsection will explain the motivation behind the study, including why a comprehensive survey of AI techniques is necessary. This could include identifying the gaps in current research, the need for a better understanding of AI techniques, or identifying specific challenges in applying AI in different fields.

DOI: 10.1201/9781003643173-15

15.5 AIM AND OBJECTIVES OF THE STUDY

The aim and objectives of this study will be presented in this section. These include: an introduction to a range of AI methods; the style and design of those techniques; actual uses of AI in industry; and new techniques alongside further possible applications in the future.

15.6 OUR CONTRIBUTION

This section aims to emphasize the originality of the research, which involves thoroughly examining AI methodologies encompassing both traditional and contemporary methods. The study's potential impact on the development of AI and its applications in various fields will also be discussed. The study's novelty can be further explained by identifying the specific contributions it will make to AI research. Overall, this section will provide readers with an introduction to the topic of AI and the motivation behind the study, setting the foundation for the remainder of the paper. The subsections will be supported by relevant literature and data, and may include citations and references from various sources such as academic articles, reports, and case studies [1].

15.7 ADVANTAGES

Based on the abstract provided, here are four advantages of the study:

1. Comprehensive overview: This paper provides a detailed account of the modification of AI, with a focus on rule-based expert systems, followed by advanced developments such as deep learning and reinforcement learning. By covering this broad spectrum, it provides valuable insights into how AI technologies have progressed and adapted over time.
2. Diverse AI techniques: It gives an idea about methods like machine learning, natural language processing (NLP), computer vision, expert systems, and fuzzy logic, upon which this study is built. This comprehensive analysis benefits researchers, practitioners, and students, showcasing the wide range of applications these techniques offer.
3. Ethical and societal considerations: The study emphasizes AI's ethical and societal impacts, advocating for trustworthy, understandable, and responsible AI systems. This focus on ethics highlights the importance of developing AI technologies that are transparent and accountable.
4. Educational resource: The study is a comprehensive survey and a valuable resource for newcomers to AI, providing a roadmap through AI's historical and current landscape. It demystifies some of the entrenched concepts regarding AI while promoting the generation of intelligent solutions that address significant issues without compromising on the two principal values of ethical AI: explainability.

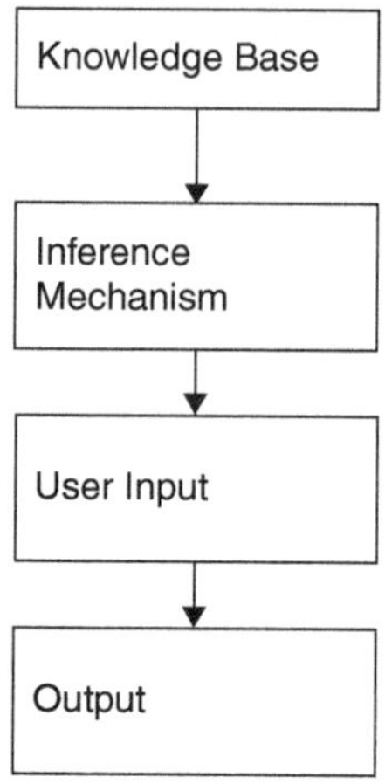

Figure 15.1 Block diagram for Knowledge base system.

15.7.1 Classic AI techniques

Classic AI techniques refer to the first generation of AI techniques developed in the 1960s and 1970s. These techniques are based on symbolic reasoning and rely on rule-based and expert systems. This section will discuss the following classic AI techniques; Figure 15.1 explains the blocks of the Knowledge base system:

15.7.2 Definition and explanation of classic AI

This subsection provides the definition and explanation of classic AI. Classic AI is a branch of AI based on symbolic reasoning, designed to mimic human intelligence using a set of predefined rules and logic.

15.7.3 Expert systems

They are highly defined information systems that have been developed the aim of emulating detailed decision making of experts in given areas. This section will discuss and describe them as relevant social interface structures, what the elements are, and how they operate in domains.

Definition and explanation: This subsection will define and explain expert systems, highlighting their architecture, design, and operation.

Architecture and design: In this section, we will explore the structure and conception of expert systems, encompassing a collection of pre-established rules, a reservoir of specialized knowledge, and a reasoning engine. The reservoir of knowledge houses domain-specific information, while the reasoning engine employs the rules to arrive at decisions and offer suggestions [2].

Real-world applications of expert systems: This subsection will highlight practical examples of expert systems utilized across different fields, including healthcare, financial services, and industrial sectors, showcasing their impact and effectiveness in solving domain-specific challenges.

Challenges and limitations of expert systems: This subsection will cover the challenges and limitations of expert systems, which include their inability to acquire knowledge and adjust to unfamiliar circumstances, the difficulty in acquiring domain-specific knowledge, and the need for constant maintenance and updates.

15.7.4 Rule-based systems

Rule-based systems are specialized computer programs developed to make decisions by following a set of predefined rules. This subsection will explore various aspects of these systems, including their structure, functionality, and applications:

Definition and explanation: This subsection will define and explain rule-based systems, highlighting their architecture, design, and operation.

Architecture and design: This subsection will discuss the structure and development of systems based on rules, encompassing a predefined set of rules, a repository of domain-specific knowledge, and an inference engine. The knowledge base houses the domain-specific information, while the inference engine employs the rules to make decisions and offer recommendations.

Examples of rule-based systems in real-world applications: This subsection will provide examples of rule-based systems in unique domains, such as healthcare, finance, and manufacturing.

Challenges and constraints of rule-based systems: This subsection will cover the challenges and limitations of rule-based systems, which include their inability to acquire knowledge and adjust to unfamiliar circumstances, the difficulty in acquiring domain-specific knowledge, and the need for constant maintenance and updates [3].

15.7.5 Fuzzy logic

Fuzzy logic is a technique that is used to deal with uncertain or ambiguous information. This subsection will cover the following aspects of fuzzy logic:

Definition and explanation: This subsection will define and explain fuzzy logic, highlighting its architecture, design, and operation.

Architecture and design: This subsection will cover the architecture and design of fuzzy logic, which involves a collection of fuzzy rules, a membership function, and a fuzzy inference engine. The membership function ascertains the extent of membership of a variable within a fuzzy set, and the inference engine uses the fuzzy rules to make decisions and provide recommendations.

Examples of fuzzy logic in real-world applications: This subsection will provide fuzzy logic in different fields, such as control systems, image processing, and pattern recognition [4].

15.7.6 Machine learning techniques

Artificial Intelligence is a big subject that encompasses three big branches of study, such as NLP, Machine learning, and Robotics. As an introduction to the main topic, this section provides the definition of the machine learning field and mentions its applications.

15.7.7 Definition and overview of machine learning

Machine learning is all about learning algorithms that can learn how to make decisions themselves without being taught exactly what to do in the case of specific problems. These techniques fall into three main categories: the three categories include supervised learning, unsupervised learning, and reinforcement learning. All of these approaches are unique in the type of data that they are applied to and the circumstances during problem solving.

15.7.7.1 Supervised learning

Supervised learning is a machine learning strategy that deals with models trained on labeled data to make predictions on outcomes. This section focuses on the discussion of the supervised types of learning encompassing decision trees, random forests, and support vector machines. It also discusses applications of supervised learning, such as image classification, speech recognition, and NLP. We also discuss two problems that are associated with supervised learning, for instance, a requirement of a large amount of labelled data and the risk of overfitting [5].

15.7.7.2 Unsupervised learning

Some of the models trained under this technique don't even require predetermined labels of data and are aimed in identification of patterns or similar data units. In this portion, the author discusses clustering and association rule mining, which are the major unsupervised learning techniques, and reviews some typical applications, including anomaly detection and customer segmentation. Additionally, the section highlights challenges in unsupervised learning, such as evaluating clustering quality and the need for domain expertise.

15.7.7.3 Reinforcement learning

Reinforcement learning is where a model learns to make a decision based on the outcomes resulting from it. This section explains the architecture and topology of reinforcement learning, including the Markov decision process and the Q-learning. It also highlights real-world applications such as gaming, robotics, and autonomous driving. Additionally, the section addresses the challenges of reinforcement learning, including the need for extensive training and the potential for negative outcomes [6].

15.7.7.4 Deep learning

Deep Learning is a subfield of machine learning, that relies on artificial neural networks with multiple layers to address the data. In this section, you will meet deep learning models like the Convolutional Neural Network (CNNs), and Recurrent Neural Networks (RNNs). We'll also study the practical scenarios of the use of deep learning, including object and speech identification, language processing, and self-driving cars. But like everything, deep learning has its drawbacks such as the requirements of large datasets, the host of computations required, and the understanding of these models' decision-making process (Figure 15.2).

15.7.8 Natural Language Processing techniques

15.7.8.1 Definition and explanation of Natural Language Processing (NLP)

Natural Language Processing (NLP) is an AI application that looks at how a computer will comprehend and process language that humans use. In a nutshell, it means processing-and-generating natural language as both rational

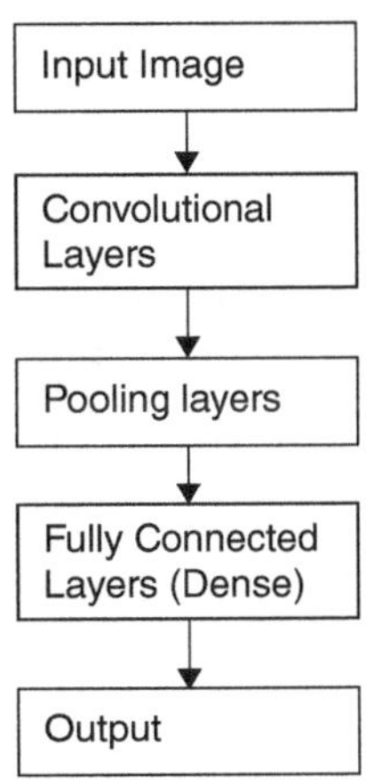

Figure 15.2 Block diagram for CN network.

and linguistic encompassment with the precise concern for effectiveness and adequacy. NLP is an interdisciplinary branch of AI that draws from computer science, linguistics, psychology, and cognitive sciences.

15.7.8.2 Information retrieval

Indeed, NLP is a system that benefited a lot from information retrieval (IR) techniques since it involves the capability of looking for and selecting relevant data from unformatted and partially formatted data, typically text documents, web pages, emails, etc. The purpose of IR is to provide a user with the answer to whatever question they have.

Definition and Explanation: Information retrieval is the process of selecting, refining, and locating information that is stored in an unstructured or a partially structured manner. These are indexing, query processing, and relevance ranking. Architecture and Design: An IR system usually has three main parts: a set of documents, an indexing module, and a query processing module. All these components enable it to search, classify, and return the information that best fits the user's query.

Examples of Information Retrieval in Real-World Applications: A familiar example of an IR system is the search engine of Google. Other applications of IR are spam filters, recommendation systems, and legal document search engines.

Challenges and Limitations of Information Retrieval: Managing the extensive volume of unstructured data available on the web is among the main difficulties in IR.

15.7.8.3 Text analytics

Text mining is also an important subfield of NLP because it is specially aimed at analysis of large volumes of texts in pursuit of useful information and patterns. It is applying statistical, machine learning, and NLP methods to text data.

Definition and explanation: Text analytics hence, all aim at converting textual data in a less structured format into one that is more structured. Includes methods such as IR, Information Extraction, Sentiment Analysis, Topic Modelling, and Text Classification. These methods aid in the discovery of patterns most relevant from large and complex text information.

Architecture and design: A typical text analytics system has three key components: These three components are Data Preprocessing Component, Feature Extraction Component, and Modeling Component or Engine, respectively. Data Preprocessing: This part is completely responsible for tasks like cleaning and normalization of raw text in order to make it suitable for analysis. Feature Extraction: Here, we also detect

features in the data, which would be relevant in the proceeding steps.
Modeling: In this phase, methodological statistical or machine learning tools are used to extract relevant information from the data sets.

Real-life uses of text analytics: Text analytics is used in many industries, including marketing, finance, and the healthcare industry. Some of the applications are the trends on social media networks, customers' feedback, or fraudulent activities.

Challenges and limitations of text analytics: The most significant issue in text analytics is the nature of natural language, being easy to misinterpret or grossly oversimplify. There are also other challenges: erroneous data, domain-specific language that can be difficult to analyze, and no access to annotated data that allows applications to be trained adequately.

15.7.9 Robotics and autonomous systems

15.7.9.1 Definition and explanation of robotics and autonomous systems

Robotics and Autonomous Systems identified as RAS, is a form of advanced mechanics created for self-operation for function- executing purposes. Imaging integrating computer vision and NLP with machine learning capability; this forms the basis for enabling machines to understand the physical environment and act on it, this is evidenced by RAS.

15.7.9.2 Robotics in manufacturing

Definition and explanation: Robotics in manufacturing refers to the use of robots to automate various tasks in the production process. These robots handle repetitive and hazardous tasks, such as welding, painting, and material handling, which are often done by humans. The integration of robotics in manufacturing can boost productivity, lower costs, and enhance safety in the workplace.

Architecture and design: The structure that defines a robotic system, especially in manufacture, is usually defined both in the physical entity as hardware and in the logical entity as software. The hardware parts are considered to be major components, such as the robot arm, the end effector or the tool that is mounted on the robot arm, the sensors, and the actuator. Altogether, these components allow for the robot to perform particular tasks with exactness and effectiveness [7]. The components include the control software, programming environment, and simulation tools. The design of a robotic system relies on the application's particular needs, like the type of manufacturing process, the size and weight of the objects to be handled, and the level of precision required.

Challenges and limitations of robotics in manufacturing: Using robotics in manufacturing presents several challenges and limitations. Among the difficulties is the high cost of robotic systems, which can hinder adoption for small and medium-sized enterprises. Another challenge requires specialized skills to design, program, and maintain robotic systems. The limitations of current robotic technology include the lack of talent and flexibility of the robot arms, which limits the range of tasks they can perform [8].

15.7.9.3 Autonomous vehicles

Definition and explanation: Self-driving cars are built to run on the freeway with no human intervention. These particular vehicles employ the use of sensors, cameras, and GPS systems to acquire information, which the vehicles then act upon. Autonomous cars have the potential to make roads safer, eliminate traffic jams as well as increase mobility for people who might otherwise have a difficult time getting around.

Architecture and design: The system of an autonomous vehicle consists of several software and hardware features that have been mentioned in the thesis. Such hardware elements are sensors such as LiDAR, cameras, radar, and mechanical mechanisms like steering and braking. The software components include a perception system, which analyzes the signal coming from the sensors; decision system, which decides what has to be done; and the control system, which actually performs the required action (Figure 15.3).

Examples of autonomous vehicles in real-world applications: Self-driving cars are under development and testing by Waymo, Tesla, Uber, and so many others. Waymo, a subsidiary of Alphabet Inc., has been conducting tests of autonomous vehicles on public roads since 2015.

Challenges and limitations of autonomous vehicles: Today, autonomous automobiles are proving much promise; however, their implementation is not without several difficulties. One of the critical issues is security – crashes that occurred with cars of automated control forced us to doubt the complete reliability of such cars and their ability to manage critical, unpredictable real-life scenarios. Another is structural because the laws and regulations governing the use of autonomous vehicles are not the same across the world, and therefore developers find it difficult to put in place acceptable standards by which auto vehicles should be designed. It is becoming increasingly clear that current autonomous vehicle technology may be problematic at the present moment in time, for example, with issues operating under poor illuminated conditions such as heavy rain or snow. Additionally, it requires high-quality mapping data to function effectively in urban environments [9].

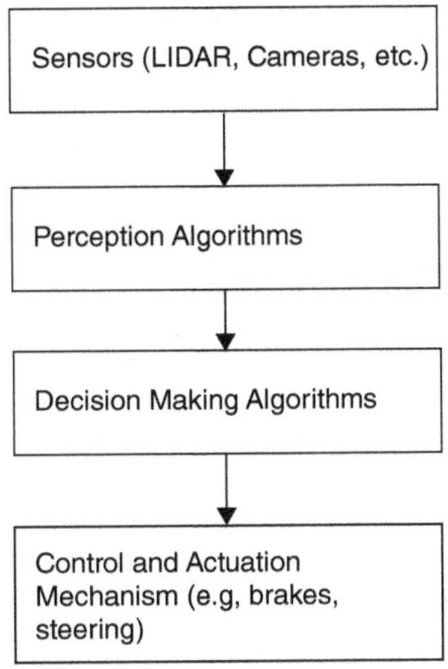

Figure 15.3 Block diagram for autonomous vehicle.

15.7.9.4 Drones

A drone is an uncrewed aerial system, programmed to be flown autonomously or by the operator through a remote-ground – control station. Recently, drones have emerged as popular due to the flexibility of use and the ease they provide in their operations. The uses of drones in military are surveillance or reconnaissance, and assassination. Amid the peaceful uses of drones, the most common and well-known uses are aerial photography and videography, delivery, aerial cinematography, mapping and surveying, search and rescue, and agriculture.

Definition and explanation: Drones are small aerial systems that may be operated autonomously or with the help of a pilot from the ground. They come with sensors and cameras for collecting data on the environment where the unmanned car is located and images, respectively. AI applications for drones include computer vision, machine learning, and path planning to enable the drones to make decisions on its movements.

Architecture and design: The architecture of a drone comprises some critical subsystems whose main function is to facilitate its flight and control. These include the structure, which is airframe, the source of power, otherwise called the propulsion system, and the directing and mapping system in order for the drone to cover the intended area. The navigational aids of this drone include GPS, accelerometers, and gyroscopes that enable the device to see where it is and in which direction it is flying, including the stability when it is flying. The control system includes the flight controller, which manages the drone's movements and controls the speed of the propellers.

Challenges and limitations of drones: However, there are also some disadvantages and drawbacks associated with the use of drone's technology. These include:

Safety concerns: The use of drones may be Eisenhower to different aircrafts, persons, and property in case of a wrong maneuver.

Privacy concerns: Aerial photography violates the privacy of people since the drones take pictures and record videos, as a result of which people's consent is not sought.

Regulatory challenges: Operations of drones are restricted in many ways, and it may be quite a decision-making process to try and arrange all these legalities and acquire the relevant licenses.

15.7.10 Emerging AI techniques

15.7.10.1 Reinforcement meta-learning

Reinforcement Meta-Learning is a relatively burgeoning approach in reinforcement learning as an AI sub-discipline. It allows an AI to train the system in a way that the learning process is adapted to the given task in the course of its functioning. In other words, Reinforcement Meta-Learning is the ability of an AI to acquire new knowledge and to do so within a reasonable time. This approach is most effective particularly, when an AI system has to learn from minimal information and, in the second case, when an AI system has to be "trained" for new tasks and circumstances. It uses reinforcement learning in an effort to attain a meta-learning that actually leads to the creation of a system that can indeed "learn learning."

The architecture of Reinforcement Meta-Learning consists of two main components: The first aspect is what I termed the meta-learner, and the second aspect is what I termed the task-learner. Meta-Learner: In this case, it becomes essential to get the approaches or methods that can be employed in learning the approaches or methods. Task-Learner: This component is the one that is fully responsible for a given task. Training in the particular process of acquiring new tasks is all about the meta-learner wanting and trying to improve the task learner to be capable of learning other tasks in the shortest time possible. They will also provide for unperturbed interactions with new conditions and the gradual enhancement of the AI system performance implemented.

Reinforcement Meta-Learning is applied in almost any discipline, such as computer vision, NLP, robotics, and many others, in order to enhance the ability of a system to learn in a way that enhances learning. For instance, an AI system trained with reinforcement learning can quickly adapt to new environments in computer vision tasks, such as changes in lighting or new objects. In NLP, an AI system can quickly learn new languages or dialects with Reinforcement Meta-Learning. In robotics, an AI system can adapt to new tasks or environments in real time, such as changes in terrain or object positions [10–12].

15.7.10.2 Generative Adversarial Networks (GANs)

GANs means Generative Adversarial Networks – an innovative, perspective direction in AI that involves two neural networks. There is one network known as a generator that is responsible for generating fake data which resembles the real data, such as image and text data. The other network, called the discriminator, distinguishes between the produced data and the real data. The flow of intense rivalry between the two systems benefits the generator in enhancing its capacity to generate more realistic data in the long run.

Architecture of GANs: GANs consist of two main components: the generator and the discriminator. Generator: This network helps you generate a varied data set from noise. Discriminator: This network decides whether what has been produced looks like data that has been obtained naturally. The generator component is adjusted here to produce data that is greater in realism than the training data, and, conversely, the discriminator component is adjusted to make a more accurate determination of the real and fake types of data. This exchange process, I suggest, results in improvement of both over time.

GANs have a variety of applications, such as: Image and video synthesis: Designing lifelike images and videos right from the ground up. Text-to-image generation: Superimposing of text descriptions onto images. Data augmentation: Creating more data for AI applications, for instance, creating synthetic medical images to amplify AI's ability to work on healthcare tasks, such as the diagnostics of diseases. GAN can be applied to produce realistic medical images for training healthcare AI and even produce the virtual environment for the simulated augmented reality.

Challenges and limitations of GANs: Another problem of GANs is managing the adversarial loss, which consists of achieving a proper balance between the generator and the discriminator. Failure in cooperation may lead to the generation of low-quality data among the parties involved in the process. A key problem is that in certain applications, it is challenging to make a determination regarding the realism and quality of the generated synthetic data.

15.8 CONCLUSION

15.8.1 Summary of key findings

This survey has provided an extensive overview of a range of AI techniques, spanning both traditional and contemporary methods. This encompasses: expert systems, rule-based systems, fuzzy logic, machine learning, NLP, and robotics. The survey also discusses newer trends in AI, such as reinforcement meta-learning and GANs. The issues examined are related to the architectural and design aspects of the techniques, as well as their usage in real-life applications and the deficiencies that must be addressed for the respective fields to evolve.

15.8.2 Contributions and novelty of the study

This study is novel in its comprehensive coverage of various AI techniques, providing a thorough analysis of each approach's architecture, design, and real-world applications. The study has also highlighted each technique's challenges and limitations, which is essential in developing future AI applications that address these limitations.

15.8.3 Study limitations

One drawback of this study is that the coverage of each AI technique is not exhaustive. Due to the vast number of AI techniques and their applications, it was impossible to cover every approach in-depth. Additionally, due to the rapidly evolving nature of AI, some of the information presented in this study may become outdated over time.

RECOMMENDATIONS FOR FUTURE RESEARCH

There is a set of urgent questions and issues associated with the development of AI technologies, which future studies should address, namely: ethical or legal implications, the problem of bias and unfair treatment of clients, and problems of security and privacy protection. There is also a need to advance AI techniques and applications to overcome these obstacles. Additionally, more exploration is required into the social and economic implications of AI to develop policies and regulations that promote its responsible and beneficial use.

In conclusion, AI techniques can potentially revolutionize various fields, from manufacturing to healthcare. It is therefore important that these challenges be dealt with in order to prevent misuse of AI and bring effectiveness in social welfare. This paper has given an idea of different types of AI, their structure, development, and their practical use in industries and commerce, along with their drawbacks. These findings are essential for new AI applications to successfully address these challenges and move toward the advancement of the field.

REFERENCES

1. Z. Ullah, F. Al-Turjman, L. Mostarda, and R. Gagliardi, "Applications of artificial intelligence and machine learning in smart cities," *Comput. Commun.*, vol. 154, pp. 313–323, 2020.
2. R. Bhatia and R. Nagpal, "Reviewing expert systems in healthcare," *J. Biomed. Inform.*, vol. 85, pp. 112–127, 2018, doi: 10.1016/j.jbi.2018.07.003

3. H. Chen, X. Liu, D. Yin, and J. Tang, "A survey on dialogue systems: Recent advances and new frontiers," *ACM Trans. Inf. Syst.*, vol. 36, no. 2, pp. 1–34, 2018, doi: 10.1145/3185591
4. I. Goodfellow et al., "Generative adversarial nets," in *Advances in Neural Information Processing Systems*, vol. 27, pp. 2672–2680, 2014.
5. S. B. Kotsiantis, "Supervised machine learning: A review of classification techniques," *Informatica*, vol. 31, no. 3, pp. 249–268, 2007.
6. G. Rajan, M. Murugappan, and S. Yaacob, "A comprehensive review on autonomous vehicles: Progress, challenges, and future directions," *J. Adv. Res. Dyn. Control Syst.*, vol. 11, no. 5, pp. 1–14, 2019.
7. H. N. Suresh and P. Karthikeyan, "A survey on machine learning techniques in fault diagnosis for mechanical systems," *Mech. Syst. Signal Process.*, vol. 85, pp. 486–504, 2017
8. J. Wang and W. Liu, "A survey on deep learning for big data," *Inf. Fusion*, vol. 42, pp. 146–157, 2017, doi: 10.1016/j.inffus.2017.10.006
9. Y. Wu et al., "Google's neural machine translation system: Bridging the gap between human and machine translation," *arXiv preprint arXiv:1609.08144*, 2016.
10. Y. Zhang and B. C. Wallace, "A sensitivity analysis of (and practitioners' guide to) convolutional neural networks for sentence classification," *arXiv preprint arXiv:1510.03820*, 2015
11. D. Ghai, S. L. Tripathi, S. Saxena, M. Chanda, and M. Alazab, *Machine Learning Algorithms for Signal and Image Processing*, Wiley-IEEE Press, 2022.
12. S. L. Tripathi and M. Mahmud, *Explainable Machine Learning Models and Architecture*, Wiley & Scrivener Publishing LLC, 2023.

Chapter 16

Sustainable computing, greener design for computer server systems and softwares

Kadambri Agarwal, Sandhya Avasthi, Aditya and Aman Pandey

16.1 INTRODUCTION

The fast-growing nature of information and communication technology (ICT) has swept the world, changing the way we live, work, and communicate. While this technological growth is going on, it is also taking us a big step back in terms of the environment. The IT industry is the leading cause of greenhouse gas emissions, use of energy, and exhaustion of limited natural resources. The ever-growing need for global computing power and digital services consumption places this sector under the process of proactively addressing the likely environmental impact.

Green computing, in general usage, synonymous with sustainable computing or the eco-friendly computing, is a term that embraces several practices and strategies with a view to lesser the ecological footprint of IT operations. From conceptualization, production, and use to final discard, this includes the circular system of use of ICTs and systems. Introduction of green computing is not only a way to reduce the organization's environmental impact but also the final result of the long-term cost savings, supporting proper social responsibility and compliance with the stricter regulations of the environment [1, 2].

This chapter focuses on two crucial aspects of green computing: harnessing solar energy as a green power source and promoting reusability of old goods that could be used to prolong the life of an electronic device and drive out the e-waste.

16.2 WHAT IS GREEN COMPUTING?

Green computing is a very wide-ranging concept that puts a point of focus on the care of the environment in terms of computers processing and handling. It implements the range of methods and considerations, which the latter are meant to keep minimal ecological footprint of the IT service side of the business regardless, as well as to improve or increase effectiveness.

DOI: 10.1201/9781003643173-16

The core principles of green computing include the following: Energy efficiency is the optimization of devices; virtualization is the use of the right resources when they are needed; and power management is the establishment of the ways that power can be controlled.

Resource conservation: The introduction of new regulations on mining excessive and uncontrolled raw inputs from the earth for the sake of computer equipment production by relying on the reusing, recycling, and responsible disposal of such materials as the new standard [3, 4].

Renewable energy utilization: In my opinion as a policy advisor, introducing renewable energy sources such as solar, wind, and hydropower is inevitable since it will help to reduce the use of coal plants and greenhouse gases.

Sustainable design: Sustainable production has certain things in it: using environmentally friendly materials, reducing the consumption of hazardous materials, and creating the whole product considering the environment at every step, including design, production, and waste disposal.

Awareness and education: Ensuring that the environmentally responsible approach is built into organizational and institutional culture, and through engaging stakeholders that include the individual population as well as policymakers in green computing practice awareness and provision of relevant information [1].

16.2.1 Applications

The given illustration is a graphical depiction of the basic definitions of Green Computing, which are computer processes and technology that are ecologically sustainable (Figure 16.1).

Green design: It indicates the necessity of taking environmental friendliness into account at every stage of computing hardware, software, and systems production. The range of green technologies requires green architectures, and eco-friendly materials and minimizes the consumption of dangerous materials.

Green manufacturing: Green manufacturing procedures are directed toward the stage of computer device and component production. This embraces implementing renewable energy sources, lessening waste and emissions, and adopting a sustainable supply chain management strategy.

Green usage: By doing so, this shows the importance of deploying energy-efficient and environment-friendly computing processes. It comprises strategies like energy management, virtualization, and optimization of resources, which eventually decrease the amount of energy and extent of carbon footprint.

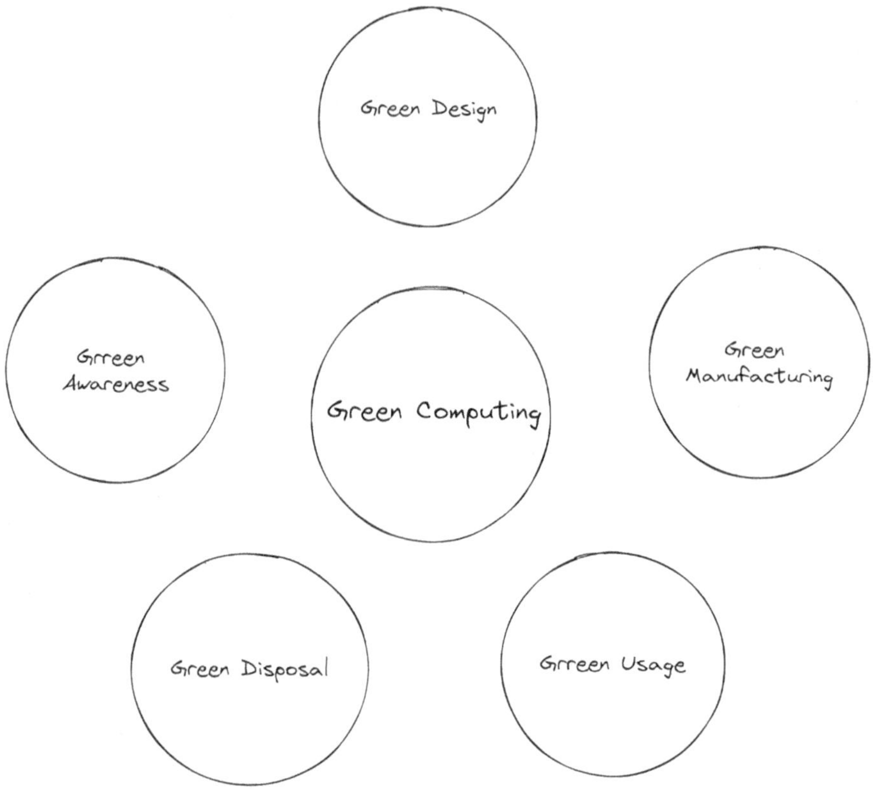

Figure 16.1 Definition of green computing.

Green disposal: Appropriate disposal and recycling of computer hardware and electronic waste make them important aspects of green computing. Thus, this unit stresses the significance of responsible waste management, electronic scrap recycling and minimizing environmental burden caused by discarded computing equipment.

Green awareness: Developing the consciousness and implementing the eco-friendly computing practices among and between individuals, organizations and society as a whole is vital for the realization of the green computing initiative. This component covers equal learning, training, and cultivating a green culture.

The interrelated factors in the painting indicate that green computing is a comprehensive method that implies the entire phases of computing devices and systems, starting from the design to the usage and the disposal. Every part of the segment is essential in decreasing the environmental impact of computing, and implementing the benefits in the ICT sector. [1, 2]

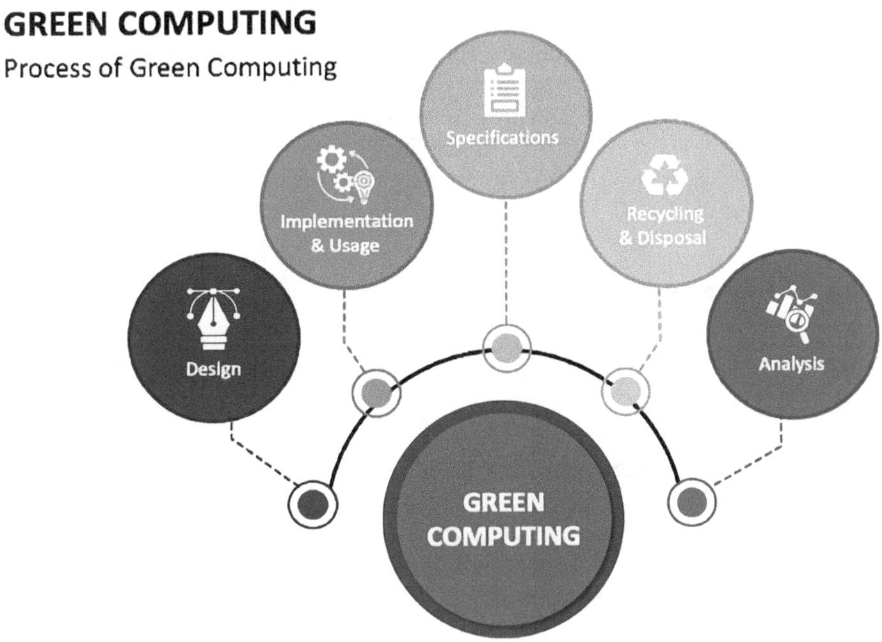

Figure 16.2 Process of green computing.

16.2.2 Process of green computing

(See Figure 16.2.)

16.2.2.1 Design

Normal design process of any software and cloud architecture includes designing independent modules that work autonomously and collaboratively to serve a complete purpose or feature that the architecture was built to provide.

However, a green computing architecture requires any particular design to adhere to some core principles of energy consumption, environmentally friendly strategies, processes of deployment, use of technologies, etc.

Assessment of environmental impact: Identification of impact caused by designed architecture and utilized technologies. For example, AWS choosing to allow shared hosting is an environmentally friendly step.

Set goals and objectives: Turning core objectives of the design decisions into an environmentally friendly step could help the process to be more aligned with green computing.

Renewable energy resources: Using renewable sources of energy to power our software, cloud, or servers is also a good step in green design, utilizing solar energy to power the servers.

Resource conservation: Designing an architecture which is optimized to save energy consumption on the basis of its momentary usage, this helps in reducing the consumption of resources when a piece of tool is not in use.

16.2.2.2 Implementation and usage

In green computing, implementations and usage mainly refer to the stage when green technologies are used for energy efficiency and sustainability. This includes:

Install power management functionality in software as well as hardware phases to minimize electricity consumption during use.

Promoting consumers to engage themselves in lifestyle-oriented power-saving habits, for instance, switching off devices when they are out of use and activating sensors which help to save energy.

Making provision for education and orientation of users on eco-friendly computing and handling of computing resources without compromising sustainability.

Based on the usage data monitoring and formation of the usage patterns, as well as identification of the points for improvement, and making of optimized resource utilization.

16.2.2.3 Specifications

The primary step in implementing Green computing standards is to design devices and programs that conform to environmental standards and sustainability. This involves:

Regulations such as implementation of energy-efficient designs, resource conservation, and minimal ecological damage are setting sustainable design requirements and standards.

Developing ecological materials and reducing environmental concerns using low-power processors, and recycling materials results in eco-friendly implementation.

Develop increased software optimization, in terms of energy efficiency and resource utilization, that will involve usage of energy-efficient algorithms and power-efficient features.

Joining forces with such parties as suppliers and other stakeholders so that the products, their components, and assembling details are all geared to green computing.

16.2.2.4 *Recycling and disposal*

With the objective of limiting generated e-waste and reusing as much as possible, recycling and disposal in green computing are mostly done. This includes:

Building products having recyclability as their key characteristics, such as using separable components that can be recycled, or the material being recyclable.

Launching collection campaigns, take-back programs, and designing the recycling systems to help dispose of the electronic waste in a pollutant-free manner.

Partnering with verified recyclers and garbage collection companies that ensure all the e-waste is disposed of the right way using environmentally friendly techniques.

Educating the Public: consumers and businesses on why recycling electronic waste is essential, how such waste should be taken care of, and we should provide guidance on how to properly dispose of worn-out devices [4].

16.3 GREEN COMPUTING IN IT

The IT industry contributes greatly to the green computing practices developed and integrated, as it is the one with the biggest environmental footprint. The following subsections explore two specific aspects of green computing in IT: use of solar power and hardware reuse.

16.3.1 Solar energy in IT

The adoption of solar energy usage in IT operations is a promising technique to decrease the environmental footprint of the industry and lower its dependency on conventional sources of energy. Clean, renewable, and endless sun rays could be utilized to run data centers, offices, and different IT infrastructure.

16.3.1.1 *Benefits of solar energy in IT*

Reduced carbon emissions: Solar energy helps IT organizations to efficiently prevent air pollution, greenhouse emissions and thereby, take measures toward environmental protection. Data centers and IT infrastructure are recognized as energy-intensive, more often than not being contributors to the emission of carbon dioxide into the atmosphere. When an organization switches to solar energy, it can partly overcome pollution and is in line with sustainable objectives. [5]

Cost savings: The cost of the installation of solar power systems can be pretty high initially, but the operating costs are generally cheaper than those of other energy sources in the long run. Solar power plants have a very low operational cost, as the generated energy is basically cost-free by the time your initial investment is returned. Moreover, the different government incentives, tax credits, and Renewable Energy Certificates (RECs) can also reduce the upfront costs, which makes solar energy investment more and more an attractive financial option. [5]

Energy resilience: The installation of solar energy systems can help provide a stable and uninterruptible power supply and concurrently decrease the chances of power outages and disaster occurrences. Such situations are especially common when it comes to IT duty that demands uninterrupted performance with secure supplies of electricity. If the solar power generation systems are integrated with the energy storage devices, organizations can maintain an uninterrupted power supply and business continuity even when there are grid failures or undesirable climate conditions. [5]

Corporate social responsibility: Embracing the use of solar energy shows the organization's devotion to environmental sustainability, which in turn will boost the reputation and brand value of the company throughout every stakeholder that supports the cause. Consumers and stakeholders are becoming more aware of environmental issues. As sustainable practices are made a priority by businesses, these companies get a competitive edge and are able to draw eco-conscious customers, employees, and investors.

16.3.1.2 Implementing solar energy in IT

On-site solar installations: The business enterprises can install some solar photovoltaic (PV) panels if there is space available on the roof or ground of their offices, including IT companies, data centers, and head offices. This power production is the direct use of electricity and the reduction of unnecessary energy loss during the processes of transmission, and that also brings about community-centralized controls on energy supply. I choose from a variety of pharmaceutical products, machines, water, and suitable tools that can be calibrated to produce the energy equivalent to my IT operations.

Off-site solar farms: In such cases, as there is insufficient space/moral conditions on-premise for roof-mounted installations, or when preconditions for implementing the system are not provided, organizations can either rent solar energy from solar farms constructed off-site or opt for different contractual partnerships called power purchase agreements (PPAs). PPAs provide an opportunity to use solar energy without a large upfront capital investment that is needed to install the solar systems. Hence, with the PPA, solar power is affordable and can be enjoyed by more and more people.

Grid-tied systems: Grid-tied solar power can do this when it is the major electricity supplier, and also, the excess electricity produced can be fed back into the grid, or the organization can also draw power when needed. These grids established a link between other resources, which made us have stable and constant energy available presently and gradually made integrating with the old infrastructure easy, and reduced the irregularity of solar energy.

Energy storage solutions: Storage methods, like batteries or thermal storage, can be loaded contingencies to achieve the target of providing a grid that is uninterrupted and stable. For this purpose, the surplus energy stored when sunlight is abundant is used for necessary demands, and hence, an uninterruptible power supply during IT operations is obtained.

Energy management and monitoring: The importance of adopting energy management systems with quality monitoring tools along the whole process of solar energy utilization cannot be overstated from the point of view of energy efficiency. The resulting energy management systems are enabling companies to see the overall energy use trends, find out waste, and eventually take improvement actions to optimize energy efficiency. The application of this energy in parallel with grid-supplied electricity can make economies gain independence.

16.3.2 Hardware reusability

The fast pace at which erasable computing devices or the prevailing need for new technologies are being used has all played a major role in creating tons of that electronic waste or e-waste.

Resuscitating hardware is one important aspect of green computing, which allows usage of old computing devices for longer and reduces the need for remanufacturing those components, which consume a lot of resources.

Benefits of hardware reusability:

Waste reduction: Some organizations are trying to rectify this by using of lesser new equipment but rather recycling of the old electronic devices' components, which is a move aimed at reducing the amount of e-waste disposed which is proven to be very unfriendly to the environment during the recycling and disposal process. At the same time, the waste is willingly or not E-contained with several hazardous substances, including lead, mercury, and cadmium, being among the critical compounds that create immediate threats to the ecosystems and human health when not treated professionally.

Resource conservation: Utilizing the already-existing computers either by repurposing them prolongs the period until replacement, increasing the one-time industry-bad process where finite natural resources are used up, destroying the environment. Fabrication of these "high tech" gadgets involves the taking or mining of metals that are scarce, like the precious metals and rare earth elements, and also plastics as the end products, which can keep the resource base finite for quite some time [3, 4].

Cost savings: In restoring commonly used projects, the process of replacing old hardware may even be cheaper than buying new equipment. This gives space for companies to save big if they exactly use what they already have as their primary technology assets, so they would not often request hardware upgrades or buy new hardware.

Environmental compliance: Additionally, several communities and particular municipalities are enjoined by the EU-based WEEE Directive to implement a certain method of managing electrical and electronic waste. Such organizations can practice environmental sustainability by cultivating proper management of machines that can be reused, hence rightly adhering to waste e-rules and regulations and thus, not being slammed with any fines or penalties that might emanate from such regulations [5].

16.4 WHY DO WE NEED GREEN AND SUSTAINABLE COMPUTING?

We need green and sustainable computing for several important reasons:

Environmental impact: The ICT sector is a major contributor to carbon emissions and energy consumption worldwide due to the utility of this technology. Green computing practices seek to achieve an environmentally friendly aspect of computer devices and infrastructure through measures like improving power efficiency, minimizing wastage in the range of facilities, and the mainstream usage of renewable energy [2].

Energy efficiency: Computational systems and data centers use a lot of energy, and these operations cost us a big deal and emit large quantities of carbon dioxide. Green computing methods like energy-efficient devices, virtualization, and efficient power management that employ conditioning, load leveling, and outages, among others, can contribute to the reduction of energy consumption and expenses [3–5].

Resource conservation: Mining, using, and processing computing equipment imposes the demand on the natural resources and their finiteness, which are rare earth metals and precious materials. Sustainable computing methods encourage resource efficiency, recycling, and the listing of circular economy principles, which contribute to the sustainable development goals.

Regulatory compliance: A lot of governments and business entities have brought numerous laws and norms into this field with a focus on energy efficiency and waste management. Adopting greener and sustainable computing ways can enable businesses to remain in compliance with these regulations and, at the same time, keep out of trouble that comes with fines and penalties.

Corporate social responsibility: Despite the different tools that raise the awareness of individuals toward environmental issues, many people have a higher expectation from corporations to behave in a more environmentally friendly manner. Green computing and sustainable computing practices can make your corporation more significant socially and externally, creating an image that your company is much future-oriented.

Cost savings: Efficient computing technologies and competent resource management can yield major savings in operation in the future because these solely lead to less power usage, waste removal, and each device's life extension [4].

Future-proofing: With the increasing global concerns on environmental issues and both national and international long-term regulations becoming stricter, adopting green and sustainable computational processes early puts organizations that are already prepared to compete and even thrive in present-day environments where such practices are not yet required.

16.5 TURNING OLD COMPUTER PARTS INTO GREEN

The fastest-growing waste products in the United States are electronics. Electronics or computer parts occupy significant landfill space, and many individuals are unaware that these components contain hazardous chemicals such as lead oxide, mercury, nickel, zinc, and cadmium, posing risks to both humans and the environment. Green computing advocates the ER3 principle: Eliminate, Reduce, Reuse, and Recycle. While some computer parts can be reused, certain components like chips requiring upgrades and motherboards for their mounting may not be. Consumers should recognize that nearly all computer parts can be recycled and reused, a process known as computer recycling. This involves refurbishing old computer parts into new, usable ones. For those looking to dispose of old computers, there are refurbishing businesses that repair and clean them. Options include resale or donation to schools and non-profit organizations, or sending them to recycling facilities where metal, plastic, and glass components can be broken down and repurposed. Presently, many individuals perceive it as more economical to replace computer parts than to repair them due to the associated labor costs [3, 4]. They also look for computer parts that are environmentally friendly. Computer parts that can be recycled are as follows:

Glass monitor
Keyboard
CD-ROM drive
Plastic case
Cathode ray tube (CRT)
Cables
Copper in the power cord
Metal from the circuit board
Printer cartridges
Batteries

16.6 SURVEY

We initially started with a quantifiable survey of target **Django and Wagtail CMS/other CMS** based sites.

Main objective of this survey is to quantify the related carbon emissions of these sites yearly. total sites surveyed: **23856** (Django only).

Total carbon emission recorded: **264,869 tCO2e/year.**

Complete survey data has been incorporated into the spreadsheet provided below, which also deals with sites of various stacks such as LAMP, MERN, MEAN, etc. [6] (Figures 16.3 and 16.4).

16.7 PRACTICAL APPROACH TO GREEN COMPUTING

This section contains a research- and analysis-based demonstration of how we can build a green computing server to serve our own apps (Figure 16.5).

We would also make use of some benchmarking softwares to analyze the difference between green practice based application-server combination vs regular system combination.

16.7.1 Overview of this endeavor

This initiative aims to tackle the increase in carbon emissions from data centers in response to growing concerns and the urgent need for sustainable technological solutions. Environmental awareness by constructing a server using recycled hardware components and solar panels as an energy source. To further reduce power consumption significantly, we commit to implementing workload enhancements and energy-efficient software solutions. As part of the project, the "Green Metric Tool" is employed for real-time monitoring of carbon emissions to track the reduction in carbon footprint. The anticipated outcomes include decreased carbon emissions, cost savings from energy utilization, and evidence showcasing the practicality and benefits of

3	https://ar.skokka.com/	518.9882813	1000	1,200,000,000	0.1772004933	212640.5919
4	https://pubmed.ncbi.nlm.nih.gov/	507.9228516	1000	1,200,000,000	0.1734223741	208106.8489
5	https://chaturbate.com/	2048.306641	1000	1,200,000,000	0.6993625102	839235.0122
6	https://br.skokka.com/	524.8583984	1000	1,200,000,000	0.1792047537	215045.7044
7	https://www.udemy.com/	2080.118164	1000	1,200,000,000	0.7102240611	852268.8733
8	https://prezi.com/	1338.289063	5000	1,200,000,000	0.4569380285	548325.6342
9	https://nextdoor.com/	3569.939453	5000	1,200,000,000	1.218900416	1462680.5
10	https://prodoctorov.ru/	1122.075195	5000	1,200,000,000	0.3831151594	459738.1913
11	https://www.zurnal24.si/	9285.957031	5000	1,200,000,000	3.170545899	3804655.079
12	https://milano.bakecaincontrii.com	548.3300781	5000	1,200,000,000	0.1872187944	224662.5533
13	https://ii.skokka.com/	511.4902344	5000	1,200,000,000	0.1746404016	209568.4819
14	https://www.slovenskenovice.si/	3421.732422	5000	1,200,000,000	1.168297426	1401956.911
15	https://www.mreader.co/	1882.829102	5000	1,200,000,000	0.6428627729	771435.3275
16	https://www.teachoo.com/	4268.892578	5000	1,200,000,000	1.457547112	1749056.535
17	https://roma.bakecaincontrii.com/	550.4082031	5000	1,200,000,000	0.1879283379	225514.0055
18	https://emojipedia.org/	1858.617188	5000	1,200,000,000	0.6345959906	761515.1888
19	https://www.jstor.org/	4665.105469	5000	1,200,000,000	1.592827854	1911393.425
20	https://siol.net/	5111.911133	5000	1,200,000,000	1.745382713	2094459.256

Figure 16.3 Yearly carbon emissions sourced by random Django sites.

1	url	rank	pageviews/year	kgCO2e/year	total_kb	gCO2e/pageview	html_kb	js_kb	css_kb	kgCO2e/year	img_kb	gCO2e/pageview	g
2										Images only			
3			Projection (7800 sites)	35,799	tCO2e/year				Projection (7800 sites)	20,753	tCO2e/year		
4			Average	5	tCO2e/year				Average	3	tCO2e/year		
5			Total (2470 sites)	11,336	tCO2e/year				Total (2470 sites)	6,572	tCO2e/year		
6													
7	https://www.nhs.uk/	1,000	1,200,000,000	192,340	469	0.16	13.18	59	225	56,846	139	0.05	
8	https://octopus.energy/	10,000	100,000,000	166,178	4,867	1.66	9.95	4,342	87	6,654	195	0.07	
9	https://www.slovenskenovice.si/	10,000	100,000,000	234,459	6,867	2.34	165.74	1,743	59	91,882	2,691	0.92	
10	https://www.thinkwithgoogle.com/	10,000	100,000,000	55,072	1,613	0.55	74.87	406	50	35,264	1,033	0.35	
11	https://www.vinmec.com/	10,000	100,000,000	405,538	11,877	4.06	32.84	1,728	203	335,905	9,838	3.36	
12	https://about.google/	100,000	50,000,000	21,106	1,236	0.42	14.34	560	133	7,113	417	0.14	
13	https://admob.google.com/	100,000	50,000,000	14,170	830	0.28	35.80	513	33	1,383	81	0.03	
14	https://arvr.google.com/	100,000	50,000,000	225,012	13,180	4.50	5.14	136	13	196,104	11,487	3.92	
15	https://autot.tori.fi/	100,000	50,000,000	45,061	2,640	0.90	38.28	896	38	28,091	1,645	0.56	
16	https://blog.google/	100,000	50,000,000	250,607	14,680	5.01	92.67	1,338	173	221,193	12,957	4.42	
17	https://blog.youtube/	100,000	50,000,000	19,117	1,120	0.38	47.86	316	25	9,334	547	0.19	
18	https://cryptokinews.com/	100,000	50,000,000	22,216	1,301	0.44	27.24	417	29	11,491	673	0.23	
19	https://dj1jklak2e.28car.com/	100,000	50,000,000	5,797	340	0.12	12.63	53	2	4,643	272	0.09	
20	https://emojigraph.org/	100,000	50,000,000	2,010	118	0.04	4.29	59	31	401	23	0.01	

Figure 16.4 Yearly carbon emissions sourced by random Wagtail sites.

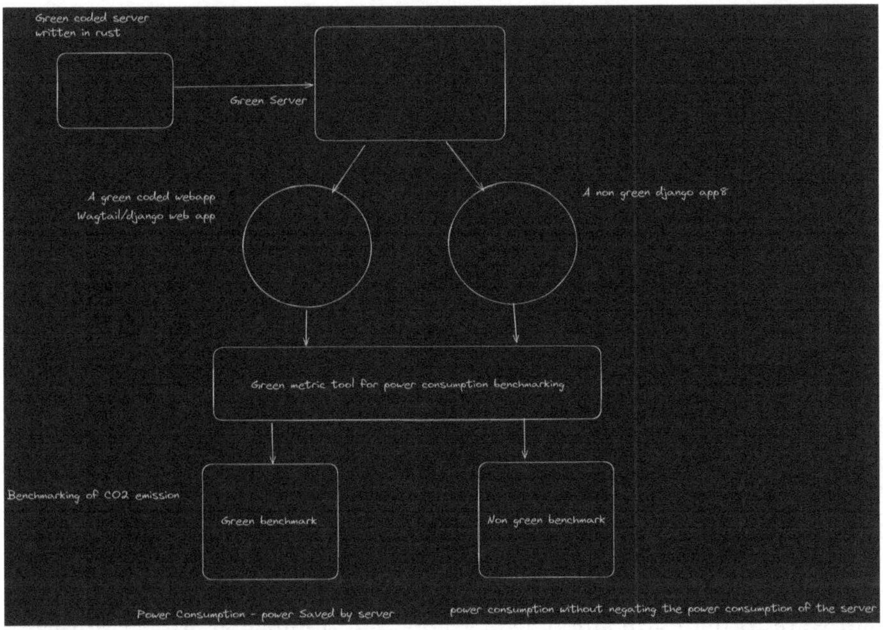

Figure 16.5 High-level diagram for benchmarking green versus non-green server-software setup.

computing that could lead to transformative impacts. Our objective is to promote the adoption of conscious practices in the tech industry by demonstrating the economic viability of sustainable computing, thus paving the way for a more sustainable and eco-friendly future [7].

16.7.2 Green metric tool and carbon components

Embedded carbon:

"Embedded carbon" refers to the total carbon footprint associated with the creation and manufacture of server hardware, including production, transportation, and assembly, akin to raw material extraction. It is a valuable indicator of the environmental impact of hardware even when not in use. Recognizing and quantifying embedded carbon is crucial in evaluating the overall sustainability of computer infrastructure. [8]

Associated carbon:

Associated carbon refers to the carbon emissions produced during the process of powering the server, showing the environmental impact of the electricity used. Servers are typically powered by energy from fossil fuels, renewable sources, or a mix, depending on the local energy mix. Measuring associated carbon helps understand the indirect environmental costs of server operation. [8]

Holistic perspective with GMT:

The Green Metric Tool (GMT) comprehensively calculates the environmental impact of computing by combining software weight and embedded carbon along with related carbon. The following components are included in GMT readings:

Software weight: GMT readings consist of Software Weight, which considers program complexity and resource needs. Heavier software tends to require more processing power. Our study proposes objectives for eco-friendly server settings to promote sustainable and energy-efficient computing practices.

Embedded carbon: The carbon impact resulting from the server hardware's manufacturing.

Associated carbon: The carbon released during the process of producing the energy needed to run the server.

The combination of these components in GMT values offers a benchmark for evaluating the environmental effectiveness of computing systems. Our research aims to assess and reduce these parameters by installing a solar panel system to power the server with sustainable energy, using recycled hardware to minimize embodied carbon, and optimizing critical software. [7]

16.7.3 Objective

Our research presents a complete set of objectives to develop sustainable and energy-efficient server settings, with the aim of advancing environmentally friendly computing practices.

Optimized green software:

One of our research goals is to create a small-scale, highly optimized application. First, optimization focuses on lowering the software's resource requirements and increasing energy efficiency. This entails choosing a coding style which emphasizes efficient algorithms and databases. Furthermore, there are considered programming languages and frameworks that are renowned to have lower ecological effect [9]. The purpose of our research is to develop the perfect green software prototype as a standard reference for more extensive energy-efficient coding standards that may be used in the wider computing environment [1].

Recycled server hardware:

The use of reused or second-hand parts in the building of a server supports the concept of embodied carbon. This objective includes choosing, reworking and testing parts from the server demolition or e-wastes. We intend to minimize the total amount of greenhouse emissions that occur during the lifetime of the servers by adopting alternative uses for equipment that could potentially damage the ecosystem. The use of this approach is also informed by the Green IT initiatives, which support the proper handling of end-of-life electronics, including recycling [3].

Solar panel system:

Using a solar panel system as a source of power for the server is a commitment to embracing environmentally friendly and renewable sources of energy. The second objective is determining the energy needs for servers; identifying solar options and adopting an energy storage plan so as to guarantee constant power supply. Motivated by studies carried out by Solar Energy Industries Association (SEIA), we seek to add to the existing understanding regarding environmentally friendly energy services for computing equipment. [10]

Holistic green server setup:

The last end result of the optimized green software program, recycled server hardware, and solar panel machine is anticipated to obtain the best inexperienced server setup. This complete technique displays the opportunity of bringing collectively numerous green computing principles in one environment-friendly conception. Our studies, therefore, offers sensible implications for agencies which can be inquisitive about enforcing a complete green computing approach.

16.7.4 Methodology

Optimized green software:

In order to create an optimized green software program, a thorough analysis of current data structures, algorithms, and coding methods is required. After that, the group will choose frameworks and programming languages with a reputation for having the least negative effects on the environment while taking runtime effectiveness and resource usage into account. Efficiency will be the top priority during the coding process, minimizing computing demands and maximizing energy conservation.

Performance metrics, such as energy and resource consumption, will be monitored and evaluated on a regular basis during development. Iterative testing and refining will be done to ensure that the final application maintains functionality and performance while achieving the highest possible level of energy economy. [11]

Recycled server hardware:

Building a server from recycled and repurposed hardware calls for an intensive identification, testing, and refurbishment technique. The team will collaborate with suppliers of decommissioned servers and recycling centers for digital waste to pick out hardware components that have the least capacity damaging environmental effect. The decided on hardware components will undergo extensive testing to make sure reliability and operation. During the refurbishment manner, parts are up to date, constant, and cleaned as wanted. The intention is to increase the existence of these hardware components and reduce the requirement for sparkling production. [5]

Solar panel system:

An interdisciplinary approach is necessary to integrate a solar panel system into the server configuration. To evaluate the server's energy needs and choose the best solar panel arrangement, the team will work with engineers and specialists in solar energy. This entails assessing seasonal fluctuations, solar exposure, and geographic location.

The solar panel system will be built to produce enough energy to keep the server running constantly. The implementation of an energy storage solution, such as capacitors or batteries, will guarantee continuous operation during times when solar input is low. By referencing standards and guidelines set by organizations such as the Solar Energy Industries Association (SEIA) and the International Renewable Energy Agency (IRENA), the methodology integrates industry best practices for solar energy integration.

16.7.5 Result and analysis

16.7.5.1 Before

The following is the benchmarking of the GMT for the Django test suite before the green optimization. Single-phase followed by total-phase energy consumption data (Figure 16.6).

The following is the benchmarking of energy consumption for a simple WordPress site.

Which is 1.3 times more optimized (for the runtime environment) in terms of the Django test suite [12] (Figure 16.7).

16.7.5.2 After

The following is the benchmarking of Wagtail CMS (simple Django CMS with green optimizations) (Figures 16.8 and 16.9).

The runtime environment of the after benchmark shows that energy consumption is at an all-time low, which provides a direct fewer carbon emissions score, as it is directly proportional to energy consumption and other predetermined values [13].

16.7.6 Final impact of this endeavor

This demonstration opens up a world of new possibilities for greener sustainable computing, but also a new level of benchmarking and standards

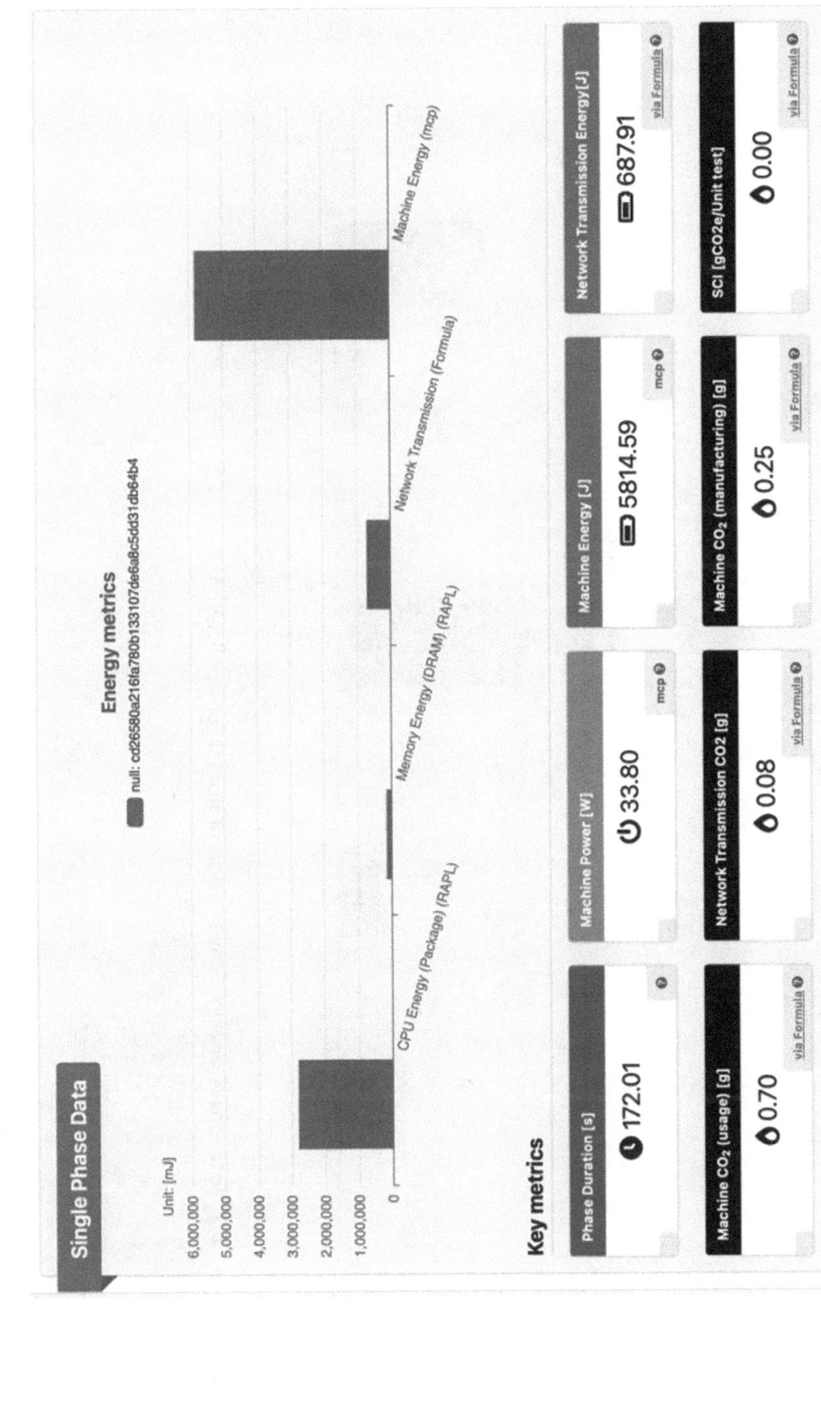

Figure 16.6 Energy metrics benchmarked by green metric tool for non-green web app.

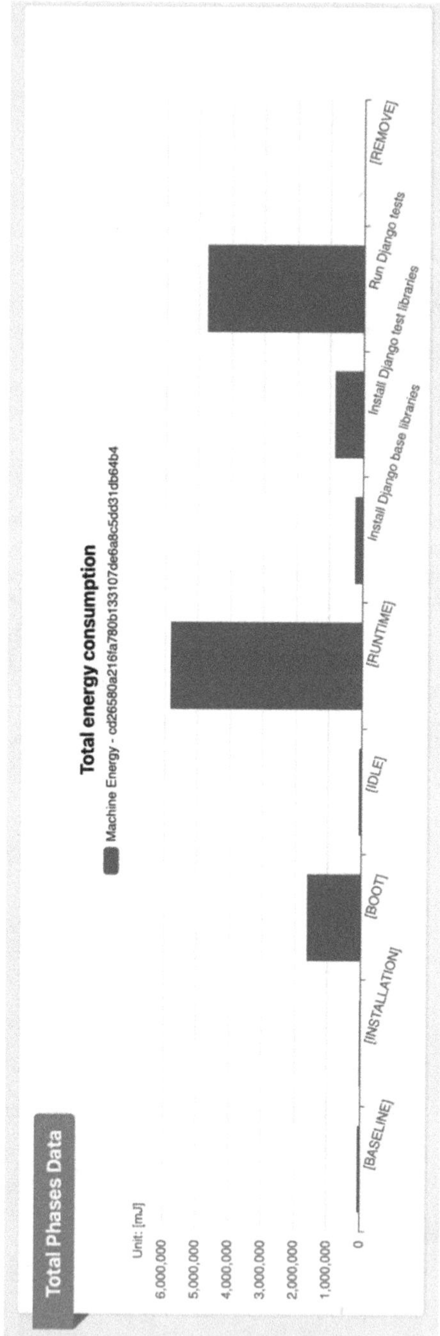

Figure 16.7 Total phase energy consumption of non-green application.

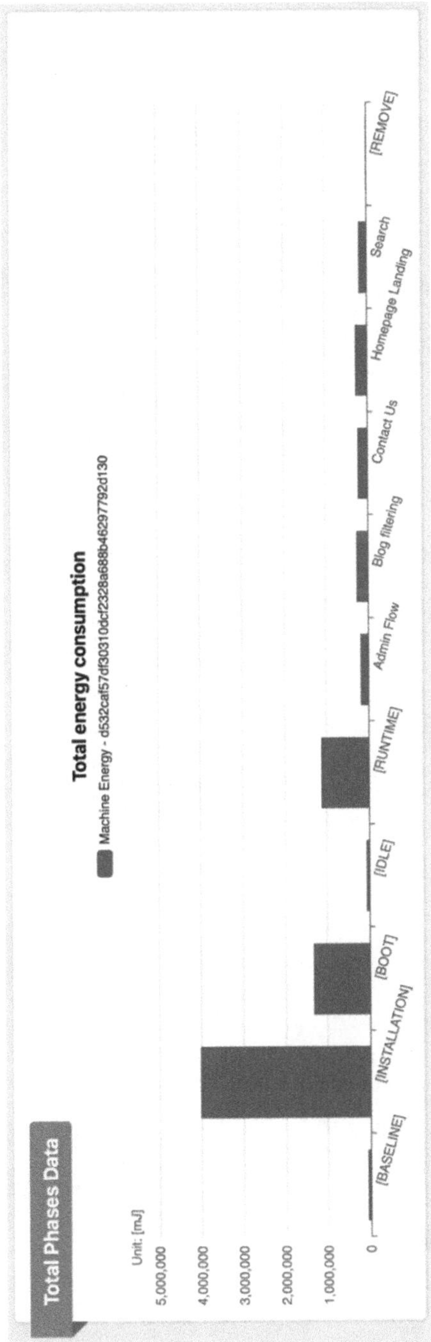

Figure 16.8 Energy metrics benchmarked by green metric tool for green web app.

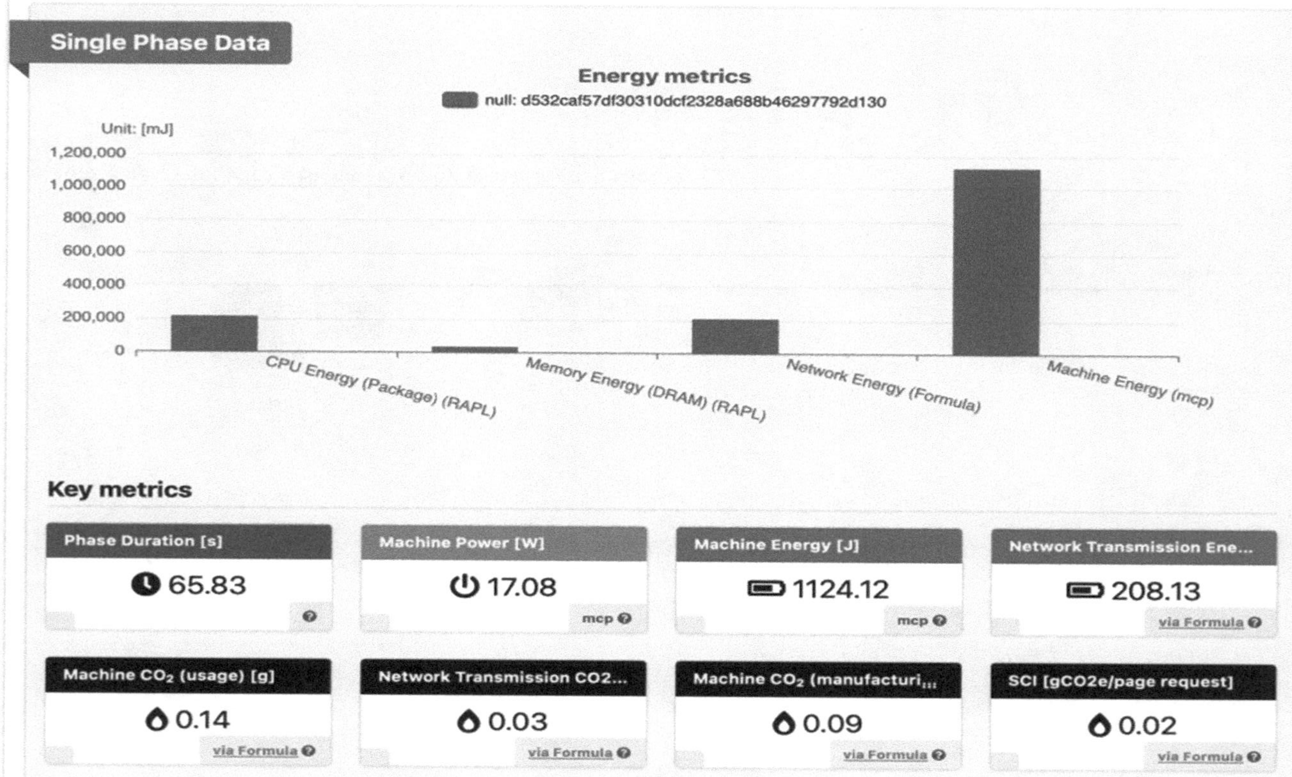

Figure 16.9 Total energy consumption of green application.

for software and hardware to adhere to. This endeavor called "Green Computing Using Solar Energy" is a new and comprehensive way of reducing the impact of information technology on the environment. The demonstration proposes to contribute greatly to sustainable computing through the coordination of optimized software, recycled hardware, and solar energy solutions. For benchmarking using the GMT, the study wants to show that this eco-friendly server setup succeeds using the tangible evidence, which is necessary in order to support the importance of the integration of green concepts into computing infrastructure. These elements are blended together in a way that addresses current environmental problems to ensure that IT becomes a green sector.

16.8 CONCLUSION

Green computing is achieved through the application of solar energy and the promotion of hardware reusability which is a very important direction in the development of the IT industry to a more sustainable and environmentally friendly one. Through introducing these approaches into their business operations, the organizations will be able to significantly enhance the overall effort to mitigate climate change by reducing greenhouse gas emissions – including those of carbon dioxide, the main cause of global warming.

By utilization of solar energy in IT-operations one will clearly target the emission associated with the use of conventional non-sustainable sources such as fossil coast. Through the extensive use of solar power that is widely available and renewable organizations can lessen their dependence on non-renewable energy thus decreasing their carbon emissions. With an ongoing trend of the IT industry to enlarge and the energy consumption growing, it becomes a strategic transition by the sector to move to solar energy in order to decrease its total carbon footprint and to be aligned with the world's climate goals.

On the other hand, the process of creating the software represents an even greater percentage of the emissions notably during data centers maintenance and decommissioning. New hardware production uses up a lot of energy and also takes away resources that are needed to reduce the greenhouse gases. One of the ways to accomplish these is to prolong the life of existing hardware through repair, reuse, and responsible e-waste management methods. This will reduce the need for manufacturing processes that are power consuming and such will likely lower carbon footprint that is traceable to these processes.

Apart from cropping up the direct ecological influence, solar energy, and hardware reuse practices help to initiate better innovations and create a culture of sustainability in the field of IT. As companies are using these sustainable practices as their top priorities, they are creating a demand for more efficient and eco-friendly technologies, which in turn, are spurring the research and development in the areas like renewable energy integration, energy storage solutions, and circular economy principles for electronic waste management.

Similarly, a better integration with the international regulatory initiatives on social responsibility of corporate organizations and sustainable development makes it more reasonable to have green computing methods which recognize these trends. Customers, shareholders, and other stakeholders today are more and more convinced about the need for environmental conservation, and businesses that achieve to show a commitment to sustainability can definitely be so well-endowed by a better reputation, get people ecologically conscious to work for their companies and be highly competitive in today's world which is much concerned about environmental issues.

To sum up, the union of solar energy and hardware reusability into IT processes is a major breakthrough towards a greener and low-carbon future for the industry. Through a process of cutting greenhouse gas emissions, use resource and do the job or responsibility to environmentally organizations are not only mitigating their ecological footprint but also leading in the innovation, linking to the environmental sustainability goals and global goals, and turning as leaders in the change towards a digital environment that is more a green and resilient.

REFERENCES

1. Saha, Biswajit. (2018). "Green Computing: Current Research Trends." *International Journal of Computer Sciences and Engineering*, vol. 6. https://doi.org/10.26438/ijcse/v6i3.467469
2. Podder, Suplab Kanti, Samanta, Debabrat. (2022). "Green Computing Practice in ICT-Based Methods: Innovation in Web-Based Learning and Teaching Technologies." *IJWLTT*, vol. 17, no. 4, pp. 1–18. http://doi.org/10.4018/IJWLTT.285568
3. André, H., Ljunggren Söderman, M., Nordelöf, A. (2019). "Resource and Environmental Impacts of Using Second-Hand Laptop Computers: A Case Study of Commercial Reuse." *Waste Management*, vol. 88, pp. 268–279. https://doi.org/10.1016/j.wasman.2019.03.050
4. Kiddee, P., Naidu, R., Wong, M.H. (2013). "Electronic Waste Management Approaches: An Overview." *Waste Management*, vol. 33, no. 5, pp. 1237–1250. https://doi.org/10.1016/j.wasman.2013.01.006
5. Albarracín, Cristian Laverde, Venkatesan, Srinath, Torres, Arnaldo Yana, Yánez-Moretta, Patricio, Vargas, Juan Carlos Juarez. (2023). "Exploration on Cloud Computing Techniques and Its Energy Concern." *MSEA*, vol. 72, no. 1, pp. 749–758.
6. Survey of Target Django and wagtailCMS/Other CMS Based Sites https://docs.google.com/spreadsheets/d/1Y3yh5k-7mIIUmXbIKX1SpWvIUUnDjL5WeKZD20q4TCQ/edit#gid=30107539
7. Green Metric Tool (GMT). https://github.com/green-coding-berlin/green-metrics-tool

8. Kamiya, G. (2020). "Factcheck: What Is the Carbon Footprint of Streaming Video on Netflix?" *Carbon Brief*. Available online: https://www.carbonbrief.org/factcheck-what-is-the-carbon-footprint-of-streaming-video-on-netflix (accessed on 7 April 2022).
9. Pandey, Aman, Colas, Thibaud, Adams, Chris (2023). "Google Summer of Code 2023 Wagtail Project Greener Coding: Backend Development." https://gist.github.com/salty-ivy/fe8616a28bf7d4a577d84b646141e8a9 by green web foundation and Wagtail.
10. Rahman, N. (2022). "Existing Green Computing Techniques: An Insight." In: Agarwal, P., Mittal, M., Ahmed, J., Idrees, S.M. (eds.) *Smart Technologies for Energy and Environmental Sustainability*. Green Energy and Technology. Springer, Cham. https://doi.org/10.1007/978-3-030-80702-3_5
11. Bharany, S., Sharma, S., Khalaf, O.I., Abdulsahib, G.M., Al Humaimeedy, A.S., Aldhyani, T.H.H., Maashi, M., Alkahtani, H. (2022). "A Systematic Survey on Energy-Efficient Techniques in Sustainable Cloud Computing." *Sustainability*, vol. 14, p. 6256. https://doi.org/10.3390/su14106256
12. Green Metric Tool Measurements for Django Test Suite. https://metrics.green-coding.io/stats.html?id=4a0a3ffd-f737-466c-a250-52b595732049#
13. Green Metric Tool Measurements for Wagtail GOLD. https://metrics.green-coding.io/stats.html?id=d2948f26-d257-4be9-a287-61a963db6f7c#

Chapter 17

IoMT

The key to affordable and accessible healthcare for all

Lavpreet Kaur, Pavan Thimmavajjala, Simarpreet Kaur and Indu Bala

17.1 INTRODUCTION: FROM IoT TO IoMT

We are currently living in a period when the Internet of Things (IoT) has become the largest area of research in technology and is also the key ingredient of the digital transformation that we are experiencing. IoT is a communication network of smart devices (sensors) that perform certain objectives [1]. Nearly every facet of human existence has been facilitated by IoT applications. The healthcare and medical professions have long been the most enticing applications for IoT, which eventually became known as the Internet of Medical Things (IoMT). The IoMT comprises a network of medical equipment, such as sensors, diagnostic and imaging apparatus, and medical servers, interconnected via Wi-Fi or the internet, facilitating machine-to-machine communication for the continuous monitoring of patient health [2]. IoT-powered health services are replacing existing medical infrastructure at a rapid rate, and it is projected that IoMT applications will cut costs, improve quality of life, and boost customer satisfaction [3].

There has been a dramatic increase in the demand for personal healthcare services. Health professionals like doctors, nurses, and guardians of patients, as well as patients themselves, may find m-health (the provision of healthcare through mobile devices) to be a lucrative and effective option in this case. Figure 17.1 shows how communication works in an IoT network. Such an interconnection includes doctors, patients, and all medical devices. Sensor nodes (in the form of medical equipment) collect data and send it to a medical server, which continuously monitors a patient's health. If there is an emergency, interventions must be immediately implemented to prevent the patient's health from deteriorating. In addition to this, the digital record-keeping facility allows doctors to access their records whenever they wish; they can also ensure that further treatments are provided while keeping in mind previous and current data. The IoT provides advanced and complete healthcare for patients, especially those who suffer from chronic diseases.

Today, medicine and healthcare are no longer confined to a hospital setting. With telemedicine, there is virtual home assistance for the elderly, serving patients throughout the world. With the development of remote patient

DOI: 10.1201/9781003643173-17

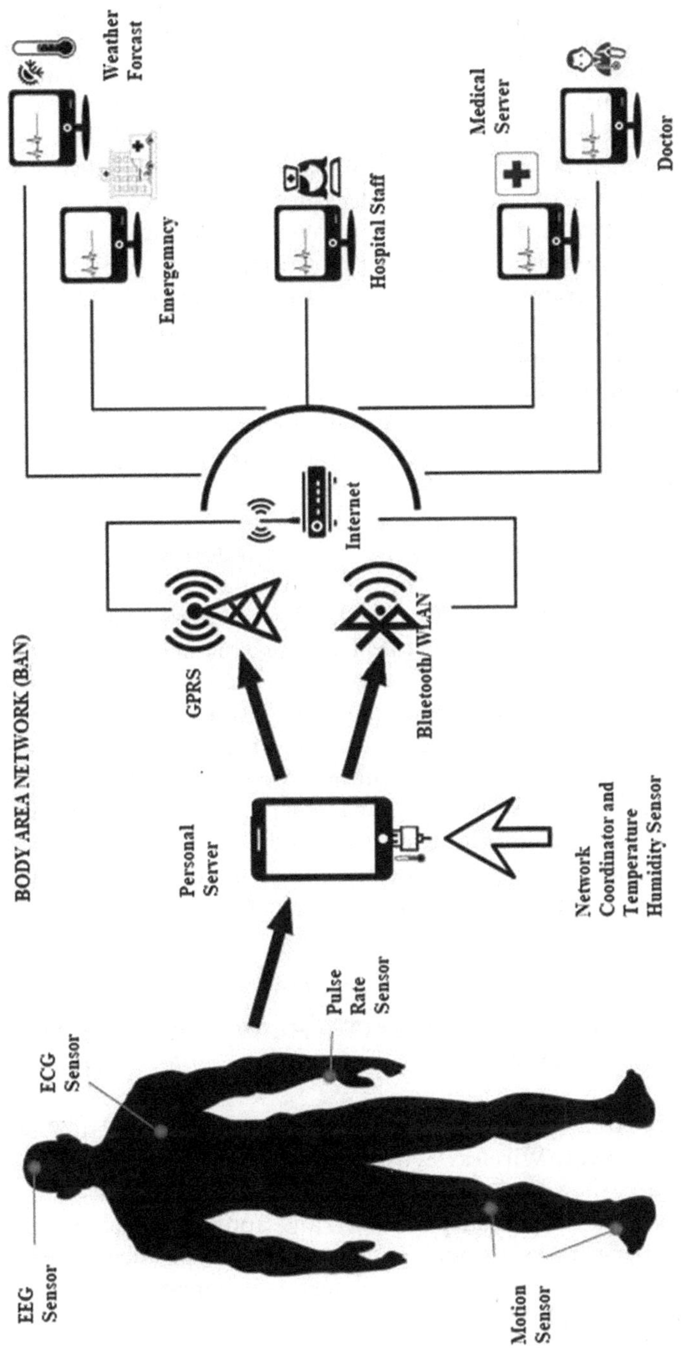

Figure 17.1 Working of IoMT through wearables.

monitoring, it is now possible to take immediate action in case of emergency when a patient's condition changes. This is specifically useful when a patient is suffering from chronic illnesses but doesn't want to be in the hospital at all times. As an additional benefit, IoMT enables monitoring medical equipment remotely for any performance problems. The idea of wearable smart devices is not a new one either. Patients can use the wearable devices to keep a close eye on their heart rate, blood pressure, and other health conditions, thanks to their attractive colors and sleek designs.

Thus, the shift from IoT to IoMT proves that technology has no geographical borders when it comes to application, and the way in which it is being used demonstrates its promise to lower healthcare costs and increase the quality of care provided to the public. The opportunities offered by IoMT would not only improve quality of healthcare, but also make health systems more efficient because doctors, medical staff, and patients would all be aware of their health, and the growing trend of wearable technology makes it possible for individuals to generate their own health reports based on variables that are factored in by the devices. The devices provide all kinds of information ranging from heartbeat levels to SpO2 and level of activity based on calories burned, etc. It is the medtech industry's role to contribute to reducing costs, improving quality, improving efficiency, and supporting the transition from volume-based to value-based care (VBC). Therefore, in this context, it is not just IoT and IoMT but also big fields such as data analytics and particularly medical data analytics that gain a lot of attention. Medtech is rapidly transforming its role and relationships within the healthcare industry due to the IoMT. More precisely, the connection between sensors and technologies is helping healthcare companies to optimize their clinical operations and workflow management, and improve patient care, even from remote places. As long as medtech businesses can convince physicians and patients of the value and benefits of linked medical devices, the pace and scale of healthcare revolution will accelerate exponentially [4] (Table 17.1).

17.2 THE SCIENCE BEHIND IoMT – HOW DOES THE TECHNOLOGY WORK?

Wireless Body Area Networks present a three-layered structure, which is an example of IoMT. Wireless Sensor Networks are the backbone of the IoT architecture; they are formed of five layers-physical layer, data connection layer, network layer, transport layer, and application layer. The objective of each layer is to carry out some function. The physical layer consists of sensors that gather data. The layer acts as a gateway for transmitting data streams over a physical medium for further processing and decision-making. The data link layer, as the name suggests, is responsible for transmitting data efficiently to the network layer. It carries out tasks like framing,

Table 17.1 Benefits of IoT

S. No.	*Benefits of IoT in medical domain*	*Ref. No.*
1.	1. Accuracy in medicine 2. Patient health reports can be shared with family and friends	[5]
2.	24/7 medical services enabled – medical advice	[6]
3.	1. Improved quality of life 2. Improved healthcare facilities and patient care 3. Improved patient outcome 4. Reduced costs 5. Improved user experience 6. Improved disease prevention and management	[7]
4.	Patient data is more manageable, and doctors can visit the records anytime	[8]
5.	Saves energy, time, and cost of healthcare services	[9]

multiplexing, and error control to ensure strong networking between the nodes. The network layer is responsible for routing, and hence routing protocols have a significant role to play in the functioning of this layer. The transport layer is responsible for delivering information across the network. In order to ensure no or minimal data loss between sender and receiver, this layer ensures end-to-end packet delivery. The application layer acts as the user's interface and has the primary responsibility of managing software.

Remote patient monitoring is one of the applications for IoMT, where portable smart devices or sensor nodes are wirelessly connected with the medical server. Sensor nodes (either on the body or implanted) collect data from their point of contact and transfer it to actuators, which collect the data from all these sensor nodes. A Personal Device, or PD, is an interface that receives data from nodes that connect to sensors and actuators, where the data can be analyzed by a Medical Server (MS), which allows for continuous monitoring of patients. The dynamic nature of this technology makes healthcare very efficient and personalized (Table 17.2).

Table 17.2 Technologies for IoMT

S. No.	*Future technologies*	*Use in IoMT applications*
1.	Big Data	Big Data accounts for the vast volume of health data generated from different sensor nodes and medical devices available in an IoMT network. Big Data leverages increased data analytics to improve the quality of health diagnosis and monitoring.
2.	Augmented reality	Augmented Reality plays a critical role in the IoMT network, especially during emergency procedures and remote patient monitoring.

(*Continued*)

Table 17.2 (Continued)

S. No.	*Future technologies*	*Use in IoMT applications*
3.	Cloud computing	IoMT consists of distributed devices (sensor nodes and medical equipment) to analyze, and communicate clinical information to the cloud in real-time. This vast stream of data is collected, stored, and evaluated for accurate and effective decision-making. Data gathering with Cloud Computing enables for constant and quick access to medical information from any connected device over the internet.
4.	Blockchain technology	The two main features of Blockchain technology make it suitable for healthcare applications: (i) It is decentralized, and (ii) It is distributed. The main application of Blockchain technology is for security, privacy, and data sharing reasons among multiple parties in the healthcare industry including Personal Health Record (PHR) Data Management, clinical data exchange, and drug supply chain data.
5.	Wearable technologies	Wearable medical devices allow for real-time health monitoring of patients along with enhanced public safety [10].
6.	Fog computing	Fog computing refers to a distributed network environment where multiple heterogeneous (geographically dispersed) devices are connected to provide reliable and scalable connectivity, computation, and storage facilities. The main features that make Fog Computing a significant part of IoMT applications are the high response time and low latency rate. This technology is a great alternative to cloud-enabled technologies that tend to create delays that are intolerable in critical medical applications.

17.3 APPLICATIONS OF IoMT

The IoMT is a critical aspect of the digital transformation of healthcare. As technology advances, new business models emerge, transforming processes, boosting productivity, improving cost-effectiveness of services, and increasing customer satisfaction. Some of the most important applications of IoMT that have been making healthcare better include

17.3.1 In hospitals and clinics

As hospitals and clinics become more advanced, healthcare is improving. The traditional medical equipment is becoming increasingly integrated into the IoMT network patients, devices/equipment, and doctors are all connected in such a way that their data can exchange smoothly. Telemetry ICU is one of the finest examples of this application, where patient data is collected with the help of a Wireless Body Area Network (or smart medical devices), and health monitoring is done on a medical server. Apart from this, biomedical equipment remote monitoring is also a remarkable application of IoMT, under which equipment can be remotely monitored for performance issues [11]. Also called Predictive and Preventive maintenance planning, this method reduces cost, energy, and ensures that the patient's health is controlled 24/7 without interruption.

17.3.2 Home

By making medical and healthcare services and applications accessible across multiple platforms, the IoMT makes medical and healthcare services a lot more accessible. Telemedicine is one such application that serves beyond borders. Patients with chronic diseases can benefit from Digital Remote Patient Monitoring (RDM), which offers them freedom while keeping constant track of their health. Patients don't have to visit the hospital again and again with portable RDM devices providing complete healthcare. Virtual Home Assistance is a remarkable application, especially for the health monitoring of elderly patients. The combination of telemedicine and virtual health assistance allows for personalized healthcare even for those who do not have a family to take care of them.

17.3.3 On the body

Sensors are the heart of IoT networks. Biological sensors are electrical devices expressly intended to transduce biological signals into easily detectable electric impulses for the aim of monitoring and analysis [12]. Sensors that are employed in remote sensing or endoscopy can be categorized into three primary categories, based on their application in either of these fields:

- Sensors embedded in apparel
- Sensors attached to the skin (on-body sensors)
- Sensors implanted in the body

Wearable smart devices are used for fitness tracking and diagnostics. Application-specific sensor technology is constantly developing. Medical sensors are a key component for various medical diagnostic instruments and equipment. These form a healthy chunk of the IoMT product marketplace. Extensive research is being carried out in this field, and this technology is expected to develop advanced personalized and preventive healthcare in the years to come. The modern world of mobile apps and wearables supports applications like fitness, disease management and care symptom tracking, and health education. The analysis of big data enables real-time decision-making, which has led to groundbreaking healthcare innovations [13].

17.4 CASE STUDIES – IoMT IN PRACTICE IN INDIA

The IoMT business is growing at a phenomenal speed and is anticipated to surpass US$158 billion by 2022 from US$41 billion in 2017 to 2018. The recent COVID-19 problem will speed the process of hypergrowth in this sector. As noted previously in Figure 17.2, its uses can be separated into two segments – Clinical and Non-Clinical, with each having its own variable dimensions and specific applications on the ground. To get a deeper understanding of how it's being adopted in India, we have to look at the practical examples in place. Many startups are emerging to assist doctors, hospitals, and medical and health infrastructure in using technology effectively, including IoMT [14]. Through support provided by multiple arms of the government, such as the Technology Development Board (TDB), NMITLI, the Department of Biotechnology (DBT), BIRAC, and Millennium Alliance, Indian MedTech companies and SMEs have increased in number during the last two decades [15].

The areas that these startups have entered into, and later specialized in, are remote health diagnostics, mental health, critical care monitoring, and remote health monitoring. The ongoing pandemic woes, for instance, have created opportunities for IoMT innovation in ambulatory, remote monitoring, and patient care assistance. Public and private healthcare providers are harnessing the power of IoMT data to improve the accuracy of diagnostics and the impact of targeted treatments [16]. A compilation of the same has

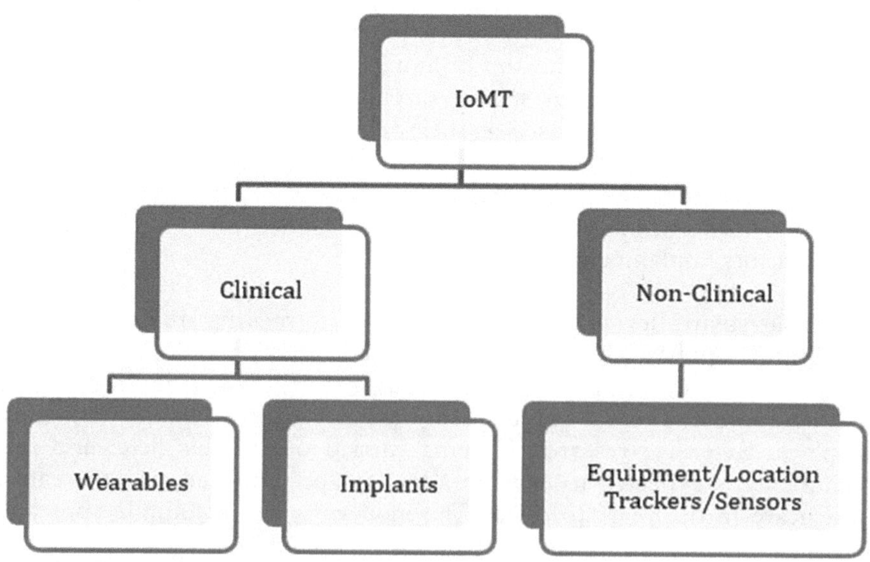

Figure 17.2 IoMT classification.

been provided in [17], which lists out the various products that are being developed to aid IoMT. In this piece, we look at a few examples of IoMT's adoption in India.

17.4.1 Case 1: Periwinkle Technologies – SmartScope and Net4Medix

This particular product, developed by Periwinkle Technologies, was an outcome of using healthcare communications as seen in [18]. Net4Medix is the software application that is connected to the cloud in real-time, while SmartScope is a handheld device that helps the doctor scan a patient for cervical cancer detection and treatment. Used for women patients between 21 and 65 years of age, it provides close examination and enlarged viewing of the cervix, hence enabling the diagnosis of anomalies or precancerous alterations at an early stage. It is beneficial in keeping track of the cervical health of an individual patient over a period of many years [19] (Figure 17.3).

In light of the fact that it may be used by any skilled health professional or nurse to monitor or study the patient's cervical health over time, it is easy to use and operate. The process of its working is as follows: If a patient with worries about cervical cancer and her cervical health visits a physician who utilizes SmartScope, the physician uses the equipment to screen the patient in a normal context, diagnoses any issues, and recommends appropriate treatment. The gadget records the level of relevant indicators, provides a report, and stores the photos through Net4Medix software. Stored photos can be examined by an expert doctor from a remote location. This decreases the demand on the constant availability of human expertise on the site of the exam, denies the subjectivity of a visual screening test when performed by

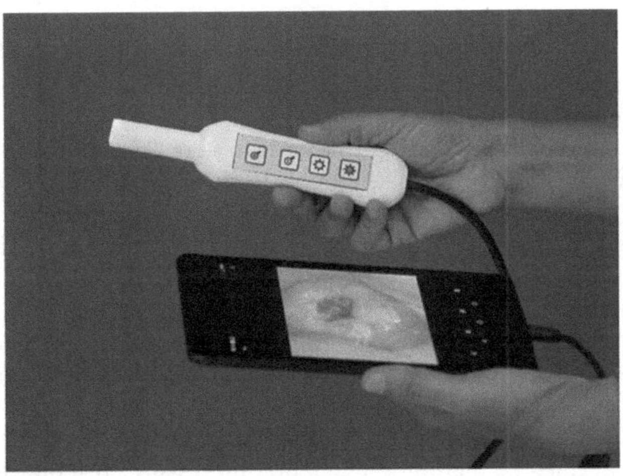

Figure 17.3 Working device of SmartScope [19].

experienced auxiliary workers, and removes the multi-step lab-based process of expensive cytopathology testing [19].

In this way, it enables early detection of cancer in precancerous stages, and enables appropriate treatment in a timely manner, and reduces the costs of multiple tests, which would have to be recommended in the case of spreading cancer further. The important thing about this is the remote usage of SmartScope and Net4Medix. In fact, by updating the patient's records and saving them, Net4Medix software helps any other doctor the patient consults get an overview of his or her condition from glancing at the report online. Thus, remote health monitoring through such hardware combined with equally efficient and effective software can go beyond its regular usage, and as seen in [18], the firm that is producing SmartScope is also looking at using it for oral cancer treatment as well. This shows a broad usage of IoMT possible by identifying the right applications.

17.4.2 Case 2: Neurosynaptic's ReMeDi-remote monitoring

This is another example of remote health monitoring using Telemedicine developed by Neurosynaptic Communications [20]. The product is titled 'ReMeDi', implying Remote Medical Diagnostic Kit. The idea was to connect doctors to patients in rural areas, and its USP is a real-time measurement of vital parameters. While it works on a similar pattern as that of SmartScope, its purpose is for general treatment and helps in one sense, the development of remote clinics or clinics on the go. The device has state-of-the-art android-based portable solutions, which would come in handy given the growing usage of mobile phones, which can help connect better through mobile data [20]. It has an inbuilt stethoscope and spirometer. It is also equipped with Low Power Bluetooth connectivity to the Tablet/Mobile Phone. The catch lies in its ability to provide both online and offline access, and is powered by a Lithium Polymer rechargeable battery, which is incredibly power efficient.

The difficulties while fostering this gadget were admittance to web availability in distant rustic regions, alongside poor data transfer capacity in regions where broadband network was available. The justification for such an innovation to have critical effect is given the detachment of specialists contacting remote and the absence of value clinical consideration at the neighborhood level. Primary Health Centers, a clinical set-up like ReMeDi would assist towns with associating with specialists in metropolitan communities and significant urban areas in adjacent regions and accelerate clinical findings (Figure 17.4).

In addition, the ReMeDi platform offers solutions to diagnose 20 various diagnostic parameters, including ECG, blood pressure, auscultation, oxygen saturation, fetal doppler, blood, and urine tests. Such a transformative technology, therefore, reduces the costs of healthcare on one end, raises

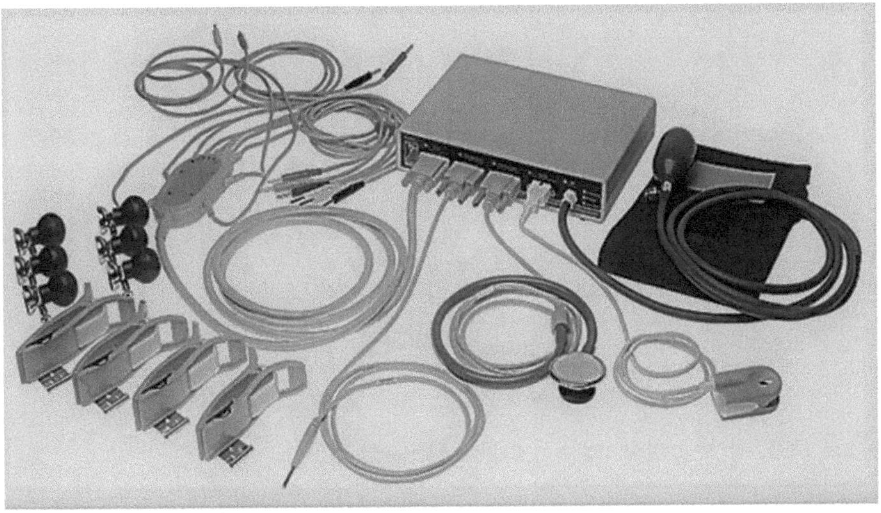

Figure 17.4 Models demonstrate ReMeDi set-up for remote monitoring [21].

awareness about such simplistic technologies at the ground level, and adds to the creation of a strong market for more such IoMT-based solutions to take shape and paves the way for indigenous solutions just like ReMeDi.

Both models demonstrate that remote monitoring of health through IoMT can not only reduce the cost of health care, but also improve the quality of health care provided through remote means, and assist with interfacing the rustic regions better and achieve a more generous effect on the ground level.

17.4.3 Case 3: Jana Care: point-of-care diagnostics – self-diagnosis device for diabetes

AINA device has been created by JanaCare Solutions and marketed to quantify blood glucose, HbA1C, lipids (HDL, LDL, and TrG), creatinine, and hemoglobin. The device can be used in hospitals for inpatient treatment, for self-monitoring by patients at home, and for general screening by health staff. This device possesses strip readers, such as those seen in conventional diabetic testing, and is used to check the sugar level. The difference between this and the typical diabetes check-up devices is that, upon contact with blood, capillary action forces passive flow over or through a strip, the flowing liquid activates the chemistry, and the chemical provides a signal that Aina can read [21].

This can be connected to our phones and, we can activate the device through a mobile application, which helps us generate the data that we need – blood sugar levels, calorie levels, and other relevant metrics that need to be captured. This way, the device is connected to a mobile phone and would help in generating instant results based on the strip reader's

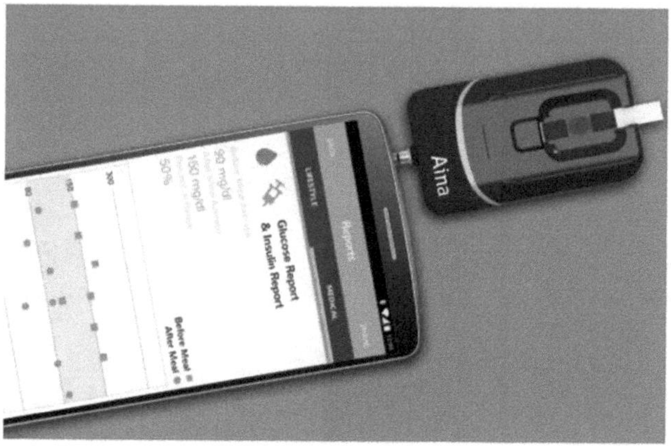

Figure 17.5 AINA handheld test device [19].

signals. The differentiation from other devices is that there is an unmet demand for an affordable diagnostic system for chronic illness management in primary and community care settings. Diagnostic testing using array technologies and microfluidics demands expensive equipment and capital-intensive manufacturing. Though these forms of diagnostics could be useful for established institutions that require testing hundreds of samples at a time, they are not designed to fulfill the needs of the community and home care [21].

Such easy-to-use point-of-care devices would greatly help in putting in place the basic medical device infrastructure in PHCs, local area hospitals, etc., and bring about cost-effectiveness in terms of technology (Figure 17.5).

17.5 RECOMMENDATIONS – BASED ON IOMT CASE STUDIES

The preceding section discusses a few advances in IoMT and has highlighted just some of the examples of various technologies in the workplace and their applications. The thought here is that IoT gives the essential specialized design through numerous layers and installed frameworks, which can be adjusted in different areas according to their needs. IoMT specifically addresses the aspects of healthcare that see the need for human intervention – be it physical monitoring of patients or hospital management in addition to point-of-care testing. The advantages of IoMT – based solutions show that they make administering adequate treatment easier, as they can function remotely, and are easy to use to deal with most of the cases, enabling the technology's easy portability.

Based on the above analysis and presentation of the technologies that are at work, there are a few aspects which may warrant a need for experts in the technological development process, to take a look and customize the technology accordingly, and for businesses to pitch in new ideas to take the concept further. The alternative uses of IoMT can be many, and so are the current applications that are in place already, and are functioning well. However, to get into more specific details, the following areas can be explored with regard to development of IoMT in other areas of healthcare.

17.5.1 Patient wellness indicators in hospitals

The key idea behind this proposal is to separate critical patient wards from the normal/regular wards and use Embedded Systems and IoMT to monitor the levels of each patient in intensive care or critical units by merging the data (ECG, SpO2, blood pressure levels, and other important metrics) patient-wise, with indicators in red and green showing real-time status of patients, giving the medical staff, the doctors, and medical practitioners, a clear picture as to when further treatment should be given. Such a technology would provide critically injured/ill patients with immediate medical attention and save their lives by keeping the concerned doctors informed of their status.

17.5.2 Integrating IoMT platforms – one stop IoMT dashboard for hospitals

This idea stems from the fact that interoperability poses a challenge to the simultaneous monitoring of IoMT devices, which later poses a logistical problem. The root of this idea is to explore customization in IoMT-based solutions by integrating them into one definite application, to monitor each of the platforms, which in turn, may be monitoring at least a dozen connected IoMT devices. This way, a hospital, for instance, can monitor the doctors' availability, matching them with patients' health, and, in addition, ensure a supply of required medicines and other equipment. In a way, it is a one-stop dashboard monitoring patients' status, doctors' consultancies, and surgeries scheduled, in addition to their availability [22].

17.6 CHALLENGES FOR IOMT IMPLEMENTATION

There are multiple demanding situations that arise while looking to put into effect IoMT in Healthcare.

17.6.1 Absence of healthcare regulatory body

As cited in [15], there may be no regulatory oversight in the field of clinical era, which places this industry at a disadvantage, as without any regulatory framework, it is difficult to maintain the security and privacy of the documents and reports. The Information Technology (IT) Act 2000 afforded early supervision to the telemedicine industry in India, but lacked any established rules for data protection, security, and confidentiality. In this backdrop, while MoHFW, Government of India, in cooperation with NITI Aayog, has come forth with Telemedicine Practice Guidelines [23], which lay out a comprehensive framework with the overriding purpose to enable registered medical practitioners to offer treatment utilizing telemedicine.

The purpose of these recommendations is to offer practical guidance to physicians, so that all services and models of care, employed by doctors and health workers, are encouraged to consider the use of telemedicine as a part of their usual medical practice. These recommendations will enable the medical practitioner to follow a good course of action, to give effective and safe medical treatment built on current knowledge, available resources, and patient requirements to guarantee patient and provider safety [24].

These provide extensive details about what telemedicine constitutes, and how various modes can be adapted to improve the quality of healthcare provided, particularly, in the National Digital Health Mission, which will be launched in the coming days. Electronic Health Record Standards, 2016, should incorporate IoMT cloud-based services, that would strengthen the legal framework in a big way and prevent misuse of healthcare data, which brings us to the next challenge that comes up as part and parcel of the shift to IoMT and cloud-based remote medical treatment – Data Security.

17.6.2 Data security and privacy

Given the repeated concerns about privacy of personal data, and after the Supreme Court of India ruled privacy to be a Fundamental Right in a landmark judgment in 2017, a similar concern echoing with regard to data safety is inevitable, and was well-founded. Data security comes into question as most of the cloud-based applications collect personal data such as addresses, names, and identification in addition to a person's health history. For this, the broad legal framework is already in place, with the Ministry of Health bringing about Electronic Health Record Standards in 2016. At present, in addition to the EHR, the IT Act 2000 helps safeguard sensitive personal data and imposes punitive action upon its misuse [25].

However, with IoMT being cloud-based, where people voluntarily submit their information and health history is recorded at their request, there should be a strict policy regarding the data generation, protection, and dissemination in line with both the IT Act and EHR, 2016. Access to health data by third parties should be tracked using blockchain technology to prevent misuse, particularly in the case of insurance claims, health difficulties, and other actions that are illegal in nature.

It remains a challenge, as even the government databases are at the risk of cyberattacks. The concern remains valid, as the example of the ransomware cyberattacks on the UK's National Health Scheme's systems show.

17.6.3 Question of affordability

While switching from old systems to IoMT can lower healthcare costs for end users in the short term, the cost of shifting or adoption can be substantially greater because it now requires investing in the correct IoMT equipment, depending on the demands of hospitals, doctors, and medical personnel. This cost can be incurred in two ways: financially and through training. Finance is concerned with obtaining the appropriate equipment, installing the necessary drivers, software, and hardware, testing it (at first in trial projects), and finally finalizing it. This is a significant expense in and of itself. So it will be a challenge to see how the lowest rungs of medical infrastructure – PHCs, zonal and district level health centers (both public and private) – will accept it by allocating enough funds [26].

The training expenditures would be the next expense to be incurred. The IoMT technology, which is rapidly changing, requires a sufficient skill set to handle it, no matter how basic, which means a large amount of training will be required at various levels. Every layer, from technical equipment handlers to data analysts to doctors and practitioners, as well as support staff, requires constant upskilling, which can significantly increase costs in the early months, but, as with any other technology, this cost can be reduced in the long run with adequate ease of interface. In this context, handheld, phone-app-based IoMT solutions can help to alleviate cost difficulties and facilitate a smoother, less expensive transition to IoMT at both the local and macro levels. Furthermore, as the world moves on to become a more health-conscious world in the post-COVID era, personal wearables that are gaining widespread use can be used better to give the correct health picture for those interested, with incentives to encourage more people to start using them.

17.6.4 Interoperability – information exchange between more than one system

Another key challenge in the medtech market is the problem of interoperability. Interoperability may be generally described as the ability of two or more systems to share information and utilize the information that has been communicated. Interoperability is the ability of two or more devices, systems, platforms, or networks to perform in conjunction. Interoperability facilitates communication across diverse devices or systems, in order to accomplish a shared aim. However, the existing devices and systems are fragmented with regard to the communication technologies, protocols, and data formats. This variety makes it tough for devices and systems in the IoT network to connect and exchange their data with one another. The usability of the IoT network is restricted by the lack of compatibility [27]. A great majority of current IoT solutions are proprietary and intended to function primarily inside a predefined hardware or infrastructure environment. The lack of IoT interoperability implies that data can't be effectively transferred across diverse, sometimes overlapping devices and systems [28].

17.7 FUTURE SCOPE OF IoMT

Healthcare and medicine will be transformed by disruptive technologies. One of IoMT's primary forecasts is that emerging mobile technologies would spur seamless and secure telehealth communication, bringing physicians and patients together at a fraction of the expense of traditional consultations. In the coming years, digital health advisors, also known as customized preventive health coaches, will become a trend. Such specialists will be able to assess and analyze data relating to health and well-being, as well as give online solutions. The application of developing technologies such as Big Data, AI, improved sensor technology, and IoMT would enhance not only patient health but also healthcare costs. It will make healthcare far more accessible and ensure that patients receive individualized and effective treatment.

Furthermore, modern technologies will raise the market for such innovations and engage local businesses more, as well as collaborations with those creating such technologies on the global market, improving treatment quality and making healthcare more accessible. IoMT networks will promote good improvements at each stage of the patient experience for a population that is growing at a consistent rate. To achieve better results, sophisticated application-specific network protocols are required to increase the privacy and security of patient data in IoMT networks.

In addition to the foregoing, we believe that the IoMT's future scope will include numerous connected areas such as wellness, fitness applications, and monitoring disease risks based on fitness levels, all of which will require wearable technology to be scaled up. Similarly, key parameters can be recorded and updated from time to time to understand how fit a person is and categorize them into age groups to study how fit a society is at the macro level, while analyzing during general checkups to improve patients' chances of early recovery with respect to certain diseases. The use of RFID in IoMT, combined with geo-tagging, can assist hospitals and government authorities in determining the prevalence of seasonal illnesses. RFID's application in IoMT has the potential to be game-changing [29]. These health apps can help a variety of people, including caregivers for patients, patients themselves, doctors, nurses, and healthy people. The advancement of m-health results in better medical services, increased efficiency, and increased effectiveness of health plans and services, lowering the cost of health maintenance [30].

Blockchain technology has the potential to help the hospital manage not only money transactions (together with the patients) but also data storage. Blockchain has a lot of potential for improving the security of IoT-based healthcare apps [31–36]. This may also assist in addressing the issue of personal data privacy and security. Any further adoption of IoMT at the ground level will be based on its cost-effectiveness, and so the key to improving IoMT adoption will be determined by the scale of application as well as the cost of production. The rising market for IoMT indicates that more and more options will be available in the future, allowing for a lot more customized solutions and a multitude of IoMT-based solutions to flourish and favorably influence the health and well-being of our citizens and the entire medical community.

REFERENCES

[1] https://www.mckinsey.com/industries/healthcare-systems-and-services/our-insights/telehealth-a-quarter-trillion-dollar-post-covid-19-reality
[2] M. Javaid, and I.H. Khan, "Internet of Things (IoT) enabled Healthcare helps to take the Challenges of Covid-19 Pandemic," *Journal of Oral Biology and Craniofacial Research*, vol. 11, no. 2, Apr.–Jun. 2021, pp. 209–214.
[3] S. Razdan, and S. Sharma, "Internet of Medical Things (IoMT): Overview, Emerging Technologies, and Case Studies," *IETE Technical Review*, May 2021. https://www.tandfonline.com/doi/full/10.1080/02564602.2021.1927863
[4] Deloitte Centre for Health Solutions, 2018. https://www2.deloitte.com/content/dam/Deloitte/global/Documents/Life-Sciences-Health-Care/gx-lshc-medtech-iomt-brochure.pdf
[5] https://behrtech.com/blog/interoperability-the-secret-to-a-scalable-iot-network/
[6] https://www.janacare.com/technology.html
[7] https://cse.iitkgp.ac.in/~smisra/theme_pages/interop/index.html
[8] Digital Strategies for Big Picture Results. http://www.tcs.com/resources/white_papers/Pages/Internet-of-Things-Medical-Devices.aspx
[9] C. H. Huang, and K. W. Cheng, "RFID Technology Combined with IoT Application in Medical Nursing System," *Bulletin of Networking, Computing, Systems, and Software*, vol. 3, no. 1, 2014, pp. 20–24.
[10] M. Ghamari, B. Janko, R. Sherratt, W. Harwin, R. Piechockic, and C. Soltanpur, "A Survey on Wireless Body Area Networks for eHealthcare Systems in Residential Environments," *Sensors*, vol. 16, no. 6, Jun. 2016, pp. 1–33. https://doi.org/10.3390/s16060831
[11] J. Maktoubian, and K. Ansari, "An IoT Architecture for Preventive Maintenance of Medical Devices in Healthcare Organizations," *Health and Technology*, vol. 9, 2019, pp. 233–243. https://doi.org/10.1007/s12553-018-00286-0
[12] J. Maktoubian, and K. Ansari, "An IoT Architecture for Preventive Maintenance of Medical Devices in Healthcare Organizations," *Health and Technology*, vol. 9, 2019, pp. 233–243.
[13] J. D. Enderle, and J. D. Bronzino, *Introduction to Biomedical Engineering* (3rd Edition). Academic Press, Elsevier, Chapter 10, 2012, pp. 609–666.
[14] D. V. Dimitrov, "Medical Internet of Things and Big Data in Healthcare," *Healthcare Informatics Research*, vol. 23, no. 3, Jul. 2016, pp. 156–163. https://www.ncbi.nlm.nih.gov/pmc/articles/PMC4981575/
[15] A. Basu, and A. Majumdar, "Reimagining the Possible in the Indian Healthcare Ecosystem with Emerging Technologies," *The Pharma Times New Bureau*, Mar. 2018.
[16] S. P. Dash, "The Impact of IoT in Healthcare: Global Technological Change & The Roadmap to a Networked Architecture in India," *Journal of Indian Institute of Science*, 100, 2020, pp. 773–785. https://doi.org/10.1007/s41745-020-00208-y

[17] S. Tah, "Will IoMT be Able to Salvage Indian Healthcare?," Jun. 2021. https://www.pcquest.com/will-iomt-able-salvage-indian-healthcare/
[18] https://www.birac.nic.in/webcontent/BIRAC_Exhibitor_PROFILE_26_9_2017.pdf
[19] https://yourstory.com/2018/01/meet-veena-moktali-making-real-time-diagnosis-cervical-cancer-reality/amp
[20] https://birac.nic.in/product-detail.php?product=169
[21] https://www.meddeviceonline.com/doc/the-internet-of-medical-things-raises-novel-compliance-challenges-0001
[22] https://www.neurosynaptic.com/remedi-telemedicine-solution/
[23] D. Menon, and R. Sharma, "Neurosynaptic Communications: A Case Study in Telemedicine," *Neurosynaptic Communications Pvt. Ltd.*
[24] S. Arora, "IoMT (Internet of Medical Things): Reducing Cost While Improving Patient Care," *IEEE Pulse*, Oct. 2020.
[25] https://www.embs.org/pulse/articles/iomt-internet-of-medical-things-reducing-cost-while-improving-patient-care/
[26] D. Dinakaran et al. "Telemedicine Practice Guidelines of India, 2020: Implications and Challenges," *Indian journal of Psychiatry* vol. 63, no. 1, 2021, pp. 97–101.
[27] Electronic Health Standards, 2016. https://main.mohfw.gov.in/sites/default/files/EMR-EHR_Standards_for_India_as_notified_by_MOHFW_2016_0.pdf
[28] M. Segura, C. M. Butler, and F. Tabibkhoei, "The Internet of Medical Things Raises Novel Compliance Challenges," *Med Device Online*, Jan. 2018.
[29] M. Fischer, and M. Lam, "From Books to Bots: Using Medical Literature to Create a Chat Bot," in *Proc. First Workshop on IoT-enabled Healthcare and Wellness Technologies and Systems*, Jun. 30, 2016, pp. 23–28.
[30] D. D. Kumar, and P. Venkateswarlu, "Secured Smart Healthcare Monitoring System based on IOT," *Imperial Journal of Interdisciplinary Research*, vol. 2, no. 10, 2016.
[31] R. Singh, "A proposal for Mobile E-Care Health Service System using IOT for Indian Scenario," *Journal of Network Communications and Emerging Technologies*, vol. 6, no. 1, 2016, pp. 21–23.
[32] C. Kotronis et al., "Evaluating Internet of Medical Things (IoMT)-Based Systems from a Human-Centric Perspective', *Internet of Things*, vol. 8, 2019, pp. 100–125.
[33] G.J. Joyia, R. M. Liaqat, A. Farooq, and S. Rehman, "Internet of Medical Things (IOMT): Applications, Benefits and Future Challenges in Healthcare Domain," *Journal of Communications*, vol. 12, no. 4, 2017, pp. 240–247.
[34] N.N. Thilakarathne, M. K. Kagita, and T. R. Gadekallu, "The Role of the Internet of Things in HealthCare: A Systematic and Comprehensive Study," *International Journal of Engineering and Management Research*, vol. 10, no. 4, Aug. 2020, pp. 145–159.
[35] D. Ghai, S. L. Tripathi, S. Saxena, M. Chanda, and M. Alazab, *Machine Learning Algorithms for Signal and Image Processing*, Wiley IEEE Press, USA, 2022. https://doi.org/10.1002/9781119861850
[36] S. L. Tripathi, and M. Mahmud, *Explainable Machine Learning Models and Architecture*, Wiley & Scrivener Publishing LLC, USA, 2023. https://doi.org/10.1002/9781394186570

Chapter 18

Next-generation cybersecurity system in integration with Artificial Intelligence and blockchain

Rajesh Singh, Anita Gehlot, Shaik Vaseem Akram, Ravindra Sharma and Praveen Kumar Malik

18.1 INTRODUCTION

Industry Revolution 4.0 (IR 4.0) is a concept intended to enhance the way modern businesses operate by utilizing cutting-edge technologies such as those used in the development of the Industrial Internet of Things (IIoT), robotics, and Big Data applications. Blockchain is one such technology that may provide trust, security, and decentralization to a variety of industrial fields [1]. The reason for this is that many IIoT, and IR 4.0 tools have been linked to networks without proper security controls, resulting in devices that are incorrectly set or inadequately secured. Securing networks requires fairly explicit skills, which are unfortunately rarely offered in technical universities [2]. Organizations must be aware of their risks and prepare for quantum-resilient systems. The evaluation is based on three variables: the time necessary to move to quantum-resistant systems, the security layer life of information assets, and the interval remaining before quantum computers violate security [3]. Information security is one of today's most pressing concerns. When most people think of cybersecurity, they think of "cybercrimes," which are becoming increasingly common. To tackle this sort of cybercrime, several governments and companies make various attempts. Many people are concerned, in addition to many cybersecurity precautions [4]. Cybersecurity (CS) is also known as information technology security. The basic goal of CS is to protect networks, programs, devices, and information against unauthorized access and manipulation. Cybercrime (CC) is classified into three types: cybercrime against entities, assets, and individuals [5]. The development and implementation of appropriate sophisticated strategies to address the complex difficulties in this environment demand the identification of all the individuals and activities that act, directly or indirectly, on the cybersecurity of digital ecosystems [6]. Security Operations Centre (SOC) is a phrase that is extensively used nowadays, particularly in the context of network security. While we exist in a virtual world full of hazards, most senior executives in modern businesses still do not completely comprehend the threat picture [7]. The phrase "Industrial Control System" (ICS) refers to various control system arrangements such as Dispersed Control Systems (DCS), Supervisory Control

DOI: 10.1201/9781003643173-18

and Data Acquisition (SCADA), and Programmable Logic Controllers (PLC). Critical infrastructure includes thermal and nuclear reactors, power production, heavy industries, water treatment plants, and distribution networks [8].

18.2 IoT AND IIoT ASSISTANCE IN CYBERCRIME

Regardless of the growing diversity of networking infrastructures, computation frameworks, and software architectures that efficiently reinforce the execution of decentralized multi-domain value chains, an integrated architecture that successfully manages all associated vulnerabilities continues lacking: reciprocal credibility of structures in barely undefined layouts, identification and preventive of complex multi-vector threats, authentication, and network consistency [9]. However, the increasing incidences of cybersecurity assaults on networking architecture and its associated frameworks, as well as system flaws, are a source of concern for both providers and users. While the invention improves prospective energy production and distribution, utilization surveillance, and data dissemination, its performance is directly connected to the system's successful integration of technologies [10]. With the increased reliance on real-life robots that operate near human beings, ensuring the safety of these systems is becoming more crucial to avoid cyberattacks that might result in privacy invasion, vital operations, and harm to individuals. Robotics, automation, and associated Artificial Intelligence (AI) technologies have grown ubiquitous, raising issues about security, safety, accuracy, and trust [11]. AI and machine learning (ML)-assisted cybersecurity may enable data-driven robotics, empowering safety measures to identify and react to attacks in an immediate response. The implementation of AI and ML to confront cybersecurity challenges is gaining prominence in business and academia, attributed primarily to widespread attack vectors on critical networks such as cloud computing platforms, government agencies, and healthcare facilities, as well as the enormous amounts of data created by these attacks [12]. Deep Learning (DL), a unique type of ML, is attracting research attention owing to its effective usage in a variety of traditional AI applications as compared to standard ML algorithms (CMLAs). Recently, DL architectures have been imaginatively modeled for a wide range of cybersecurity applications [13]. CS training and exercises are essential for acquiring a sufficient amount of proficiency in the virtual environment. The old military adage "You Fight As You Train" applies in the cyber domain. Cyber breaches can have a wide range of physical repercussions. A cyberattack or invasion against a power infrastructure, for example, might affect how the healthcare system operates [14]. The complexity, dynamism, resilience, efficiency, interfaces, and interactivity of Internet of Things (IoT) devices must be thoroughly re-examined. As a result of this occurrence, network communication will be frictionless, and the IoT will be able to provide support [15].

18.3 BLOCKCHAIN TECHNOLOGY ASSISTANCE FOR THE PREVENTION OF CYBERCRIME

AI and ML techniques have improved dramatically as a result of blockchain technology, which enables global and immutable repositories that assure non-repudiation and accountability of recorded information [16]. Blockchain technology is a software system that facilitates the safe transmission of unique value instances (such as funds, assets, agreements, and identifying attributes) across the Internet, deprived of the interference of an external mediator like a financial institution or agency. Bitcoin and Ethereum are illustrations of unreliable and unsupervised public Blockchains, whereas private Blockchains are trusted and supervised [17]. Blockchain technology appears as a feasible option for delivering prominent traits such as continuous authentication, confidentiality, and robustness. To cope with various attacks on IoT networks, it is necessary to conduct research on next-generation IoT security frameworks and build cutting-edge confidentiality protection solutions [18]. Blockchain applications incorporate essential attributes such as self-validation and autonomy to eradicate the requirement for reliable external access. Blockchain applications, including cyber-physical systems (CPS), are crucial for modernizing obsolete industrial processes, technical procedures, and business models [19]. Financial services, insurance companies, cybersecurity, prediction, healthcare, cryptocurrency, and other industries benefit from technological convergence. As more virtual systems are utilized and services are delivered by these entities, the greater the probability that these systems may be manipulated. Blockchain is an exceptional information security system that generates networks that combine old entries retained by peers and fresh data sequentially by exchanges between nodes [20]. In terms of connection, data throughput, and latency, 5G currently outperforms 3G/4G/4G, shifting from the traditional people-to-people communication paradigm to the modern people-to-things and things-to-things communication paradigm [21]. The emphasis on the development of next-generation cellular networks, or 5G, has shifted to meet expanding data rate expectations and boost the capacity of the microcell wave spectrum. Next-generation networks must achieve extremely rapid data rates, minimal latency, and enormous data processing capacities [22]. Robots will soon become a realistic choice given the rapid advancement of smart devices with wireless sensors that function in hostile conditions. Unfortunately, as technology evolves, safety concerns grow, especially when it comes to cooperation between humans and robots [23]. The IoT in robotics is a new research and development industry that combines ubiquitous sensors with self-driving mechanisms. The integration of robotics and IoT technologies will broaden the potential of present IoT and robotics [24] (Figures 18.1 and 18.2).

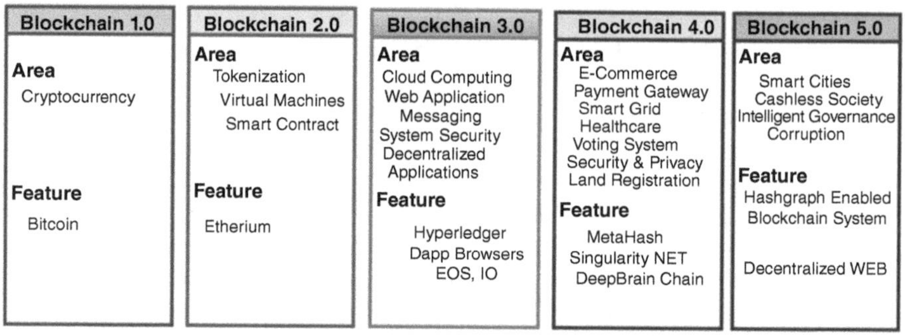

Figure 18.1 Different layers of blockchain technology assistance for providing a prevention of cybercrime in terms of area and features.

Recommendations:

- Cyber-Physical Systems (CPSs) enable one-of-a-kind computer framework designs, as well as the integration of ubiquitously embedded sensors and communication services via better networking technologies [25].
- Modern big data analytics strategies may reduce computing and processing expenditures for next-generation networks.
- Additional research is needed to enhance Quality of Control (QoC) and Quality of Service (QoS) parameters. Attackers are attracted to the communication routes of robotic devices [26].
- Big data analytics exposes hidden understanding inside enormous amounts of data. To guarantee that data cannot be tampered with or altered, it should be adequately encrypted during the storage, administration, and processing phases.
- Deep learning and federated learning are unknown areas that researchers must address immediately.
- Future research might use a novel framework that combines semi-supervised approaches with online incremental learning to find latent patterns in network traffic of various types.
- Blockchain integration with edge computing, which uses a distributed architecture and distributed services to address future network, system security, performance, and scalability objectives.
- Existing communication models' data transmission security should be reinforced, and a new set of technologies with higher degrees of dependability should be developed.

18.4 CONCLUSION

We evaluated and contrasted the most contemporary works on the convergence of big data, data science, and network science disciplines in this study, allowing researchers to grasp from a technical perspective. The literature

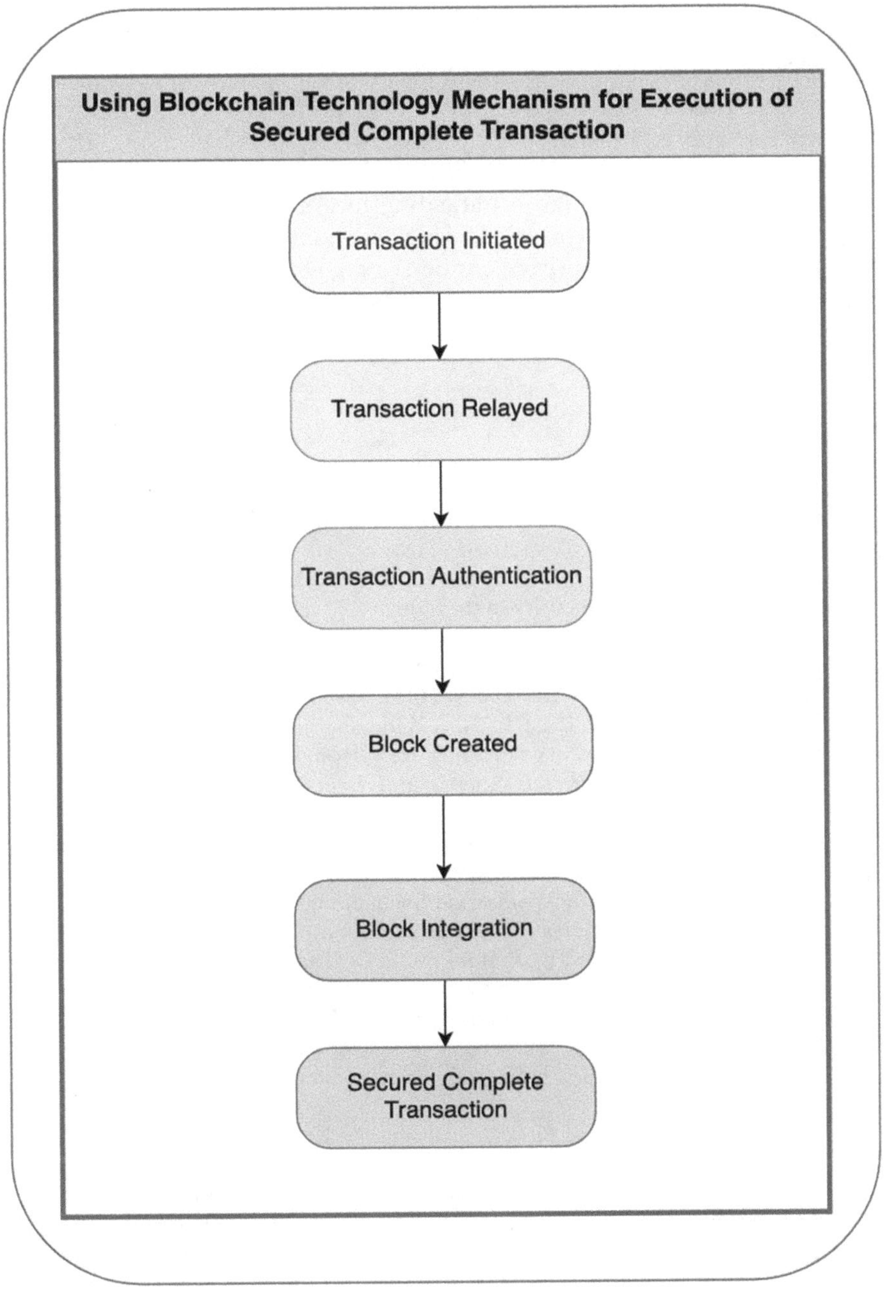

Figure 18.2 Blockchain technology mechanism for execution of secured complete transaction.

review describes a cybersecurity course that will investigate cutting-edge skill sets for leveraging AI and ML for malware analysis. It also describes a course meant to build an environment for planning against AI vulnerabilities in pervasive robotic systems. To combat cybercrime assaults, an innovative, trustworthy, and highly effective cybersecurity system based on ML principles should be researched. The findings indicate that a comprehensive security architecture is beneficial in mitigating future cybersecurity threats for innovative research problems in ensuring data confidentiality. Several industrial agencies, public and private entities, and governments pose substantial vulnerabilities to vital networks, and a cybersecurity role in all of its components will be critical to future expansion, creativity, and competitiveness.

REFERENCES

1. Fernandez-Carames, T. M., & Fraga-Lamas, P. (2019). A review on the application of blockchain to the next generation of cybersecure industry 4.0 smart factories. *IEEE Access*, 7, 45201–45218.
2. Fernández-Caramés, T. M., & Fraga-Lamas, P. (2020). Use case based blended teaching of IIoT cybersecurity in the industry 4.0 era. *Applied Sciences*, 10(16), 5607.
3. Mosca, M. (2018). Cybersecurity in an era with quantum computers: Will we be ready? *IEEE Security & Privacy*, 16(5), 38–41.
4. Rajasekharaiah, K. M., Dule, C. S., & Sudarshan, E. (2020, December). Cyber security challenges and its emerging trends on latest technologies. In *IOP Conference Series: Materials Science and Engineering* (Vol. 981, No. 2, p. 022062). IOP Publishing.
5. Barde, S. (2021). Blockchain-based cyber security. In *Transforming Cybersecurity Solutions using Blockchain*, pp. 55–69.
6. Kianpour, M. (2020). Knowledge and Skills Needed to Craft Successful Cybersecurity Strategies. In *Norsk IKT-konferanse for forskning og utdanning* (No. 3).
7. Bikov, T., Radev, D., Iliev, T., & Stankovski, D. (2021, November). Threat hunting as cyber security baseline in the next-generation security operations center. In *2021 29th Telecommunications Forum (TELFOR)* (pp. 1–4). IEEE.
8. Bhamare, D., Zolanvari, M., Erbad, A., Jain, R., Khan, K., & Meskin, N. (2020). Cybersecurity for industrial control systems: A survey. *Computers & Security*, 89, 101677.
9. Repetto, M., Striccoli, D., Piro, G., Carrega, A., Boggia, G., & Bolla, R. (2021). An autonomous cybersecurity framework for next-generation digital service chains. *Journal of Network and Systems Management*, 29(4), 37.
10. Tweneboah-Koduah, S., Tsetse, A. K., Azasoo, J., & Endicott-Popovsky, B. (2018). Evaluation of cybersecurity threats on smart metering system. In *Information Technology-New Generations: 14th International Conference on Information Technology* (pp. 199–207). Springer International Publishing.
11. Mittal, S., & Chen, J. (2023). AI security threats against pervasive robotic systems: A course for next generation cybersecurity workforce. *arXiv preprint arXiv:2302.07953*.

12. Gupta, M., Mittal, S., & Abdelsalam, M. (2020). AI assisted malware analysis: A course for next generation cybersecurity workforce. *arXiv preprint arXiv:2009.11101*.
13. Alazab, M., Srinivasan, S., Venkatraman, S., & Pham, Q. V. (2021). Deep learning for cyber-security applications: A comprehensive survey.
14. Karjalainen, M., & Kokkonen, T. (2020, September). Comprehensive cyber arena; the next generation cyber range. In *2020 IEEE European Symposium on Security and Privacy Workshops (EuroS&PW)* (pp. 11–16). IEEE.
15. Amin, F., Abbasi, R., Mateen, A., Ali Abid, M., & Khan, S. (2022). A step toward next-generation advancements in the Internet of Things technologies. *Sensors*, 22(20), 8072.
16. Mattos, D. M. F., Krief, F., & Rueda, S. J. (2020). Blockchain and artificial intelligence for network security. *Annals of Telecommunications*, 75, 101–102.
17. Swan, M. (2018). Blockchain for business: Next-generation enterprise artificial intelligence systems. In *Advances in Computers* (Vol. 111, pp. 121–162). Elsevier. https://www.sciencedirect.com/science/article/abs/pii/S0065245818300287
18. Medhane, D. V., Sangaiah, A. K., Hossain, M. S., Muhammad, G., & Wang, J. (2020). Blockchain-enabled distributed security framework for next-generation IoT: An edge cloud and software-defined network-integrated approach. *IEEE Internet of Things Journal*, 7(7), 6143–6149.
19. Deebak, B. D., & Fadi, A. T. (2021). Privacy-preserving in smart contracts using blockchain and artificial intelligence for cyber risk measurements. *Journal of Information Security and Applications*, 58, 102749.
20. Muheidat, F., & Tawalbeh, L. A. (2021). Artificial intelligence and blockchain for cybersecurity applications. In *Artificial Intelligence and Blockchain for Future Cybersecurity Applications* (pp. 3–29). Springer International Publishing.
21. Xue, K., Liu, J., Zhu, L., Wei, D. S., & Lu, R. (2022). Guest editorial: Recent advances in mobile communication network security. *China Communications*, 19(6), iii–v.
22. Sultan, K., Ali, H., & Zhang, Z. (2018). Big data perspective and challenges in next generation networks. *Future Internet*, 10(7), 56.
23. Dutta, V., & Zielińska, T. (2021). Cybersecurity of robotic systems: Leading Challenges and robotic system design methodology. *Electronics*, 10(22), 2850.
24. Rai, A., Sharma, D., Rai, S., Singh, A., & Singh, K. K. (2021). IoT-aided robotics development and applications with AI. In *Emergence of Cyber Physical System and IoT in Smart Automation and Robotics: Computer Engineering in Automation* (pp. 1–14). Springer International Publishing.
25. Demertzis, K., Tziritas, N., Kikiras, P., Sanchez, S. L., & Iliadis, L. (2019). The next generation cognitive security operations center: Adaptive analytic lambda architecture for efficient defense against adversarial attacks. *Big Data and Cognitive Computing*, 3(1), 6.
26. Hazra, A., Alkhayyat, A., & Adhikari, M. (2022). Blockchain-aided integrated edge framework of cybersecurity for Internet of Things. *IEEE Consumer Electronics Magazine*, p. 1-1. https://doi.org/10.1109/MCE.2022.3141068

Index

Pages in *italics* refer to figures and pages in **bold** refer to tables.

FSC
www.fsc.org
MIX
Papier | Fördert
gute Waldnutzung
FSC® C083411

Kolibri 360 GmbH
Ferdinand-Jühlke-Straße 7
99095 Erfurt, Deutschland
produktsicherheit@kolibri360.de